VOLUME FORTY SIX

PROFILES OF DRUG SUBSTANCES, EXCIPIENTS, AND RELATED METHODOLOGY

VOLUME FORTY SIX

PROFILES OF
DRUG SUBSTANCES, EXCIPIENTS, AND RELATED METHODOLOGY

Edited by

ABDULRAHMAN A. AL-MAJED
Department of Pharmaceutical Chemistry, College of Pharmacy, King Saud University, Riyadh, Kingdom of Saudi Arabia

ELSEVIER

ACADEMIC PRESS
An imprint of Elsevier

Academic Press is an imprint of Elsevier
50 Hampshire Street, 5th Floor, Cambridge, MA 02139, United States
525 B Street, Suite 1650, San Diego, CA 92101, United States
The Boulevard, Langford Lane, Kidlington, Oxford OX5 1GB, United Kingdom
125 London Wall, London, EC2Y 5AS, United Kingdom

First edition 2021

ISBN: 978-0-12-824127-1
ISSN: 1871-5125 (Series)

For information on all Academic Press publications
visit our website at https://www.elsevier.com/books-and-journals

Publisher: Zoe Kruze
Editorial Project Manager: Leticia Lima
Production Project Manager: Abdulla Sait
Cover Designer: Alan Studholme

Typeset by SPi Global, India

Contents

Contributors

Alaa A.-M. Abdel-Aziz
Department of Pharmaceutical Chemistry, College of Pharmacy, King Saud University, Riyadh, Kingdom of Saudi Arabia

Zahi Alanazi
Department of Pharmaceutical Chemistry, College of Pharmacy, King Saud University, Riyadh, Kingdom of Saudi Arabia

Salem Albraiki
Department of Pharmaceutical Chemistry, College of Pharmacy, King Saud University, Riyadh, Kingdom of Saudi Arabia

Raisuddin Ali
College of Pharmacy, King Saud University, Riyadh, Kingdom of Saudi Arabia

Hamad M. Al-Kahtani
Department of Pharmaceutical Chemistry, College of Pharmacy, King Saud University, Riyadh, Kingdom of Saudi Arabia

Abdulrahman A. Al-Majed
Department of Pharmaceutical Chemistry, College of Pharmacy, King Saud University, Riyadh, Kingdom of Saudi Arabia

Nawaf A. Alsaif
Department of Pharmaceutical Chemistry, College of Pharmacy, King Saud University, Riyadh, Kingdom of Saudi Arabia

Ali D. Alshahrani
Department of Pharmaceutical Chemistry, College of Pharmacy, King Saud University, Riyadh, Kingdom of Saudi Arabia

Majed J. Al-wadei
Department of Pharmaceutical Chemistry, College of Pharmacy, King Saud University, Riyadh, Kingdom of Saudi Arabia

Abdullah Alzaid
Department of Pharmaceutical Chemistry, College of Pharmacy, King Saud University, Riyadh, Kingdom of Saudi Arabia

Ahmed H. Bakheit
Department of Pharmaceutical Chemistry, College of Pharmacy, King Saud University, Riyadh, Kingdom of Saudi Arabia; Department of Chemistry, Faculty of Science and Technology, Al-Neelain University, Khartoum, Sudan

Hany W. Darwish
Department of Pharmaceutical Chemistry, College of Pharmacy, King Saud University, Riyadh, Kingdom of Saudi Arabia; Department of Analytical Chemistry, Faculty of Pharmacy, Cairo University, Cairo, Egypt

Ibrahim A. Darwish
Department of Pharmaceutical Chemistry, College of Pharmacy, King Saud University, Riyadh, Kingdom of Saudi Arabia; Department of Pharmaceutical Analytical Chemistry, Faculty of Pharmacy, Assiut University, Assiut, Egypt

Adel S. El-Azab
Department of Pharmaceutical Chemistry, College of Pharmacy, King Saud University, Riyadh, Kingdom of Saudi Arabia

Rashed N. Herqash
Medicinal Aromatic and Poisonous Plant Research Center, College of Pharmacy, King Saud University, Riyadh, Kingdom of Saudi Arabia

Gunawan Indrayanto
Faculty of Pharmacy, Surabaya University, Surabaya, Indonesia

Galih Satrio Putra
Faculty of Pharmacy, Surabaya University, Surabaya, Indonesia

Farida Suhud
Faculty of Pharmacy, Surabaya University, Surabaya, Indonesia

Tanveer A. Wani
Department of Pharmaceutical Chemistry, College of Pharmacy, King Saud University, Riyadh, Kingdom of Saudi Arabia

Preface

In 1970, a group of scientists from Pharmaceutical Analysis Section of the Academy of Pharmaceutical Sciences in the USA suggested to compile and publish a series to be named *Analytical Profiles of Drug Substances*. The aim of producing this series was to provide all phases of the pharmaceutical development with information that are not available in the official compendial monographs of drug substances, such as the physical and chemical charactersitics, routes of drug preparation, and pathways of drug degradation and metabolism. *Klaus Florey* from the Squibb Institute for Medical Research, New Brunswick, New Jersey, USA, was appointed as the first editor of the series, and he has completed the editing of the first 20 volumes of the *Analytical Profiles of Drug Substances* series in the years between 1972 and 1991.

With the publication of Volume 21, the editorship of the series has been assumed by *Harry G. Brittain* from Center for Pharmaceutical Physics, Milford, New Jersey, USA. The scope of the series was expanded to include profiles of excipient materials. This expansion reflects the realization that all aspects of a drug formulation need to be fully specified. Therefore, the title of the series was slightly modified to become: *Analytical Profiles of Drug Substances and Excipients*. With the production of Volume 30, the series was again expanded to cover critical review chapters related to the characterization of drug substances and excipients. Therefore, the title of the series has been again modified and became: *Profiles of Drug Substances, Excipients and Related Technology*.

In 2020, Elsevier has appointed me as the editor for the: *Profiles of Drug Substances, Excipients and Related Technology* series. The scope of the series will remain unchanged. So the chapters that are published in the future volumes should fall into one or more of the following main categories:

1. Physical Profiles of Drug Substances and Excipients.
2. Analytical Profiles of Drug Substances and Excipients.
3. ADME Profiles of Drug Substances and Excipients.
4. Methodology Related to the Characterization of Drug Substances and Excipients.
5. Methods of Chemical Synthesis.

Authors may choose to develop a comprehensive profile that contains all of the five categories described above, or they may select an individual section(s) and develop a chapter whose scope falls entirely within one or more categories. Preferably, new drug compounds can be profiled as they become marketed, but also the important older drug compounds could also be developed. Updates on drug compounds that had been profiled earlier in the series are also accepted, whenever a sufficient body of new information becomes available.

Approximately 580 drug compounds, in addition to various reviews of methodology useful for drug substances and excipients, were published in all the volumes of this book series. The present volume contains comprehensive profiles of five drug compounds:

- Darunavir
- Bisoprolol
- Betaxolol
- Rabeprazole
- Irbesartan

The current volume also contains a chapter reviewing the Validation of an *in vitro* Bioassay Methods and Their Applications in Herbal Drug Research.

The success of the *Profiles of Drug Substances, Excipients and Related Technology* series will continue to be based on the contributions of the chapter authors, as well as the quality of the work. It is almost ironic to note that the need for this kind of *Profiles* has grown to be overwhelmingly important in the current climate of drug development. The goal is to cover all drug substances with comprehensive monographs is still a distance one. It is up to our perseverance authors to make it a reality. We hope that this series of book will become a source for valuable information useful in the everyday work for pharmaceutical community, scientific, and industrial researchers, pharmacists, and pharmacy students.

Many thanks and gratitude goes to *Zoe Kruze*, Publisher at Elsevier, and to *Letícia Lima*, Editorial Project Manager, for their great effort during the preparation of this volume. Also my sincere thanks go to the authors and their co-authors for the valuable contributions in this series.

I welcome communications from anyone in the pharmaceutical community who might want to provide an opinion or a contribution.

ABDULRAHMAN A. AL-MAJED
Editor

CHAPTER ONE

Darunavir: A comprehensive profile

Ibrahim A. Darwish[a,b,*], Abdulrahman A. Al-Majed[a], Nawaf A. Alsaif[a], Ahmed H. Bakheit[a,c], Rashed N. Herqash[d], and Abdullah Alzaid[a]

[a]Department of Pharmaceutical Chemistry, College of Pharmacy, King Saud University, Riyadh, Kingdom of Saudi Arabia
[b]Department of Pharmaceutical Analytical Chemistry, Faculty of Pharmacy, Assiut University, Assiut, Egypt
[c]Department of Chemistry, Faculty of Science and Technology, Al-Neelain University, Khartoum, Sudan
[d]Medicinal Aromatic and Poisonous Plant Research Center, College of Pharmacy, King Saud University, Riyadh, Kingdom of Saudi Arabia
[*]Corresponding author: e-mail address: idarwish@ksu.edu.sa

Contents

Profiles of Drug Substances, Excipients, and Related Methodology, Volume 46
ISSN 1871-5125
https://doi.org/10.1016/bs.podrm.2020.07.001

1. Description

1.1 Nomenclature

1.1.1 Systemic chemical names

(3*R*,3a*S*,6a*R*)-Hexahydrofuro[2,3-*b*]furan-3-yl [(2*S*,3*R*)-4-{[(4-aminophenyl)sulfonyl] (isobutyl)amino}-3-hydroxy-1-phenyl-2-butanyl] carbamate [ACD/IUPAC]

(3*R*,3a*S*,6a*R*)-Hexahydrofuro[2,3-*b*]furan-3-yl-[(2*S*,3*R*)-4-{[(4-aminophenyl)sulfonyl] (isobutyl)amino}-3-hydroxy-1-phenyl-2-butanyl] carbamat [German] [ACD/IUPAC]

[(2*S*,3*R*)-4-{[(4-Aminophényl)sulfonyl](isobutyl)amino}-3-hydroxy-1-phényl-2-butanyl] carbamate de (3*R*,3a*S*,6a*R*)-hexahydrofuro[2,3-*b*] furan-3-yle [French] [ACD/IUPAC]

(3*R*,3a*S*,6a*R*)-Hexahydrofuro[2,3-*b*]furan-3-yl [(2*S*,3*R*)-4-{[(4-aminophenyl)sulfonyl] (isobutyl)amino}-3-hydroxy-1-phenylbutan-2-yl] carbamate

[(1*R*,5*S*,6*R*)-2,8-dioxabicyclo[3.3.0]oct-6-yl] N-[(2*S*,3*R*)-4-[(4-amino phenyl)sulfonyl- (2-methylpropyl)amino]-3-hydroxy-1-phenyl- butan-2-yl] [1].

1.1.2 Nonproprietary names

Darunavir (German, French, Spanish), Darunavirum (INN-Latin), Darunavir ethanolate, Darunavir propylene glycolate, TMC-114, UIC-94017, UNII-YO603Y8113 [1,2].

1.1.3 Proprietary names

Darunavir, Darunavir Alvogen, Darunavir CF, Darunavir DOC, Darunavir KrKa, Darunavir Dr. Reddy's, Darunavir Mylan, Darunavir Pliva, Darunavir Sandoz, Darunavir Stada, Darunavir Teva, Darunavir Zentiva, Darunavir-Mepha, Darunavir, Prezista, Prezcobix (+Cobicistat), Rezolsta, Symtuza (+ Cobicistat + Emtrcitabine + Tenofovir), Virontar (+Ritonavir), Prezistanaive, Resisvir, Davarino [1–4].

1.2 Formulae

1.2.1 Darunavir

Empirical Formula	Molecular Weight	CAS Number
$C_{27}H_{37}N_3O_7S$	547.7	206361–99-1

1.2.2 Darunavir ethanolate

Empirical Formula	Molecular Weight	CAS Number
$C_{27}H_{37}N_3O_7S\text{-}C_2H_5OH$	593.7	635728–49-3

1.2.3 Structural formula

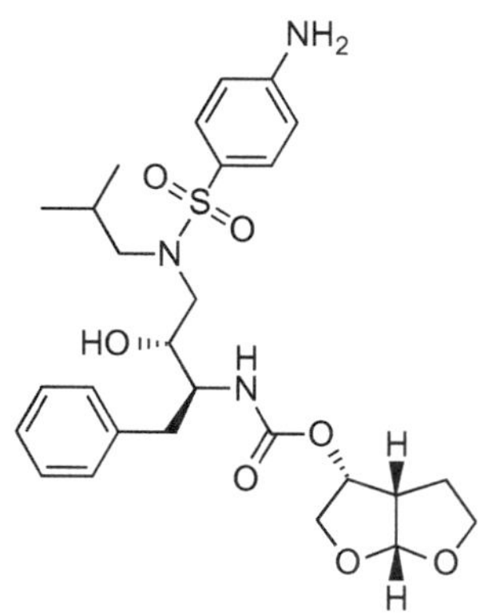

1.3 Elemental composition

The elemental composition of darunavir is as follows [3,4].

Carbon	Hydrogen	Nitrogen	Oxygen	Sulfur
59.21%	6.81%	7.67%	20.45%	5.85%

1.4 Appearance

Darunavir is a white to off-white powder [4,5].

2. Uses and applications

Darunavir is a non-peptide protease inhibitor with antiviral activity against human immunodeficiency virus (HIV). It is used in the treatment of infection and treatment of acquired immunodeficiency syndrome (AIDS) caused by in HIV infection. Darunavir is boosted with low-dose ritonavir, which acts as a pharmacokinetic enhancer. It is given orally as 600 mg (with ritonavir 100 mg) twice daily with food [6,7].

3. Methods of preparation

Mizhiritskii and Marom [8] prepared darunavir **1** using azido epoxide **2** as started material, which was reacted with isobutylamine **3** in 2-propanol at 80 °C for 12 h to afford azidoalcohol **4**. Treatment of the azidoalcohol **4** with *p*-nitrobenzenesulfonyl chloride **5** in the presence of aqueous $NaHCO_3$, provided the corresponding azide, which was hydrogenated over 10% Pd/C in ethyl acetate to afford the amine (yield of 75–78%). This amine was transformed to darunavir **1** upon reaction with hexahydrofuro2, 3-blfuran-3-yl **6** derivative in methylene chloride **7** in the presence of three equivalents of triethylamine 7 at 23 °C for 12 h with overall yield of 60–65% which is illustrated in Scheme 1.

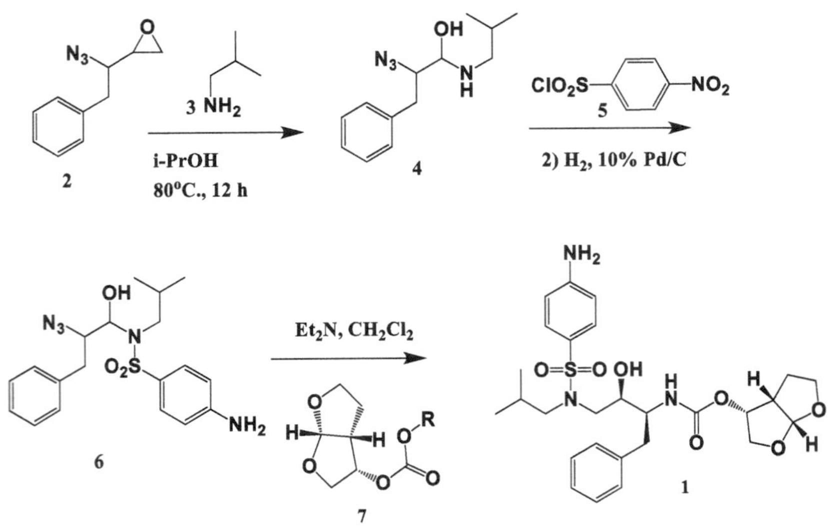

Scheme 1 Preparation of darunavir from azido epoxide.

Mizhiritskii and Marom [8] prepared darunavir **1** by amidation of (1-oxiranyl-2 phenyl-ethyl)-carbamic acid *tert*-butyl ester **2** with isopropylamine **3**. Introducing *p*-nitrophenylsulfonyl **5** with *tert*-butyl (3-hydroxy-4-(isobutylamino)-1-phenylbutan-2-yl)carbamate **4** reducing the nitro moiety of the *tert*-butyl (3-hydroxy-4-((*N*-isobutyl-4-nitrophenyl)sulfonamido)-1-phenylbutan-2-yl)carbamate **6** deprotecting of the (4-(*N*-(3-((*tert*-butoxycarbonyl)amino)-2-hydroxy-4-phenylbutyl)-*N*-isobutylsulfamoyl) phenyl) sodium 7 by hydrolysis of compound **7** to form (4-(*N*-(3-amino-2-hydroxy-4-phenylbutyl)-*N*-isobutylsulfamoyl)phenyl)sodium **8** which was coupled with (3*R*,3a*S*,6a*R*)-hexahydrofuro2.3-blfuran-3-yl 9 derivative, to give darunavir **1** as shown in Scheme 2.

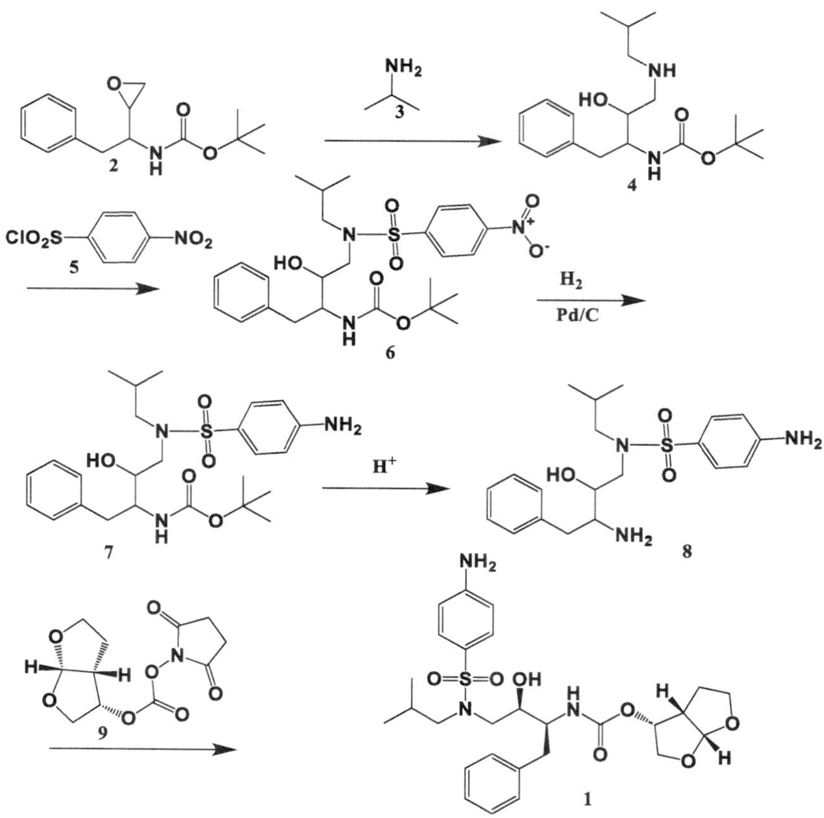

Scheme 2 Preparation of darunavir from *tert*-butyl (1-(oxiran-2-yl)-2-phenylethyl) carbamate.

Ghosh et al. [9,10] prepared darunavir **1** by coupling mixed carbonate **4** with a benzene sulfonamide derivative of 1,3-diamino-4-phenylbutan-2-ol **11**. The method of synthesis of the darunavir **1** based on converted the optical active bis-THF-ol **2** to carbonate **4** through interaction with *N*, *N′*-disuccinimidyl carbonate **2** in the presence of triethylamine in methylene chloride. The synthesis method of target compound started by reaction between commercially available (*S*)-1-((*S*)-oxiran-2-yl)-2-phenylethan-1-amine **6** with isobutylamine **5** in refluxing isopropanol to provide amino alcohol **7**. Followed by the reaction of the resulting amino alcohol with *p*-nitrobenzenesulfonyl chloride **8** in the presence of aqueous $NaHCO_3$ furnished the sulfonamide derivative **9**. Compound **9** was hydrogenation in catalytic with over 10% Pd/C in ethyl acetate that reaction reduced the nitro group to the corresponding amine **10**. Compound **11** was formed by remove the BOC group of the amine **10** with the use of trifluoroacetic acid. Reaction of the diamine with the mixed carbonate **4** in the presence of triethylamine provided darunavir **1** in high yield (Scheme 3).

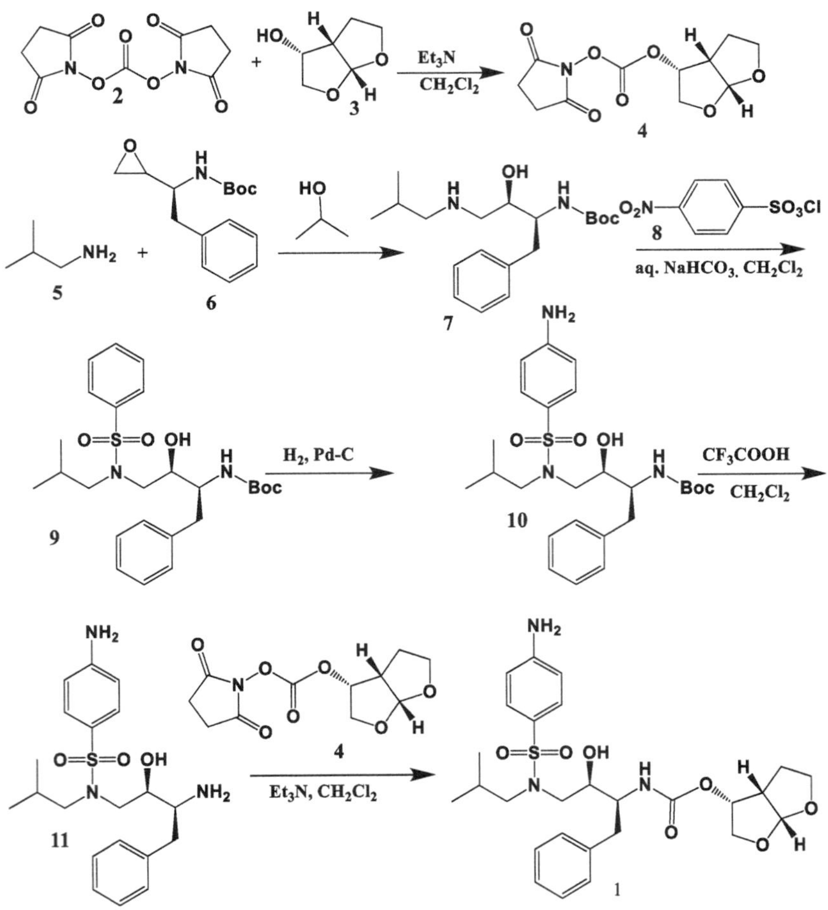

Scheme 3 Preparation of darunavir by conversion the optical active bis-THF-ol to carbonate.

Moore et al. [11] described a method for preparation of darunavir from isocitrate as illustrated in Scheme 4. The key intermediate in the synthesis of darunavir which is (3*R*,3a*S*,6a*R*)-hexahydrofuro[2,3-*b*]furan-3-ol was synthesized from monopotassium isocitrate. The isocitric acid salt, obtained from a high-yielding fermentation fed by sunflower oil, was converted in several steps to a tertiary amide. This amide, along with the compound's ester functionalities, was reduced with lithium aluminum hydride to give, on acidic workup, a transient aminal-triol. This was converted in situ to the intermediate, the bicyclic acetal furofuranol side chain of darunavir. To this intermediate (dissolved in dichloromethane), *N,N′*-disuccinimidyl carbonate and pyridine were added, and the reaction proceeded to give 2,5-dioxopyrrolidin-1-yl((3*R*,3a*S*,6a*R*)-hexahydrofuro[2,3-*b*]furan-3-yl)carbonate (8). This compound and 4-amino-*N*-((2*R*,3*S*)-3-amino-2-hydroxy-4-phenylbutyl)-*N*-isobutylbenzene sulfonamide were subsequently reacted to yield darunavir.

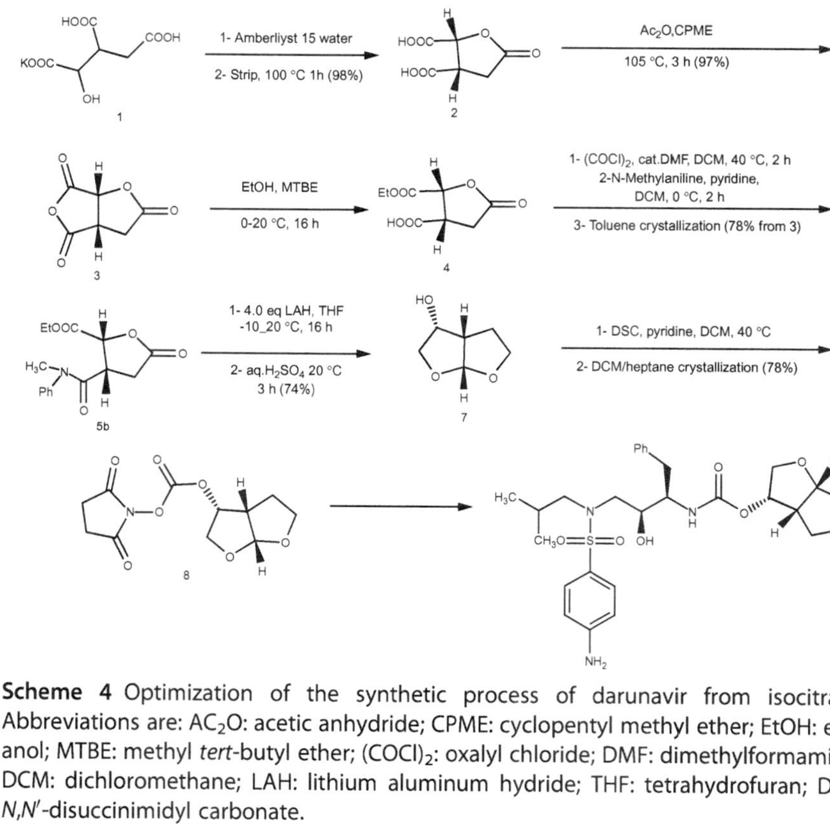

Scheme 4 Optimization of the synthetic process of darunavir from isocitrate. Abbreviations are: AC_2O: acetic anhydride; CPME: cyclopentyl methyl ether; EtOH: ethanol; MTBE: methyl *tert*-butyl ether; $(COCl)_2$: oxalyl chloride; DMF: dimethylformamide; DCM: dichloromethane; LAH: lithium aluminum hydride; THF: tetrahydrofuran; DSC: *N,N′*-disuccinimidyl carbonate.

Vellenki et al. [12] invented an improved process for the preparation of darunavir and its solvates or pharmaceutically acceptable salts with improved yield and quality (Scheme 5). The invention also relates to provide processes for the preparation of amorphous darunavir. It also relates to pharmaceutical compound of crystalline or amorphous darunavir, its solvates or pharmaceutically acceptable salts having the difuranyl impurity less than 0.1%. Darunavir was prepared via ring-opening of (2*S*,3*S*)-1,2-epoxy-3-(butoxy carbonyl)amino-4-phenylbutane (1) with isobutylamine followed by sulfonylation with *p*-nitrobenzenesulfonyl chloride. The resulting (1*S*,2*R*)-[1-benzyl-2-hydroxy-3-[isobutyl-(4-nitrobenzenesulfonyl)amino]propyl] carbamic acid *tert*-Bu ester (2) underwent reduction to give 4-amino-*N*-[(2*R*,3*S*)-(3-amino-2-hydroxy-4-phenylbutyl)]-*N*-isobutylbenzenesulfonamide (3) which underwent amidation with (3*R*,3a*S*,6a*R*)-hexahydrofuro[2,3-*b*]furan-3-yl 4-nitrophenylcarbonate (4) to give darunavir ethanolate, which underwent solvent-removing to give the title compound (5). On the other

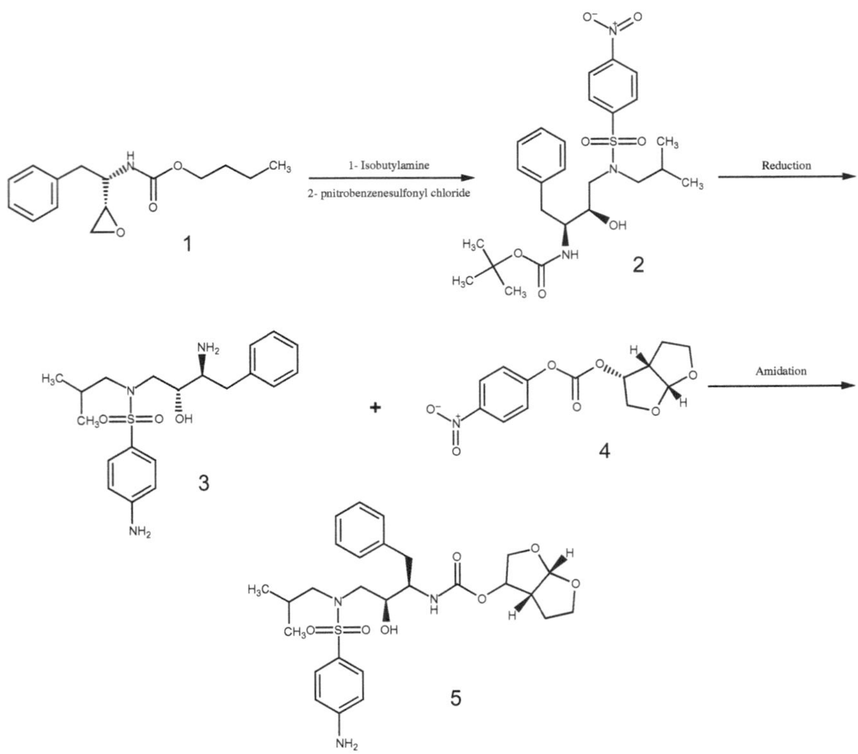

Scheme 5 Improved process for the preparation of darunavir and its solvates.

hand, darunavir was also prepared via condensation of [*N*-(*t*-butoxycarbonyl)-L-phenylalanine]chlorohydrin with isobutylamine; the resulting amine underwent sulfonylation with p-nitrobenzenesulfonyl chloride to give (1*S*,2*R*)-[1-benzyl-2-hydroxy-3-[isobutyl-(4-nitrobenzenesulfonyl)amino] propyl]carbamic acid *tert*-Bu-ester, which underwent the above procedures to give the title compound.

Stappers et al. [13] invented an improved simplified procedure for preparation of darunavir ethanolate by reacting carbonic acid 2,5-dioxo-1-pyrrolidinyl [(3*R*,3a*S*,6a*R*)-hexahydrofuro[2,3-*b*]furan-3-yl] ester (1) with 4-amino-*N*-[(2*R*,3*S*)-3-amino-2-hydroxy-4-phenylbutyl]-*N*-(2-methyl propyl)-benzenesulfonamide (2) in ethanol as solvent. Furthermore, the process allowed the isolation of darunavir (3) immediately in its ethanolate form (darunavir monoethanolate). The procedure is illustrated in Scheme 6.

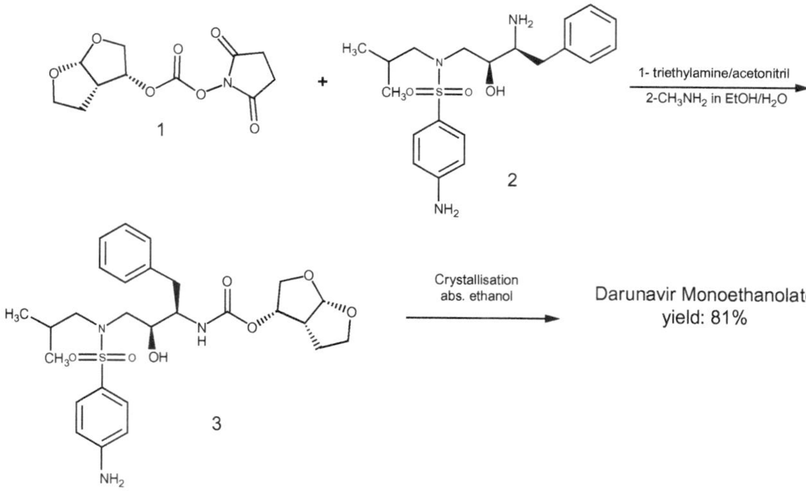

Scheme 6 Improved simplified procedure for preparation of darunavir ethanolate starting from carbonic acid 2,5-dioxo-1-pyrrolidinyl [(3*R*,3a*S*,6a*R*)-hexahydrofuro[2,3-*b*] furan-3-yl] ester.

Jiang et al. [14] invented a method for preparation of darunavir comprising: (1) preparation of 20–30 mg of *N*-dibenzyl-L-phenylalanine methyl ester, 40–90 mg of *N*-hydroxy compound, 20–40 mg of 1 equivalent of hydrogenated methane, 100–150 mg of Bu lithium, 50–80 mg of tetrahydrofuran solvent, 70–110 mg of amines, 60–100 mg of (2*S*,3*R*)-2-amino-4-gas-1-phenylbutan-3-ol, 40–60 mg of halogenated aromatic ring, 10–20 mg of triethylamine and di-methane, 5–20 mg of 1.3 equivalent of di-*tert*-Bu-dicarbonate, and 20–40 mg of metal base. The preparation method further comprises: (2) placing *N*-dibenzyl-L-phenylalanine methyl ester, *N*-hydroxy compound, 1 equivalent of hydrogenated methane, Bu lithium, tetrahydrofuran solvent, and amines at −20 °C; and (3) placing (2*S*,3*R*)-2-amino-4-gas-1-phenylbutan-3-ol, halogenated aromatic ring, triethylamine and di-methane, 1.3 equivalent of di-*tert*-Bu dicarbonate and metal base at room temperature. The method also comprises: (4) heating the mixture, adding halogenated aromatic ring, and stirring to stand in a dry and ventilated place; and (5) collecting the mixture, placing it in a beaker and stirring, stop heating, collecting the dried mixture, sieving the light yellow powder, and collecting the finished product. This darunavir preparation method had lower requirements on the preparation environment, eliminated the complicated operation in the traditional method, did not easily waste consumables, and was convenient to use.

Xu et al. [15,16] invented a method for preparing darunavir. According to this method, *N*-dibenzyl-L-phenylalanine methyl ester, halogenated aromatic ring, triethylamine, cobalt, ferrous oxide, zinc oxide, zinc, ferric nitrite solution, metal base, hydroxyl compound, aluminum oxide, silicon dioxide, Bu lithium, tetrahydrofuran solvent, ethanol solution, etc. were employed in preparation of fine powder of delanavir. The method comprised the following steps of taking 20–30 parts of cobalt, 10–20 parts of triethylamine, 10–20 parts of ferrous oxide, 10–20 parts of zinc oxide, 3–5 parts of 1.2 equivalent of methyl compound, 20–30 parts of zinc, 50–80 parts of ferric nitrite solution, 30–40 parts of metal base, 40–50 parts of aluminum oxide, 10–20 parts of silicon dioxide, 3–5 parts of *p*-acetamidophenylsulfonyl chloride, 3–5 parts of precipitant, 50–80 parts of ethanol solution. and 20–30 parts of nickel for later use. The preparation method comprises the steps of: (1) preparation of the raw materials; (2) adding nickel, cobalt and zinc to ferric nitrite solution, and stirring for reaction to obtain mixed solution 1; (3) adding aluminum oxide and silicon dioxide to the mixed solution 1, mixing, adding the precipitant, adding ferrous oxide and zinc oxide after precipitation generation, reacting at 30 °C, and standing to obtain mixed solution 2; (4) drying the mixed solution 2 in infrared-driving oven, reducing under hydrogen atmosphere at high temperature, and heating to obtain matrix 1; (5) adding triethylamine, the methyl compound, metal alkali and *p*-acetamidophenylsulfonyl chloride in a container, heating for reaction, cooling, adding the ethanol solution, sealing and heating for reaction, cooling to obtain matrix 2; (6) mixing the matrix 1 and matrix 2, adding the catalyst, and carrying out coupling reaction. The method had the advantages of simple process, high efficiency, low environmental request and low cost.

4. Physical characteristics

4.1 Solubility

Darunavir has a solubility of 8.7 mg/L in water at 25 °C and its ethanolate has a solubility of approximately 0.15 mg/mL in water at 20 °C. The solubility of darunavir is pH dependent and it is slightly increased at extreme values of pH [17,18].

4.2 Partition coefficient

Octanol/water partition coefficient; Log Kow = 1.88 [18].

4.3 X-ray powder diffraction pattern

The structural characterization of darunavir powder was done by high resolution X-ray diffraction using Rigaku Ultima IV diffractometer (Rigaku Corporation, Tokyo, Japan) with scintillation detector with reflection mode, Cu K{acute over (α)} radiation (1.5406 Å), scanning range: 3–60°2θ, Step size: 0.02°2θ and Time per step: 1 s. monochromatized with a graphite crystal. The pattern was collected at 40 kV of tube voltage and 40 mA of tube current in step scan mode. The peaks (reflections) of darunavir powder are not obtained because darunavir amorphous sample as shown in Fig. 1.

4.4 Thermal analysis

4.4.1 Melting behavior

The melting point of darunavir is 74 °C [18].

4.4.2 Differential scanning calorimetry

Differential scanning calorimetry data of darunavir was obtained from a PerkinElmer DSC8000 differential scanning calorimeter with a TAC 71DX thermal analysis controller (PerkinElmer Inc., Norwalk, CT, USA). Samples were accurately weighed (4.5 mg) into aluminum pans and thermograms obtained at a heating rate of 10 °C/min over a temperature range of 40–300 °C. Pyris software 2.04 (PerkinElmer Inc., Norwalk, CT, USA) was used for analysis. A heating rate of 10 °C/min was selected. Differential scanning calorimetric curve of darunavir showed an endothermic process around 70.86 °C (Fig. 2).

4.4.3 Thermogravimetric analysis

Thermogravimetric analysis of darunavir was obtained using a PerkinElmer Thermal analyzer (PerkinElmer Inc., Norwalk, CT, USA). The sample

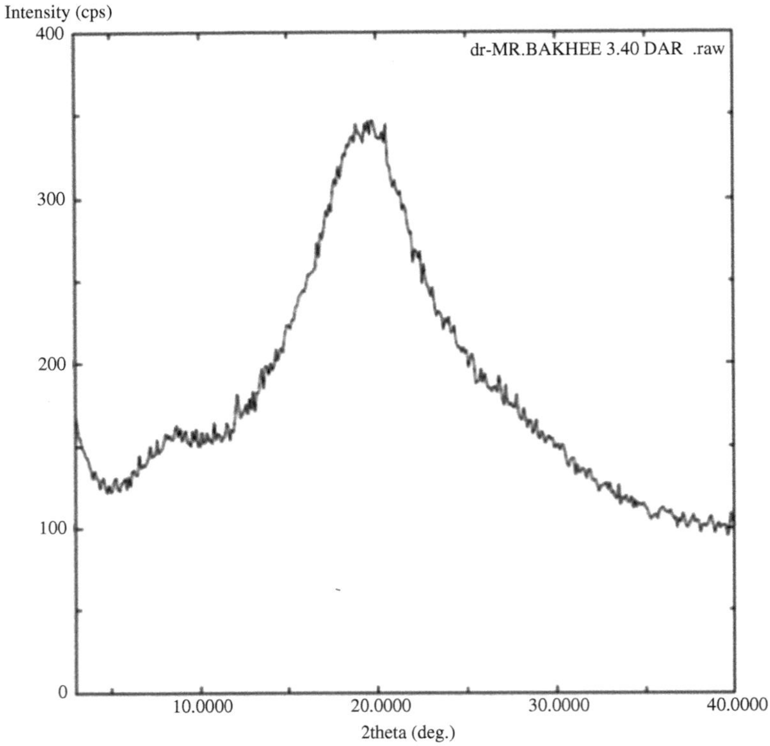

Fig. 1 X-ray diffraction of darunavir.

(4.5 mg) was placed in an aluminum pan that had been pierced prior to the analysis. The sample was heated over the temperature range of 50–400 °C at a rate of 10 °C/min under a nitrogen purge (50 mL/min). The thermogram consisted of a gradual weight loss starting at about 6.9%. The thermogram demonstrated that sublimation of darunavir base started at 260.21 °C (confirmed by hot stage microscopy) and the sudden drop in the thermogram at 327.50 °C corresponds to the thermal decomposition of darunavir that takes place during the melting process (Fig. 3).

4.5 Spectroscopic analysis

4.5.1 Ultraviolet-visible Spectroscopy

The ultraviolet (UV) absorption spectrum of methanolic solution of darunavir (20 μg/mL) in methanol was recorded using a Shimadzu UV-spectrophotometer, model UV-1800 (Shimadzu Corporation, Tokyo, Japan) with 1 cm matched quartz cells. The absorption spectrum was recorded over the range of 200–400 nm. The UV spectrum of darunavir shows an absorption maximum at 266 nm (Fig. 4).

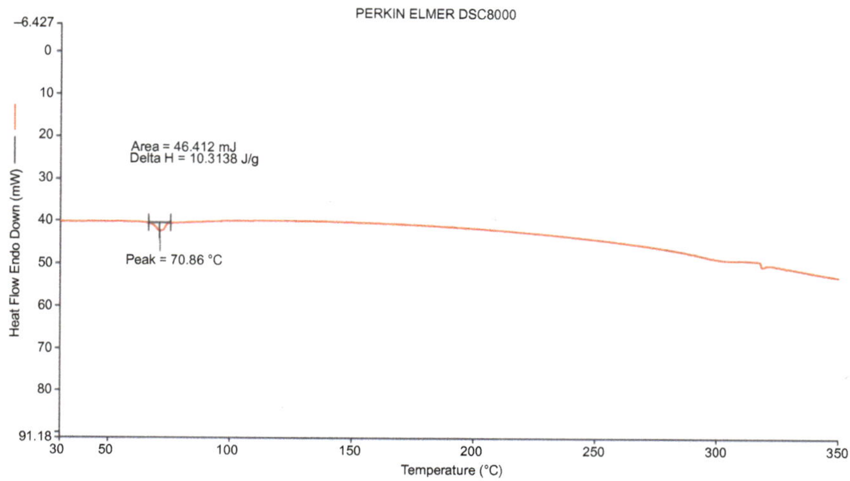

Fig. 2 Differential scanning calorimetry of darunavir.

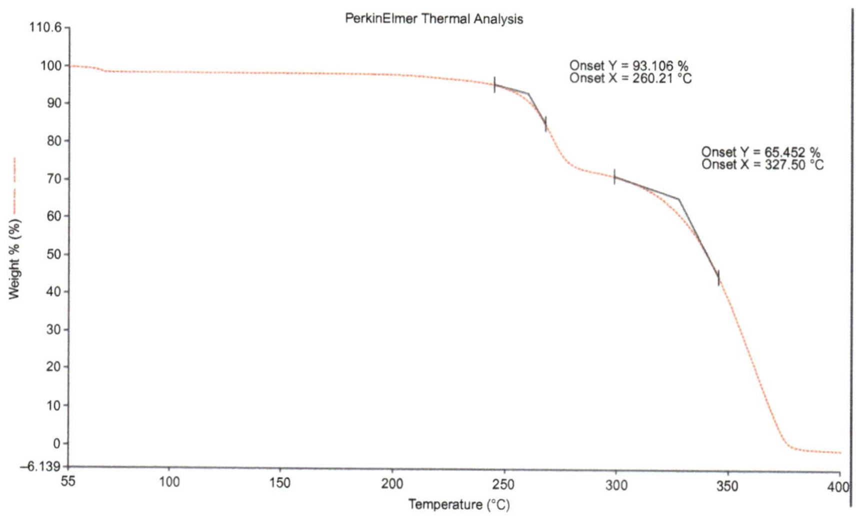

Fig. 3 Thermogram of darunavir.

4.5.2 Fourier-transform infrared absorption spectroscopy

Fourier-transform infrared (FT-IR) absorption spectrum of darunavir (as KBr disc) was recorded using the PerkinElmer FT-IR Spectrum BX apparatus (PerkinElmer, Norwalk, CT, USA); the FT-IR spectrum is given in Fig. 5. The characteristic IR absorption band of carboxylic acid appeared at

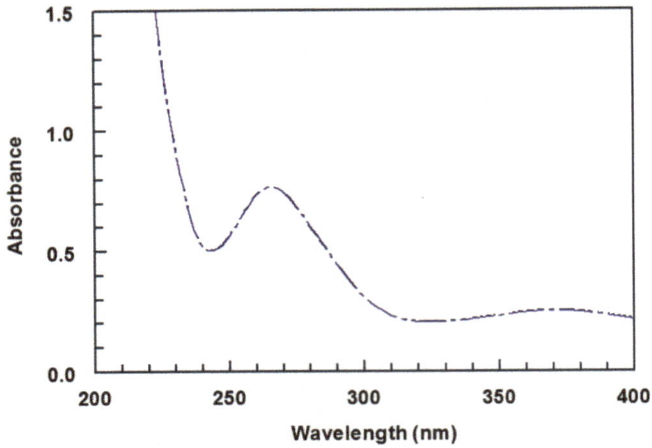

Fig. 4 UV spectrum of darunavir.

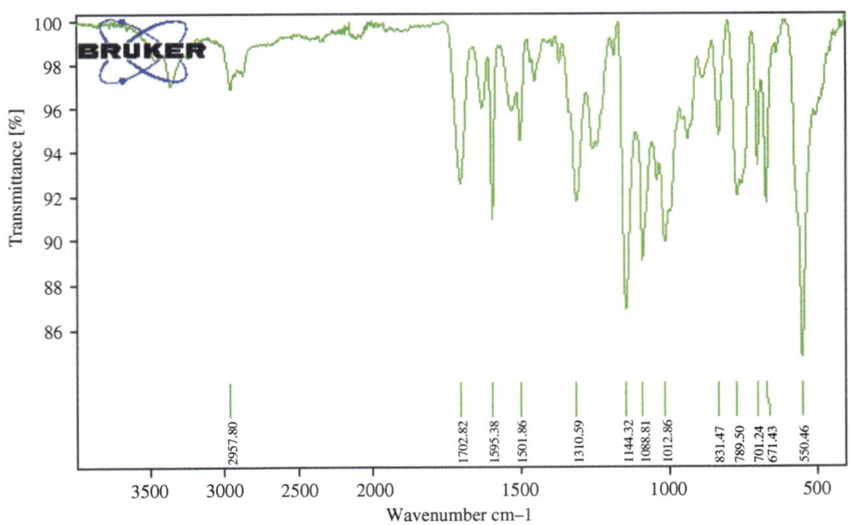

Fig. 5 Fourier-transform infrared (FT-IR) absorption spectrum of darunavir.

2957 cm^{-1}; conjugated acid C=O stretching appeared at 1702 cm^{-1}; amine N—H bending appeared at 1595 cm^{-1}; nitro compound N—O stretching appeared at 1501 cm^{-1}; sulfone S=O stretching at 1310 cm^{-1}; secondary alcohol C—O stretching at 1088 cm^{-1} and C—O stretching aliphatic ether appeared at 1144.32 cm^{-1}.

4.5.3 Nuclear magnetic resonance spectrometry

4.5.3.1 ^{1}H NMR Spectrum

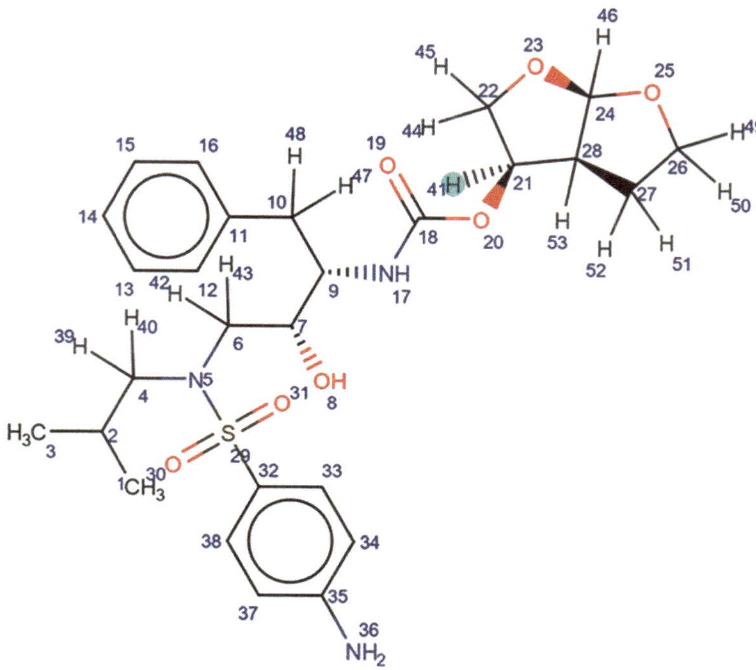

^{1}H NMR spectrum of darunavir was scanned in DMSO-d_6 on a Brucker NMR spectrometer (Bruker Co., Cremlingen, Germany) operating at 500 MHz. Chemical shifts are expressed in δ-values (ppm) relative to tetramethylsilane (TMS) as an internal standard. Coupling constants (J) are expressed in Hz (Table 1) (Figs. 6 and 1).

Table 1 ^{1}H NMR of darunavir (DMSO-d_6).

Signal	Atom number	Location (δ)	Shape	Integration
1	1, 1, 1, 3, 3, 3	0.90	m	6H
2	2	1.37	d, J= 95.1 Hz	1H
3	51	1.62	d, J= 24.2 Hz	1H
4	52	1.88	s	1H
5	53	2.71	m	1H
6	42, 43, 47	2.91	m	3H
7	48	3.06	m	1H

Continued

Table 1 ^{1}H NMR of darunavir (DMSO-d_6).—cont'd

Signal	Atom number	Location (δ)	Shape	Integration
8	39, 40	3.19	d, J=22.6 Hz	2H
9	44, 49	3.72	dd, J=9.5, 5.8 Hz	2H
10	50	3.81	dd, J=11.7, 5.4 Hz	1H
11	7, 9, 45	3.90	t, J=29.8 Hz	3H
13	46	5.05	d, J= 28.9 Hz	1H
14	36, 53	5.63	m	2H
15	14, 37, 34	7.23	s	3H
16	12, 13, 15, 16	7.28	ddd, J=26.5, 6.8, 3.6 Hz	4H
19	46	7.46	m	1H
20	41	7.61	s	1H
21	36, 36	7.7	s	2H
22	17	7.77	dd, J= 24.5, 15.8 Hz	1H

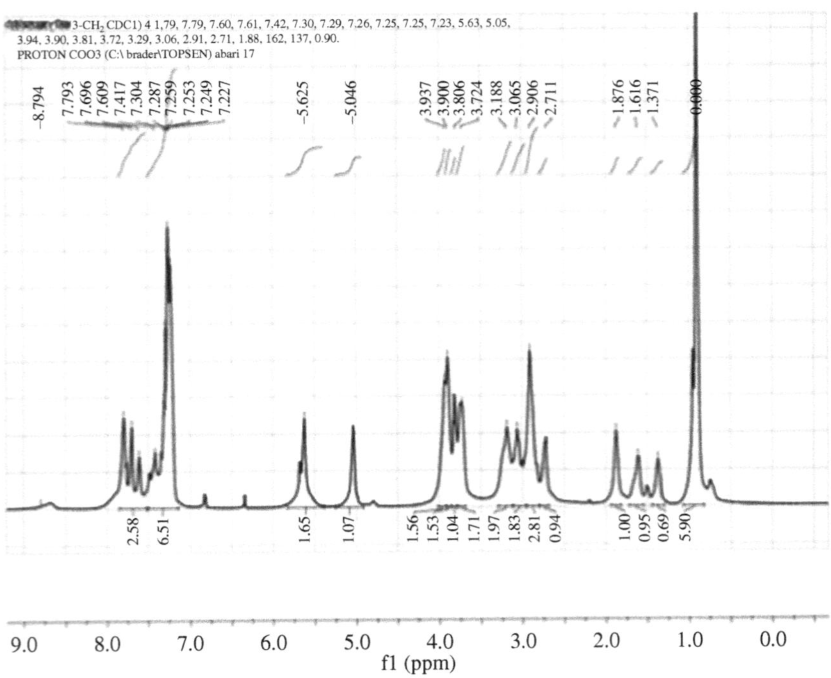

Fig. 6 ^{1}H NMR spectrum of darunavir.

4.5.3.2 ^{13}C NMR spectrum

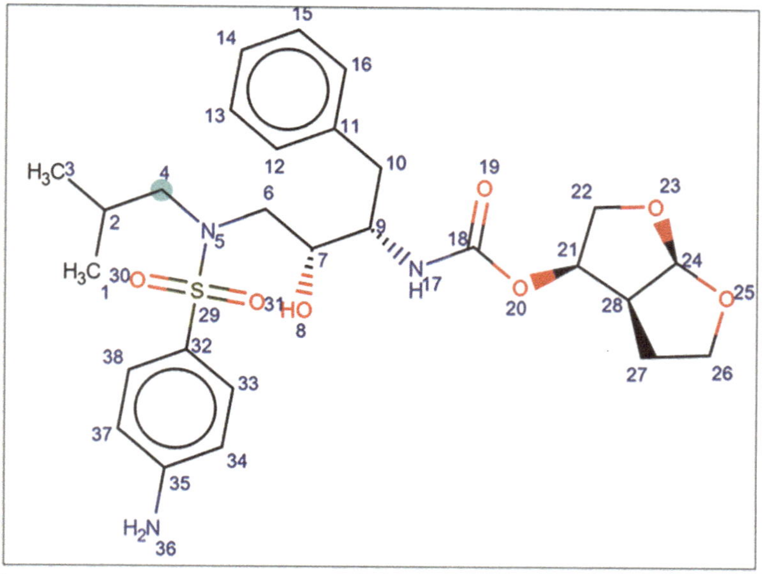

^{13}C NMR spectrum of darunavir was scanned in DMSO-d_6 on a Brucker NMR spectrometer (Bruker Co., Cremlingen, Germany) operating at 125 MHz. Chemical shifts are expressed in δ-values (ppm) relative to TMS as an internal standard (Fig. 7 and Table 2).

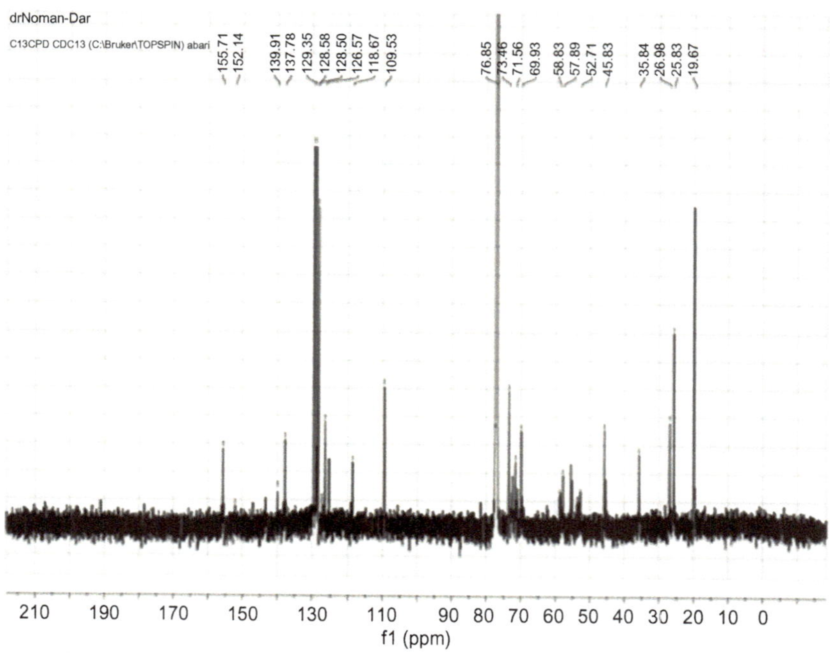

Fig. 7 ^{13}C NMR spectrum of darunavir.

Table 2 ^{13}C NMR of darunavir (DMSO-d_6).

Signal	Atom number	ppm	Signal	Atom number	ppm
1	1,3	19.67	13	21	76.85
2	27	25.83	14	24	109.53
3	2	26.98	15	34, 37	118.67
4	10	35.84	16	14	126.57
5	28	45.83	19	33, 38	128.50
6	6	52.71	20	13, 15	128.58
7	9	57.89	21	12, 16	129.35
8	4	58.83	22	11	137.78
9	26	69.93	23	32	139.91
10	22	71.56	24	35	152.14
11	7	73.46	25	18	155.71
13	21	76.85			

4.5.4 Mass spectrometry

The mass spectrum of darunavir was obtained using an Agilent 6320 ion trap mass spectrometer (Agilent technologies, Santa Clara, CA, USA) equipped with an electrospray ionization interface. A connector was used instead of column. A mobile phase composed of acetonitrile: water (50,50, v/v) was used. Darunavir sample stock solution (1 mg/mL) was prepared in methanol and diluted with the mobile phase. Test solution was prepared by diluting the stock solutions to 10–30 μg/mL depending on the ions intensities—with mobile phase. Flow rate was 0.4 mL/min and the run time was 5 min. Mass parameters were optimized for each compound. The scan was ultra-scan mode. MS2 scans were performed in the mass range of m/z 50–600. The electrospray ionization was operated in positive mode. The source temperature was set to 350 °C nebulizer gas pressure of 55.00 psi; dry gas flow rate of 12.00 L/min. Fig. 8 shows the molecular ion peak at $m/z = 548.1$ $[M+1]^+$, minor peak of molecular ion sodium salt 571.1 $[M+Na]^+$ and showed a major ion peak at m/z 586.0 $[M+K]^+$. The electrospray ionization negative ion mode was also conducted; the important fragments (m/z) are at 113.2, 153.1, 202.0, 241.3, 392.3, and 385 (Fig. 9).

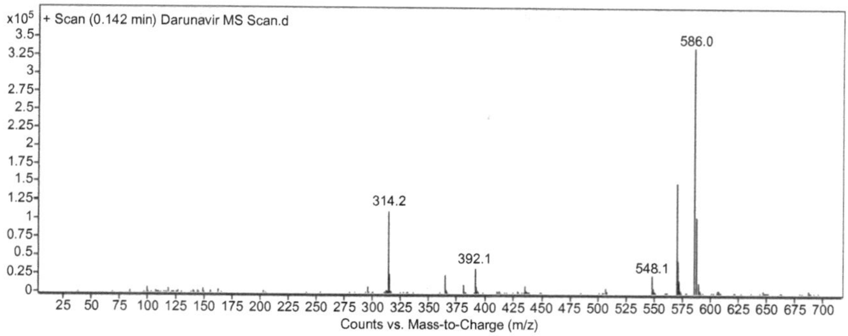

Fig. 8 Positive ion spray mass spectrum of darunavir, parent ion at *m/z* of 548.1.

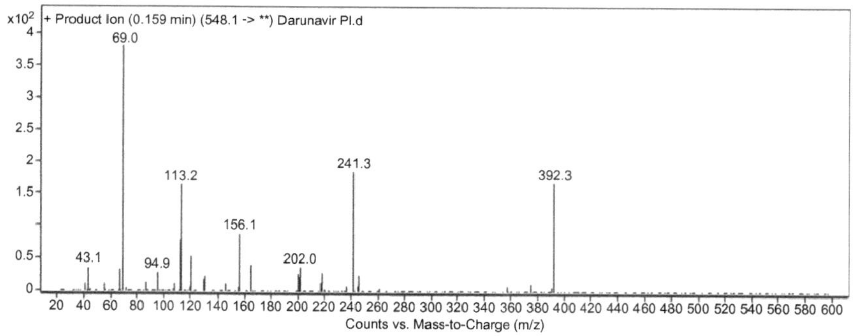

Fig. 9 Positive ion spray mass spectrum of darunavir.

5. Methods of analysis

5.1 Compendial method

Darunavir has no compendial method for its analysis.

5.2 Spectroscopic methods

5.2.1 Infrared spectroscopy

Kogawa and Salgado [19] developed an infrared spectroscopic method for the quantitation of darunavir in prezista® tablets. The method involved analyzing the spectra of carbonyl band between 1757 and 1671 cm^{-1}. The method had linear range of 1.5–3.5 mg/mL with correlation coefficients greater than 0.9991. The limit of detection (LOD) and limit of quantification (LOQ) were 0.12 and 0.36 mg, respectively.

5.2.2 Ultraviolet-visible spectrophotometry

Abdulsaleem et al. [20] developed and validated an UV spectrophotometric method for estimation of darunavir in bulk drug and tablet formulations with good accuracy and precision. An absorption maximum was determined at 254 nm. Beer's law was obeyed in the concentration range of 10–35 μg/mL with correlation coefficient of 0.998. The percent assay for darunavir ranged from 98.4–102.1% in tablet dosage form. The developed method could be used for routine analysis of darunavir in pharmaceutical formulations.

Mastanamma et al. [21] developed three UV spectrophotometric methods (A, B and C) for determination of darunavir in bulk and pharmaceutical dosage form using hydrotropic solubilization technique (8 M urea). Solubility of darunavir increased by using 8 M urea as a hydrotropic agent. Daurnavir1showed the maximum absorbance at 263.91 nm in method A, 254–274 nm in method B and 264 nm in method C. At these wavelengths, hydrotropic agent and other tablet excipients did not show any significant interference in the spectrophotometric assay. The developed methods were found to be linear in the range of 5–40 μg/mL for method A, method B and C with correlation coefficients of 0.997, 0.995 and 0.997, respectively. The mean percent label claim of tablets of darunavir in formulation estimated by the methods was found to be 107%. The methods were validated according to the International Conference on Harmonization (ICH) guidelines and values of accuracy, precision and other statistical parameters were found to be good accordance with the prescribe values. As hydrotropic agent was used in the proposed methods, these methods were eco-friendly, and they could be used in routine quantitative analysis of darunavir in bulk and dosage form in industries.

Fathima et al. [22] developed and validated two simple, accurate, precise, reproducible and economical UV spectroscopic methods (A and B) for simultaneous estimation of darunavir and rilpivirine HCl in tablet formulation. Method A employed solving of simultaneous equations based on the measurement of absorbance at two wavelengths, 265 and 290 nm which were the λ_{max} values of darunavir and rilpivirine respectively. Darunavir and rilpivirine HCl showed linearity at all the selected wavelengths and obeyed Beer's law in the concentration range of 10–35 and 10–80 μg/mL, respectively. Recovery studies for darunavir and rilpivirine HCl were performed and the percentage recovery for both the drugs was obtained in the range of 98.1–99.7% (method A) and 98.0–100.4% (method B) confirming the accuracy of the methods. Both methods showed good reproducibility and recovery with relative standard deviations (RSD) values of less than 2%.

Statistical validation of the data showed that the methods could be successfully applied for the routine analysis of drugs in commercial tablets.

Eswarudu et al. [23] developed two simple, precise and economical UV spectrophotometric methods (I and II) for the estimation of darunavir ethanolate in bulk and its pharmaceutical dosage form. The two methods were developed based on measurement of absorption at maximum wavelengths of 272.1 and 272.4 nm for methods I and II, respectively. Linearity was observed in the concentration range of 2–10 μg/mL for the both methods. The methods were validated with respect to linearity, accuracy (recovery), precision and specificity. The accuracy of the methods was assessed by recovery studies and was found to be 98.3% and 99% for methods I and II, respectively. The results were validated statistically as per ICH Q2 R1 guidelines and were found to be satisfactory. The methods were successfully applied for the determination of darunavir in pharmaceutical dosage form.

Rao et al. [24] developed and validated a simple, accurate, rapid, precise and economical spectrophotometric method for estimation of darunavir ethanolate in pharmaceutical dosage form. Darunavir ethanolate exhibits λ_{max} at 461 nm with 3- methtyl-2-benzothiazolinone hydrazone hydrochloride (MBTH) in presence of ferric chloride as oxidizing agent. It obeyed linearity in concentrations rang of 10–50 μg/mL with correlation coefficient of 0.9995.

Mrinalini and Ajinkya [25] developed two simple and rapid UV spectrophotometric methods for the estimation of darunavir and ritonavir in combination dosage form. The first method (absorbance ratio method) was based on measurement of the absorption at a maximum absorption wavelength of 239 nm and the isosbestic point at 251 nm. The second method (absorbance correction method) involved the measurement of darunavir at 267 nm. The calibration curve was linear in the range of 10–50 μg/mL for both drugs in both methods.

Ghante et al. [26] developed two simple UV spectrophotometric methods for the estimation of darunavir ethanolate in bulk and tablet dosage form. The first method (absorbance maxima method) was based on measurement of the absorption at a maximum wavelength of 266 nm. The second method (area under the curve method) was based on measurement of the area under curve in the wavelength range of 255–275 nm. Linearity was observed in the concentration range of 3–18 μg/mL for both methods. The accuracies of the methods were 100.07% and 99.58% for the first and second methods, respectively.

Vanukuri et al. [27] developed and validated three spectrophotometric methods for the quantification of darunavir in its bulk and dosage form.

These methods were: absorption maximum, first order derivative, and area under the curve. Linearity for all three methods was found in the range of 2–24 μg/mL ($r^2 = 0.999$). Tablet formulation was analyzed and the % assay for absorption max, first order derivative and area under curve methods were found to be 100.72%, 99.09% and 99.06% respectively. These methods were validated as per the International Conference on Harmonization (ICH) guidelines. Validation studies demonstrated that proposed method is simple, precise, accurate, specific, rapid, reliable and reproducible.

Corrêa et al. [28] developed and validated a first derivative spectrophotometric method for quantification of darunavir in tablets. The derivative spectrophotometric analysis was carried out to eliminate the high interference of the tablet excipients that occurred in determination of darunavir by a direct UV absorption measurement. The method involved the measurement of absorbance at 276 nm and the first-derivation of spectrum of the drug was measured between 200 and 400 nm. Beer's law was obeyed in the concentration range of 11–21 μg/mL. The intra-day and inter-day precisions, expressed as relative standard deviations (RSD), were 0.06% and 3.75%, respectively with mean recovery of 99.84%.

The method was able to quantify darunavir as raw material and tablets and can be used routinely by any laboratory applying a spectrophotometer with a derivative accessory. The method was free of placebo interferences as well as simple, fast and low cost.

Nimje et al. [29] developed two sensitive spectrophotometric methods (simultaneous equation method) and (Q-analysis method) for simultaneous determination of darunavir ethanolate and ritonavir in combined dosage form. The first method involved the measurement of absorption at 267.50 nm for darunavir ethanolate. The second method involved the use of absorbance equation at 244 nm (isoabsorptive point) and at 267.50 nm (the maximum absorption of darunavir ethanolate). The linearity was found to be 8–35 μg/mL for darunavir ethanolate and 5–70 μg/mL for ritonavir by both methods. The recovery values of the first method in tablet were found to be $102.99 \pm 1.22\%$ and $101.79 \pm 1.46\%$ for darunavir ethanolate and ritonavir, respectively. For the second method, the recovery values were found to be $101.93 \pm 0.79\%$ and 99.04 ± 0.98 for darunavir ethanolate and ritonavir, respectively. Both the methods were validated as per ICH guidelines.

Rao [30] developed an accurate and economical spectrophotometric method for determination of darunavir in pure and pharmaceutical dosage forms. The method was based on the formation of colored charge-transfer complex between darunavir as an *n*-electron donor and *p*-chloranilic acid as an electron acceptor. The absorbance of the colored complex was measured at 528 nm. The linear range of the method was 10–50 μg/mL. The values obtained by this method and reference method for formulations were compared statistically with F- and *t*-tests and found not to be different significantly. This method has been extended to pharmaceutical formulation containing darunavir.

Rao [31] developed six simple, reliable and economical spectrophotometric methods were developed for quantification of darunavir in pure and dosage forms. The methods were based on the formation of highly stable colored products upon the reaction of darunavir via its amino group with resorcinol, phloroglucinol, *p*-dimethylaminobenzaldehyde, *p*-dimethylaminocinnam aldehyde, and vanillin. The colored products were measured at 600, 510, 482, 350, and 440 nm for the reactions with resorcinol, phloroglucinol, *p*-dimethylaminobenzaldehyde, *p*-dimethylaminocinnamaldehyde, and vanillin, respectively. An excellent correlation was obtained between the absorbance and the concentrations of darunavir in linear ranges of 8–62.5 μg/mL and the values of limits of detection (LOD) were in the range of 0.013–0.032 μg/mL.

Vijayalakshmi et al. [32] developed two simple and accurate colorimetric methods for the estimation of darunavir in bulk and pharmaceutical dosage form via its condensation with *p*-dimethylaminobenzaldehyde (method A) and vanillin (method B). These condensation reactions formed colored products (Schiff's bases) measured at 452 and 406 nm for methods A and B, respectively. The linear ranges of the methods were 50–350 and 50–300 μg/mL for methods A and B, respectively. The LOD values were 6.24 and 18.93 μg/mL for methods A and B, respectively. The limits of quantitation (LOQ) values were 18.93 and 13.04 μg/mL for methods A and B, respectively. The methods were extensively validated as per ICH guidelines and all the parameters were within the acceptance criteria with a correlation of 0.9998 and 0.9999 and the relative standard deviation (RSD) was less than 2%. The results of the accuracy studies were nearer to 100%. The methods were proven to be more accurate, simple, precise and rapid by statistical validation.

Reddy and Ramireddy [33] developed and validated a spectrophotometric method for determination of darunavir. The method was based on bromination of darunavir with an excess of brominating mixture (potassium bromate and potassium bromide) in acidic medium yielding a yellow colored product measured at 350 nm. Beer's law was obeyed in the concentration range of 40–200 μg/mL. The proposed methods were simple, rapid, and validated and can be used successfully for routine analysis of darunavir in a pure and tablet dosage form.

Vijayalakshmi et al. [34] developed three simple and extraction free visible methods (A, B and C) for the estimation of darunavir ethanolate in bulk and tablet formulations. The methods involved diazotization of darunavir with nitrous acid followed by coupling with chromotropic acid (method A), Bratton Marshall reagent (method B) and α-naphthol (method C) to form colored products. These colored products were reddish-pink, dark violet and dark-greenish yellow with chromotropic acid, Bratton Marshall reagent and α-naphthol, respectively. These products were measured at 520, 544 and 464 nm for methods A, B and C, respectively. The linear relationship was observed between absorbance and the corresponding concentrations of darunavir in the range of 100–350, 10–100 and 10–60 μg/mL for methods A, B, and C, respectively. These methods were extensively validated as per the ICH guidelines. The developed methods were proven to be more accurate and precise by the statistical analysis.

Vijayalakshmi et al. [35] developed two simple, fast and extractive colorimetric methods (A and B) for the estimation of darunavir ethanolate in both bulk and tablet formulations using bromo cresol green (method A) and bromothymol blue (method B) reagents. The reactions gave yellow colored products measured at 418 and 411 nm for methods A and B, respectively. The linear relationship was observed between absorbance and the corresponding concentrations of drug in the range of 20–140 and 40–140 μg/mL for methods A and B, respectively. The methods were extensively validated as per ICH guidelines and all the parameters were within the acceptance criteria, with the correlation coefficient of 0.9999 and relative standard deviation values of less than 2% for both the methods. The results of the accuracy studies were nearer to 100%. The methods were proved to be more accurate, simple, precise and rapid by statistical validation as well as recovery studies and could be used for routine analysis.

Godambe et al. [36] developed a visible spectrophotometric method for estimation of darunavir using QBD approach and coupling agent O-phthalaldehyde reagent. The method was applied by varying different parameters. The reaction product of darunavir-phthalaldehyde in methanol with 0.1 N HCl showed linearity in a concentration range of 2–22 μg/mL with regression coefficient (r^2) = 0.998 at 355 nm. This method was found to be rugged and robust in different testing criteria with RSD values of less than 2%. The limit of detection and limit of quantification were found to be 0.2 and 0.8 μg/mL, respectively. The method was found to be precise and the percent recovery was 101.04%.

Das and Vidyasagar [37] developed and validated a simple and rapid spectrophotometric method for the determination of darunavir and ropivacain in tablets and as bulk drug. The method was based on the reaction of the drugs with 1,2-naphthoquinone-4-sulphonate reagent. Darunavir was solubilized in an alkaline medium with the reagent to form an orange-colored product with λ_{max} at 488 nm while ropivacain gave blue colored chromogen with λ_{max} at 565 nm. The chromogen obeyed Beer's law in the concentration range of 5–60 μg/mL. The results of the analysis have been validated statistically and by recovery studies.

5.2.3 Spectrofluorometry

Godambe et al. [36] developed spectrofluorometric method for estimation of darunavir using QBD approach and coupling agent *O*-pthaladehyde reagent. The method was applied by varying different parameters. The reaction product of darunavir-phthalaldehyde in methanol with 0.1 N HCl showed linearity in a concentration range of 0.5–5 ng/mL with regression coefficient (r^2) = 0.999. This method was found to be rugged and robust in different testing criteria with relative standard deviation values of less than 2%. The limits of detection and quantification were found to be 0.12 and 0.43 μg/mL, respectively. The method was found to be precise and the percent recovery was 98.15%.

5.3 Chromatographic methods

5.3.1 Capillary electrophoresis

Kogawa et al. [38] developed and validated a stability-indicating capillary electrophoretic method for the determination of darunavir in tablet and compared the method with the infrared absorption spectroscopic method. The electrophoretic separation was achieved in less than 1 min using fused uncoated silica capillary (internal diameter of 50 μm and total length of 21 cm), voltage of +20 kV, and 25 mM sodium borate buffer of pH 8.5 as a running electrolyte. The capillary electrophoresis behavior was performed at ambient temperature (25 °C), and the sample was injected by the hydrodynamic mode. The detector wavelength was adjusted at 200 nm. The method had linear concentration range of 50–200 μg/mL with correlation coefficient of 0.9998. The LOD and LOQ of the method were 7.29 and 22.09 μg/mL, respectively. The drug was subjected to acid, base, oxidation and photolysis degradation. Electrophoretic separation of degradation products indicated the stability-indicating ability of the method. The validated method was useful and appropriate for the routine quality control of darunavir in tablets.

Leonard et al. [39] developed a capillary electrophoretic method for the separation of diastereoisomers of darunavir. In total 16 isomers of darunavir have been synthesized (8 pairs of enantiomers). The eight diastereoisomers, but no enantiomers could be separated. Because of the high similarity and water-insolubility of these isomers, the separation was a real challenge. Different capillary electrophoresis modes were tried out: capillary zone electrophoresis, nonaqueous capillary electrophoresis, micellar electrokinetic capillary chromatography, and microemulsion electrokinetic capillary chromatography. Only microemulsion electrokinetic capillary chromatography offered resolution of these compounds.

5.3.2 Thin layer chromatography

Kogawa and Salgado [40] devolved a qualitative thin layer chromatographic (TLC) method for the analysis of darunavir tablets. The elution was performed on silica gel plates 60F using mobile phase consisting of methanol: water (70:30, v/v) adjusted to pH 3.0 with glacial acetic acid. The spots were localized by UV detection at 365 nm.

Kogawa et al. [41] developed a stability-indicating thin layer chromatographic (TLC) method for determination of darunavir in complex darunavir-β-cyclodextrin in the presence of its degradation products. The method employed aluminum plates precoated with silica gel 60F-254 as the stationary phase and purified water: methanol (70:30, v/v) adjusted to pH 2.4 with glacial acetic acid as the solvent system to provide spots for darunavir (Rf=0.66) and its degradation products in acidic (Rf=0.73 and 0.76), basic (Rf=0.53) and oxidative (Rf=0.71, 0,75, and 0.84) media. The chromatogram was visualized in an UV chamber at 365 nm. High performance liquid chromatography (HPLC) analysis was performed on a Waters HPLC system, Phenomenex CN Luna (250 × 4.6 mm) column and mobile phase consisting of water + 0.1% glacial acetic acid and acetonitrile +0.1% glacial acetic acid in the ratio 60:40 (v/v) at a flow rate of 1.0 mL/min and 268 nm for the separation of darunavir (retention time = 7.3 min) and its degradation products in acidic (retention time = 5.1 and 6.7 min), basic (retention time = 7.8 min) and oxidative (retention time = 5.1, 5.5, and 6.7 min) media. Darunavir-β-cyclodextrin was subjected to acid and alkali hydrolysis and oxidation and analyzed by the proposed methods in the presence of its degradation products, which were identified by liquid chromatography with mass detector. As the methods could separate darunavir from the degradation products, these techniques can be employed as indicative stability methods and can be effectively applied in quality control of darunavir complexed to β-cyclodextrin.

Patel et al. [42] developed and validated a quantitative high performance thin layer chromatography (HPTLC) method for estimation of darunavir ethanolate in tablets. The separation was achieved on silica gel 60F254 HPTLC plates with toluene:ethylacetate:methanol (7.0:2.0:1.0, v/v) as mobile phase. The plates were developed to a distance of 8 cm and the quantification was performed at a wavelength of 267 nm. The calibration curve was linear over a range of 250–1750 ng/band with correlation coefficient of 0.9994. The slope and intercept values were 0.4253 and 44.81 ng/band, respectively. The LOD and LOQ were 15.28 and 45.84 ng/band, respectively. The method was selective, sensitive, and specific, with potential application in pharmaceutical analysis.

Ramesh et al. [43] developed and validated a rapid and sensitive high performance thin layer chromatography coupled with electrospray ionization mass spectrometry (HPTLC-ESI/MS) for the quantification of darunavir in rat plasma and its application in pharmacokinetic studies. Darunavir was extracted from plasma using protein precipitation method. A mixture of toluene:acetone:methanol (6:2:2, v/v) was used as the mobile phase. Densitometric quantification of darunavir was performed at 262 nm by reflectance scanning and its mass ion was at m/z 569.80 $[M+Na]^+$ being acquired directly from the rat plasma sample bands spiked with darunavir by an elution-based interface. The method was linear in the range of 5–150 ng/μL. The LOD and LOQ were 1.24 and 3.85 ng/μL, respectively. The intra-day and inter-day precisions, expressed as relative standard deviation (RSD), were in the range of 2.15–2.77 ng/mL (n=3) and 2.13–2.47 ng/mL (n=3), respectively. Additionally, selectivity of the method was confirmed by mass spectrometry. The mass spectra showed darunavir ion at m/z 569.80 $[M+Na]^+$ being acquired directly from the rat plasma sample bands spiked with darunavir by an elution-based interface. The method was applied for the determination of plasma levels as well as pharmacokinetic study of darunavir administered orally to rats.

5.3.3 Liquid chromatographic methods

Numerous liquid chromatographic methods were developed for determination of darunavir in raw bulk material, dosage forms, and biological fluids. The methods reported before 2014 [44–66] have been reviewed and described by Kogawa and Salgado [67], Eswarudu et al. [68], Kashish et al. [69] and Herqash [70]. The other methods and those reported thereafter were reviewed [71–94] and summarized in Table 3 (methods for determination of darunavir in raw material and dosage form) and Table 4

Table 3 Liquid chromatographic methods reported after 2014 for determination of darunavir in raw material and/or dosage forms.[a]

Matrix	Separation conditions	Detection	Working range, precision; accuracy, LOD, LOQ, LLOQ	References
Raw material and tablets	Column: Shiseido C8 (250 × 4.6 mm, 5 μm); elution: isocratic; mobile phase: acetonitrile: phosphate buffer, pH 3 (60:40, v/v); flow rate: 1.0 mL/min; retention time: 4.30 min.	UV: 270 nm	Range: 5–15 μg/mL; 0.22% (RSD repeatability); 99.46–100.17% (recovery of tablet)	[71]
Raw material and tablets	Column: Phenomenex C18 (250 mm × 4.6 mm, 5 μm); elution: isocratic; mobile phase: acetonitrile: water (80:20, v/v); flow rate: 1.0 mL/min; retention time: 3.22 min	UV: 267 nm	Range: 5–35 μg/mL; 1.77% (RSD of intra-day); 0.46% (RSD of inter-day); 98–102% (recovery of tablet); 0.65 μg/mL (LOD); 1.98 μg/mL (LOQ)	[72]
Raw material and tablets	Column: Sunshell C18 (100 × 4.6 mm, 2.6 μm); elution: isocratic; mobile phase: water: acetonitrile (60:40, v/v); flow rate: 1.0 mL/min; retention time: 2.4 min	UV: 265 nm	Range: 1–25 mg/mL; 0.34% (RSD repeatability); 98.5–102% (recovery of tablet)	[73]
Raw material and tablets	Column: Enable C18 (250 × 4.6 mm, 5 μm), elution: isocratic; mobile phase: acetonitrile:0.01 M potassium acetate buffer, pH 5.1 (75:25, v/v); flow rate: 1 mL/min; retention time: 3.85 min	UV: 268 nm	Range: 40–90 μg/mL; 0.1686% (RSD of intra-day); 0.5966% (RSD of inter-day); 96.92–101% (recovery of tablet); 0.234 μg/mL (LOD); 0.734 μg/mL (LOQ)	[74]

Table 3 Liquid chromatographic methods reported after 2014 for determination of darunavir in raw material and/or dosage forms.[a]—cont'd

Matrix	Separation conditions	Detection	Working range, precision; accuracy, LOD, LOQ, LLOQ	References
Raw material and tablets	Column: C18 (250 × 4.6 mm, 5 μm), elution: isocratic; mobile phase: water: methanol, pH 3 (30:70, v/v); flow rate: 1 mL/min; retention time: 5.2 min	UV: 262 nm	Range: 5–50 μg/mL; 0.69% (RSD of intra-day); 1.3% (RSD of inter-day); 98.2–101.2% (recovery of tablet); 0.48 μg/mL (LOD); 1.5 μg/mL (LOQ)	[75]
Raw material and tablets	Column: Zodiac C18 (250 × 4.6 mm, 5 μm), elution: isocratic; mobile phase: triethylamine buffer: acetonitrile, pH 4.5 (60:40, v/v); flow rate: 1 mL/min; retention time: 2.753 min	UV: 260 nm	Range: 20–100 μg/mL; 0.889265% (RSD of intra-day); 0.8892% (RSD of inter-day); 99.71–100.01% (recovery); 2.32 μg/mL (LOD); 7.06 μg/mL (LOQ)	[76]
Raw material and tablets	Column: Kromosil C18 (250 mm × 4.6 mm, 5 μm), elution: isocratic; mobile phase: acetonitrile: methanol (90:10, v/v); flow rate: 1.0 mL/min; retention time: 2.59 min	UV: 271 nm	Range: 25–100 μg/mL; 0.48% (RSD repeatability); 99.83–100.2% (recovery); 0.64 μg/mL (LOD); 0.20 μg/mL (LOQ)	[77]
Combined dosage form	Column: Phenomenex C18 (150 × 4.6 mm, 5 μm), elution: isocratic; mobile phase: 0.1 M NaH_2PO_4:methanol (70:30, v/v); flow rate: 1.0 mL/min; retention time: 5.2 min	UV: 260 nm	Range: 80–240 μg/mL; 0.245% (RSD repeatability); 100.10–100.31% (recovery); 0.220 μg/mL (LOD); 0.733 μg/mL (LOQ)	[78]

Continued

Table 3 Liquid chromatographic methods reported after 2014 for determination of darunavir in raw material and/or dosage forms.[a]—cont'd

Matrix	Separation conditions	Detection	Working range, precision; accuracy, LOD, LOQ, LLOQ	References
Combined dosage form	Column: Tracer Excel 120 ODSB (15 × 0.4.6 cm), elution: isocratic; mobile phase: 0.037 M sodium dihydrogen phosphate buffer: acetonitrile:methanol (40:50:10, v/v/v); flow rate: 2.0 mL/min	UV: 254 nm	Range: 5–100 μg/mL; 0.25% (RSD of intra-day); 4.42% (RSD of inter-day); 4.33–3.88% (accuracy)	[79]
Combined dosage form	Column: ODS-3 V (250 × 4.6 mm, 5 μm), elution: gradient: A: potassium dihydrogen phosphate, B: acetonitrile: methanol (450:150, v/v); flow rate: 1.0 mL/min; retention time: 2.385 min	UV: 260 nm	Range: 72–216 μg/mL; 0.69% (RSD repeatability); 99.86–100.01% (recovery)	[80]
API	Column: Zodiac C18 (150 × 4.6 mm); elution: isocratic; mobile phase: acetonitrile: water (80:20, v/v); flow rate: 0.8 mL/min; retention time: 2.3 min	PDA: 290 nm	Range: 10–50 μg/mL; mean recovery: 99.13%	[81]
Combined dosage form	Column: Durashell C18 (250 × 4.6 mm, 5 μm); mobile phase: 0.1% orthophosphoric acid: methanol:acetonitrile (60:10:30, v/v); flow rate: 1 mL/min; retention time: 4.35 min	UV: 210 nm	Range: 103–149 μg/mL; LOD: 2.46 μg/mL; LOQ: 7.38 μg/mL	[82]

Table 3 Liquid chromatographic methods reported after 2014 for determination of darunavir in raw material and/or dosage forms.[a]—cont'd

Matrix	Separation conditions	Detection	Working range, precision; accuracy, LOD, LOQ, LLOQ	References
Bulk and dosage forms	Column: Symmetry C18; elution: isocratic; mobile phase: phosphate buffer (pH 2.5): acetonitrile (70:30, v/v); flow rate: 1.5 mL/min	UV: 255 nm	Range: 5.0–25 μg/mL; recovery: 99.22–99.83%	[83]
Dosage form	Column: BDS (250 × 4.6 mm, 5 μm); mobile phase: buffer: acetonitrie (40:50, v/v); flow rate: 1 mL/min; retention time: 3.984 min	UV: 210 nm	Recovery: 100.2%; LOD: 0.326 ppm; LOQ: 1.009 ppm	[84]
Dosage forms	Column: Xterra C18 (250 × 4.6 mm, 5 μm); mobile phase: phosphate buffer (0.05 M) pH4.6: acetonitrile (55:45, v/v); flow rate: 1 mL/min	PDA: 255 nm	Range: 100–500 μg; recovery: 100.5%; repeatability RSD: intermediate precision RSD: 0.1%	[85]
Bulk and dosage forms	Column: Phenomenex C18 (250 × 4.6 mm, 5 μm); mobile phase: acetonitrile: water (80:20, v/v); flow rate: 1.0 mL/min	UV: 267 nm	Range: 5–35 μg/mL; label claim: 99.15–101.88%; precision RSD: <2%	[72]

[a]Abbreviations are:
API: Authentic pharmaceutical ingredient UV: ultraviolet; PDA: photodiode array; LOD: limit of detection; LOQ: limit of quantitation; RSD: relative standard deviation.

(methods for determination of darunavir in biological matrices). Different stability-indicating liquid chromatographic methods were reported for determination of darunavir and its degradation products or impurities [98,102–110]. These methods are summarized in Table 5.

Table 4 Liquid chromatographic methods reported after 2014 for determination of darunavir in biological matrices.[a]

Matrix	Separation conditions	Detection	Working range, precision; accuracy, LOD, LOQ, LLOQ	References
Human plasma	Column: Kromasil C18 (150 mm × 4.6 mm, 5 μm), elution:isocratic:0.06 M sodium dodecyl sulfate/2.5% 1-pentanol, pH 7; flow rate: 1 mL/min; retention time: 8.2 min	UV (214 nm)	Range: 0.25–25 μg/mL; 4.6% (RSD of intra-day); 4.2% (RSD of inter-day); 89.3–103.2% (accuracy); 0.090 μg/mL (LOD); 0.250 μg/mL (LOQ)	[86]
Human plasma	Column: C18 (250 × 4.6 mm, 5 μm), elution: isocratic: acetonitrile: water (60:40, v/v); flow rate: 1 mL/min	UV (267 nm)	Range: 0.6–19.2 μg/mL; >15% (RSD repeatability); 0.6 μg/mL (LLOQ)	[87]
Human plasma	Column: UPLC C18 (50 × 2.1 mm, 1.7 μm), elution: gradient: A: 10 Mm ammonium formate, B: acetonitrile; flow rate: 0.300 mL/min; retention time: 1.02 min	MS/MS: MRM, positive ionization $[M\text{-}H]^+$	Range: 5.0–5000 ng/mL; 1.76% (RSD repeatability); 98.82–100.86% (accuracy)	[88]
Human plasma	Column: BEH C18 (50 mm × 2.1 mm, 1.7 μm), elution: gradient: A: acetonitrile/ methanol (80:20, v/v), B: (5.0 mM ammonium acetate +0.01% formic acid); flow rate: 0.4 mL/min; retention time: 1.46 min	MS/MS: TQ-MRM, ESI positive ionization $[M\text{-}H]^+$	Range: 20–12,000 ng/mL; 4.16% (RSD of intra-day); 3.61% (RSD of inter-day); 90.3–100.3% (accuracy); 20 ng/mL (LLOQ)	[89]

Human plasma and saliva	Column: C18 (1.5 × 50 mm, 5 μm), elution: isocratic:5 mM formic acid-35% (v/v) acetonitrile; flow rate: 0.2 mL/min; retention time: 4–4.5 min	MS/MS: TQ-MRM, ESI positive ionization $[M\text{-}H]^+$	Range: 1–10,000 ng/mL; 1.2–13.1% (RSD repeatability); −14.5–18.1% (accuracy); 1 ng/mL (LLOQ)	[90]
Rat plasma	Column: Luna-HILIC (1.5 × 50 mm, 5 μm), elution: isocratic: 0.1% of formic acid in water: acetonitrile (5:95, v/v); flow rate: 1 mL/min; retention time: 3.3 min	MS/MS ion-trap-SL: MRM, ESI positive ionization $[M\text{-}H]^+$	Range: 0.2–5000 ng/mL; 0.38–8.66% (intra- and inter-RSD %); 95.0–107.6% (accuracy); 0.2 ng/mL (LLOQ)	[91]
Rat serum and urine	Column: Agilent C18 (250 × 4.6 mm, 5 μm), elution: isocratic:20 mM ammonium acetate: methanol, pH 3.5 (40:60, v/v); flow rate: 1 mL/min; retention time: 9.74 min	MS-Q-TOF: ESI positive ionization $[M\text{-}H]^+$	Range: 5–5000 ng/mL; 2.54–8.92% (RSD repeatability); 0–5% (accuracy); 3.63–5.24 ng/mL (LLOQ)	[92]
Human mononuclear cell extracts	Column: UHPLC ACE C18 (3 × 100 mm), elution: isocratic: acetonitrile:water:formic acid (60:40:0.1, v/v/v); flow rate: 500 μL/min	MS/MS: TQ-MRM, ESI positive and negative ion modes	Range: 0.05–25.0 fmol/μL; 8.9% (RSD repeatability); 0.0500 fmol/μL (LLOQ)	[93]
Peripheral blood mononuclear cells	Column: XBridge™ C18 (2.1 × 75 mm, 2.5 μm), elution: gradient: A: water, B: methanol, (both containing 10 mM formic acid); flow rate: 1 mL/min; retention time: 10.2 min	MS/MS: Quattro micro, MRM, *Z*-spray positive ion mode	Range: 1.25–125 ng/mL; 4.7% (RSD of intra-day); 5.8% (RSD of inter-day); 80–120% (recovery); 0.60 ng/mL (LOD); 1.25 ng/mL (LLOQ)	[94]

Continued

Table 4 Liquid chromatographic methods reported after 2014 for determination of darunavir in biological matrices.[a]—cont'd

Matrix	Separation conditions	Detection	Working range, precision; accuracy, LOD, LOQ, LLOQ	References
Rat plasma	Column: Agilent C18 (150 × 4.6 mm, 5 μm); elution: isocratic; mobile phase: acetonitrile:triethylamine:0.025 M potassium buffer pH 2.3 (70:30, v/v); flow rate 1.0 mL/min; retention time: 2.24 min	PDA: 210 nm	Range: 1.5–3.5 μg/mL; LOD: 0.06 μg/mL; LOQ: 0.193 μg/mL. Inter-day and intra-day RSD: 2.07% and 2.001%, respectively; recovery: >90%	[95]
Human plasma	Column: Kinetex phenyl-hexyl analytical column (100 × 3 mm; 5 μm); mobile phase: water-0.05% formic acid:methanol-0.05% formic acid (55:45, v/v); flow rate: 0.5 mL/min; retention time: 6.1 min	MS/MS: TQ-MRM, Heated-ESI; *m/z*: 548.156 → 392.190	Range: 60–15,000 ng/mL; LLOQ: 60 ng/mL; precision (inter- and intra-day) RSD: <12.3%; accuracy (inter- and intra-day): −9.9% and 10%	[96]
Human plasma	Column: Acclaim RSLC C18 (2.1 × 100 mm, 2.2 μm); elution: gradient; mobile phase A and B consisted of 0.1% (v/v) formic acid in water and acetonitrile, respectively; flow rate: 0.45 mL/min; retention time: 3.63 min	MS/MS: TSQ ALTIS-ESI, positive mode; *m/z*: *m/z*: 548.2 → 392.1	Range: 0–5000 ng/mL; LLOQ: 2.5 ng/mL; intra- and inter-day precision RSD: 1.5% and 4.9%; accuracy: 0.1–5.1%	[97]
Blood plasma	XBridge C18 (150 × 4.6 mm, 5 μm); mobile phase:K_2HPO_4 buffer:methanol:acetonitrile (488:162:350, v/v); flow rate: 1.3 mL/min; retention time: ~ 14 min	UV: 262 nm	Range: 20–60 μg/mL; precision and intermediate precision RSD: 0.33 and 0.22, respectively; recovery: 99.81–100.82%	[98]

In vitro intestinal fluid	Column: Princeton sphere C18 (250 × 4.6 mm, 5 μm) and Unisphere C18 (250 × 4.6 mm, 5 μm); elution: isocratic; mobile phase: 20 mM KH_2PO_4, pH 3: acetonitrile (48:52, v/v); flow rate: 1 mL/min; retention time: 6.28 min	PDA: 266 nm	Range: 1.84–118 μg/mL; precision RSD: <15%; recover: ≥96.14%; LOD: 1.04 μg/mL; LOQ: 3.15 μg/mL	[99]
Human peripheral blood mononuclear cells	Column: YMC-Pack Pro ODS C18 (250 × 4.6 mm, 5 μm); mobile phase: 20 mM potassium phosphate buffer (pH 4.3):acetonitrile (57/43, v/v); flow rate: 1.0 mL/min; retention time: 13.8 min	FD: 235 nm for excitation and 337 nm for emission	Range: 5–100 ng/10^6 cells; LLOQ: 5 ng/10^6 cells; intra- and inter-assay precision and accuracy were <15%.	[100]
Human plasma	Column: Hypurity C18 (250 × 4.6 mm, 5 μm); mobile phase: 5 mM ammonium acetate-0.1% formic acid: acetonitrile (75:25, v/v); flow rate: 0.7 mL/min; retention time: 1.2 min	MS/MS: ESI, positive mode; MRM: *m/z* 548.1 → 194.1	Range: 20–3501 ng/mL; LLOQ: 20.2 ng/mL; precision and accuracy at LLOQ: 7.65% and 104%, respectively.	[101]

[a]Abbreviations are:

ESI: electrospray ionization, MS: mass detector, MS-Q-TOF: quadrupole time of flight mass spectrometry, MRM: multiple reaction monitoring, RSD: relative standard deviation, TQ: triple-quadrupole, UPLC: ultra-performance liquid chromatography, UHPLC: ultra-high performance liquid chromatography, HILIC: hydrophilic interaction chromatography, LLOQ: lower limit of quantification, UV: ultraviolet, PDA: photodiode array, FD: fluorescence detector.

Table 5 Stability indicating liquid chromatographic methods reported for determination of darunavir, degradation products and/or impurities.[a]

Separation conditions	Detection	Working range, precision; accuracy, LOD, LOQ, LLOQ	References
Column: Hiber, LiChrospher 60, RP-select B, C8 (250 mm × 4.6 mm, 5 μm); mobile phase:10 mM ammonium acetate: acetonitrile (52:48, v/v); elution: isocratic; retention time: 8.99 min	MS/MS: Q-TOF, ESI positive mode	Range: 50–250 μg/mL; LOD: 0.033; LOQ: 0.099; intra- and inter-day precision RSD: 0.21–0.79 and 0.24–0.8%, respectively; recovery: 99.74–100.26%	[102]
Column: Phenomenx RP-C18 (250 × 4.6 mm, 5 μm); mobile phase: acetonitrile: methanol: water (60:30:10, v/v/v); flow rate: 1.0 mL/min; retention time: 2.45 min	UV: 265 nm	Range: 1–5 μg/mL; The LOD and LOQ: 0.017 and 0.057 μg/mL, respectively; intra- and inter-assay precision RSD: 0.1137% and 0.212%, respectively; recovery: 99.2–99.8%	[103]
Column: C4 stationary phase of particle size 3.6 μm; elution: linear gradient; mobile phase: buffer:acetonitrile: tetrahydrofuran; flow rate: 0.5 mL/min	PDA: 240 nm.	Range: 0.16–0.24 mg/mL; repeatability and intermediate precision RSD: 0.24% and 0.29%, respectively; recovery: 99.77–100.17% and 92.63–100.4%, respectively	[104]
Column: Chiral PAK AD-H which is amylose tris-(3,5-dimethylphenylcarbamate) phase coated on silica matrix; mobile phase: n-hexane:ethanol:n-butanol (83:15:2, v/v/v); flow rate: 0.8 mL/min	UV: 265 nm	Range: 0.016–0.1.2 μg/mL; LOQ: 0.127 μg/mL	[105]
Column: Hi-Q Sil C18 (4.6 × 250 mm, 5 μm); mobile phase: acetonitrile: water (90:10, v/v); flow rate: 1 mL/min	PDA: 266 nm and ESI-MS/MS	Range: 15–90 μg/mL; LOD and LOQ: 5.18 and 13.70 μg/mL, respectively; precision RSD: 0.75–1.37%; recovery: 99.65 ± 0.38%	[106]

Table 5 Stability indicating liquid chromatographic methods reported for determination of darunavir, degradation products and/or impurities.[a]—cont'd

Separation conditions	Detection	Working range, precision; accuracy, LOD, LOQ, LLOQ	References
Column: Hypersil ODS (100 × 4.6 mm, 5 μm); mobile phase: phosphate buffer (pH 3.5):acetonitrile and methanol (5:1) (30:70); flow rate: 1 mL/min; retention time: 2.813 min	UV: 220 nm	Range: 120–600 ng/mL; LOD and LOQ: 3.1 and 10.10 ng/mL, respectively; precision RSD: 0.3–0.5%; recovery: 100.06%	[107]
Column: Atlantis C18 (50 × 4.6 mm, 5 μm); mobile phase: acetonitrile: water (60:40, v/v); flow rate: 0.8 mL/min; retention time: 3.15 min	UV: 230 nm	Range: 40–120 μg/mL; precision RSD: <2%; recovery: 99.8–100.01%	[108]
Column: Acquity UPLC BEH C18 (50 × 2.1 mm, 1.7 μm); mobile phase: acetonitrile: methanol (80:20, v/v) and 5.0 mM ammonium acetate containing 0.01% formic acid; flow rate: 0.4 mL/min; elution: gradient	MS/MS: ESI, MRM: positive and negative	Range: LOQ—250% for darunavir impurities; LOQ (for impurities): 0.2–0.3 ppm with respect to 5.0 mg/mL of darunavir; accuracy: 89.90–104.60%	[109]
Column: Acquity HSS-T3, C18 (100 × 2.1 mm, 1.7 μm); mobile phase: 0.1% orthophosphoric acid in water adjusted the pH 3.0 with triethylamine and acetonitrile; elution: gradient; flow rate: 0.2 mL/min	PDA: 265 nm	Range: 0.02–0.598; LOD: 0.0.007%; precision RSD: 0.3%;	[110]

[a]Abbreviations are:
MS/MS: tandem mass detector, Q-TOF: quadrupole time of flight mass spectrometry, ESI: electrospray ionization, UV: ultraviolet, PDA: photodiode array, MRM: multiple reaction monitoring, RSD: relative standard deviation, LOD: limit of detection, LOQ: limit of quantitation.

5.4 Enzyme-linked immunosorbent assay

Almehizia et al. [111] developed and validated a highly sensitive enzyme-linked immunosorbent assay for determination of darunavir in plasma samples. The assay utilized a polyclonal antibody that can recognizes darunavir with high affinity and specificity, and a darunavir conjugated with bovine serum albumin for coating onto 96-well assay plate. The assay was based on a competitive binding reaction between darunavir, in plasma samples, and the coated darunavir-bovine serum albumin conjugate for the binding to a limited amount of the anti-darunavir antibody. The antibody bound to the assay plate wells was quantified with secondary antibody labeled with horseradish peroxidase enzyme and its chromogenic substrate 3,3′,5,5′-tetramethylbenzidine. Darunavir concentrations in its samples were quantified by their ability to inhibit the binding of the anti-darunavir antibody to the plate-coated darunavir-bovine serum albumin conjugate and subsequently the color intensity in the assay plate wells. The variables affecting the assay performance were optimized and the optimum procedures were established. The working range of the assay at relative standard deviations (RSD) of $\leq$10% was 20–2000 pg/mL. The assay LOD and LOQ were 15 and 30 pg/mL, respectively. Analytical recoveries of darunavir from spiked plasma were in the ranges of 98.4–113.0 and 86.0–99.1% for intra-assay and inter-assay runs, respectively. The precision of the assay was satisfactory; relative standard deviation values were 1.87–5.76 and 3.97–7.92% for intra- and inter-assay precisions, respectively. This assay was characterized with high throughput facilitating processing of large samples. The assay is expected to significantly contribute to routine analysis of darunavir in its pharmacokinetic studies and therapeutic monitoring.

6. Pharmacological properties

6.1 Antiviral activity and mechanism of action

Darunavir is an oral non-peptidic HIV-1 protease inhibitor that selectively inhibits the cleavage of HIV gag and gag-pol polyproteins in virus-infected cells. The selective inhibition leads to prevention further infection by inhibiting virus maturation and infection of CD4$^+$ T-cells. Darunavir is highly potent against laboratory strains and clinical isolates of wild-type and multidrug-resistant HIV with limited cytotoxicity [112]. Darunavir also inhibits the dimerization of HIV-1 protease leading to inhibition of proteolytic activity and subsequent HIV-1 replication [113,114]. In vitro studies against wild-type HIV-1 and HIV-2 in acutely infected T-cells, peripheral blood and macrophages demonstrated that the 50% effective inhibitory concentration (IC_{50}) value was 0.003 μmol/L. The 50% cytotoxic concentration was found to be >100 μmol/L, yielding a selectivity index of >20,000 for wild-type HIV [115]. Furthermore, in T-cells, the potency of darunavir against HIV-1 was found to be greater than that of saquinavir, amprenavir, nelfinavir, indinavir, lopinavir and ritonavir [116]. This potent anti-HIV activity of darunavir was further supported by the results of conformational analysis, which demonstrated that darunavir can form a highly stable complex with protease, largely due to conformational flexibility and backbone interactions [117].

6.2 Viral Resistance

Drug resistance is simply defined as the ability of disease-causing germs (e.g., bacteria or viruses) to continue multiplying despite the presence of drugs that usually kill them. With HIV, drug resistance is caused by changes (mutations) in the virus's genetic structure. These mutations lead to changes in certain HIV proteins and enzymes (e.g., protease enzyme) which help HIV to replicate. Mutations are very common in HIV because it replicates at an extremely rapid rate and does not contain the proteins needed to correct the mistakes that occur during copying process. Mutations in HIV enzymes give HIV a survival advantage when antiretroviral drugs are used because these mutations can block drugs from working against the HIV enzymes they are designed to target (e.g., protease enzyme inhibitors including darunavir) and cause drug resistance. HIV drug-resistance mutations can occur both before and during HIV treatment [118]. Numerous studies have demonstrated a high barrier to resistance of HIV against darunavir [119–121]. An analysis of multiple clinical studies confirmed that the development of darunavir resistance-associated mutations was very rare [122]. In order to minimize the development of likelihood of drug resistance, antiretroviral therapy regimens generally include a combination of two to three antiretroviral agents from at least two different drug classes [123].

6.3 Pharmacokinetics

The recommended dosage of darunavir is 600 mg twice daily. Darunavir is rapidly absorbed after oral administration, reaching peak plasma concentrations after 2.5–4 h. Absorption is followed by a fast distribution/elimination phase and a subsequent slower elimination phase with a terminal elimination half-life of 15 h. Darunavir is approximately 95% plasma protein bound, mainly to α_1-acid glycoprotein. Systemic exposure is increased by 30% when darunavir is taken with a meal. Darunavir is extensively and almost exclusively metabolized by cytochrome P450 [124]. Pharmacokinetic studies showed that darunavir has a bioavailability of ~37%; however, the bioavailability is increased to 82% when darunavir is administrated with cytochrome P3A inhibitor (ritonavir) [125]. Darunavir can also be detected in the cerebrospinal fluid and cervicovaginal fluid of HIV-1 infected patients who receiving darunavir-based regimens in treatment [126,127]. Darunavir has a terminal elimination half-life of ≈15 h. Elimination of darunavir is mainly via the feces (≈79.5%) and urine (13.9%) after administration of a single dose of 400 mg, with ≈41.2% and 7.7% of the darunavir dose recovered as unchanged parent drug via these routes. The clearance of intravenous darunavir 150 mg is 5.9 L/h [128].

7. Dosing information

7.1 Dosage forms and administration

Darunavir is formulated as tablets (75, 150, 600 and 800 mg), and oral suspension (100 mg/mL). Treatment-naïve adult patients and treatment-experienced adult patients with no darunavir resistance-associated substitutions take 800 mg (two 400 mg tablets) with ritonavir 100 mg once daily and with food. Treatment-experienced adult patients with at least one darunavir resistance associated substitution take 600 mg (one 600 mg tablet) with ritonavir 100 mg twice daily and with food. Pediatric patients (3 to less than 18 years of age and weighing at least 10 kg) take darunavir and ritonavir based on body weight and should not exceed the treatment-experienced adult dose. Darunavir once daily dosing is not used in pediatric patients; darunavir/with ritonavir should be taken twice daily with food. Darunavir/ritonavir is not recommended for patients with severe hepatic impairment [129].

7.2 Contraindications

Darunavir co-administration is contraindicated with alfuzosin, dihydroergotamine, ergonovine, ergotamine, methyl ergonovine, cisapride, pimozide, oral midazolam, triazolam, St. John's Wort, lovastatin, simvastatin, rifampin and sildenafil (for treatment of pulmonary arterial hypertension). Due to the need for co-administration of darunavir with ritonavir, contraindications of ritonavir prescribing should be considered which include drugs that are highly dependent on CYP3A for clearance and for which elevated plasma concentrations are associated with serious and/or life-threatening events (narrow therapeutic index) [130,131].

7.3 Adverse reactions

Increased total cholesterol, increased triglycerides, diarrhea, pain, nausea, vomiting, acute pancreatitis, dyspepsia, flatulence, hepatotoxicity, hypersensitivity reactions, myalgia, osteonecrosis, abnormal dreams, angioedema, Stevens-Johnson syndrome, pruritus and urticarial. The most common clinical adverse drug reactions to darunavir/ritonavir (incidence greater than or equal to 5%) of at least moderate intensity (greater than or equal to Grade 2) were diarrhea, nausea, rash, headache, abdominal pain and vomiting [129–131]. Other possible side effects of darunavir include diabetes and high blood sugar (hyperglycemia), changes in body fat (lipodystrophy syndrome), changes in the immune system (called immune reconstitution inflammatory syndrome), increased bleeding in people with hemophilia [132].

7.4 Cautions

Cautions should be taken when darunavir therapy is indicated for elderly patients, patients with hepatic impairment, patients with known sulfonamide allergy, and patients with hemophilia as they may develop increased bleeding events [132].

7.5 Hepatotoxicity

Some degree of serum aminotransferase elevations occur in a high proportion of patients taking darunavir containing antiretroviral regimens. Moderate-to-severe elevations in serum aminotransferase levels (above five times the upper limit of normal) are found in 3–10% of patients overall, and rates are higher in patients having coinfection with HIV and HCV (hepatitis C virus). In clinical trials of darunavir elevations in serum ALT (alanine aminotransferase) above five times upper limit of normal occurred in 2–3% of patients, but no subject developed clinically apparent liver injury with jaundice. The serum enzyme elevations during therapy are usually asymptomatic and self-limited and can resolve even with continuation of the medication. Clinically apparent acute liver injury due to darunavir has been reported since its approval and more widescale use, but none have been well characterized for clinical features. The liver injury generally arises after 1–8 weeks of therapy and the pattern of serum enzyme elevations is usually, but not always, hepatocellular. Signs of hypersensitivity (fever, rash, eosinophilia) are rare, as is autoantibody formation. The acute liver injury is usually self-limited and resolves within a few weeks of stopping darunavir. However, fatal instances have been reported, at least to the sponsor and monitoring of liver enzymes during therapy is recommended. Finally, initiation of darunavir based highly active antiretroviral therapy can lead to exacerbation of an underlying chronic hepatitis B or C in coinfected individuals, typically arising 2–12 months after starting therapy and associated with a hepatocellular pattern of serum enzyme elevations and increases in serum levels of hepatitis B or C virus. Darunavir therapy has not been clearly linked to lactic acidosis and acute fatty liver that is reported in association with several nucleoside analogue reverse transcriptase inhibitors [18].

7.6 Use in specific population

Darunavir is used during pregnancy only if the potential benefit justifies the potential risk. Mothers should be instructed not to breastfeed during treatment with darunavir due to the potential for HIV transmission and the potential for serious adverse reactions in nursing infants. Darunavir/ritonavir is not administered once daily in pediatric patients, and is not recommended for use in patients with severe hepatic impairment [129].

7.7 Overdose and antidote

Human experience of acute overdose with darunavir/ritonavir is limited. Single doses up to 3200 mg of the oral solution of darunavir alone and up to 1600 mg of the tablet formulation of darunavir in combination with ritonavir have been administered to healthy volunteers without untoward symptomatic effects. No specific antidote is available for overdose with darunavir. Treatment of overdose with darunavir consists of general supportive measures including monitoring of vital signs and observation of the clinical status of the patient. If indicated, elimination of unabsorbed active substance is to be achieved by emesis or gastric lavage. Administration of activated charcoal may also be used to aid in removal of unabsorbed active substance. Since darunavir is highly protein bound, dialysis is unlikely to be beneficial in significant removal of the active substance [129].

References

[1] Website: http://www.chemspider.com/Chemical-Structure.184733.html?rid=0c6ee0e8-4eaf-4f28-ac42-69140e40cc85 (Accessed April 1, 2020).
[2] S.C. Sweetman, Martindale: The Complete Drug Reference, 36th ed., Pharmaceutical Press, London, 2009.
[3] A.C. Moffat, M.D. Osselton, B. Widdop, J. Watts, Clarke's analysis of drugs and poisons, in: Pharmaceuticals, Body Fluids and Postmortem Material, Pharmaceutical Press, London, 2011.
[4] Website: https://www.drugbank.ca/drugs/DB01264 (Accessed April 1, 2020).
[5] M.J. O'Neil, P.E. Heckelman, C.B. Koch, K.J. Roman (Eds.), The Merck Index: An Encyclopedia of Chemicals, Drugs, and Biologicals, Merck, John Wiley & Sons, Inc., Hoboken, New Jersey, 2007
[6] R.M. Lascar, P. Benn, Role of darunavir in the management of HIV infection, HIV/AIDS (Auckl.) 1 (2009) 31–39.
[7] A. Colbers, R. Greupink, C. Litjens, D. Burger, F.G.M. Russel, Physiologically based modelling of darunavir/ritonavir pharmacokinetics during pregnancy, Clin. Pharmacokin. 55 (2016) 381–396.
[8] M. Mizhiritskii, E. Marom, Process for the Preparation of Darunavir and Darunavir Intermediates, European Patent EP2528923 (issued on July 30, 2014), 2014.
[9] A.K. Ghosh, S. Leshchenko, M. Noetzel, Stereoselective photochemical 1, 3-dioxolane addition to 5-alkoxymethyl-2 (5 H)-furanone: synthesis of bis-tetrahydrofuranyl ligand for HIV protease inhibitor UIC-94017 (TMC-114), J. Org. Chem. 69 (2004) 7822–7829.
[10] A.K. Ghosh, J.F. Kincaida, W. Choa, W.E. Walters, K. Krishnan, K.A. Hussain, Y. Koo, H. Choa, C. Rudall, L. Holland, J. Buthod, Potent HIV protease inhibitors incorporating high-affinity P2-ligands and (R)-(hydroxyethylamino) sulfonamide isostere, Bioorg. Med. Chem. Lett. 8 (1998) 687–690.
[11] G.L. Moore, R.W. Stringham, D.S. Teager, T-Y. Yue. Practical synthesis of the bicyclic darunavir side chain: (3R,3aS,6aR)-hexahydrofuro[2,3-b]furan-3-ol from monopotassium isocitrate. Org. Process Res. Dev. 21 (2017) 98–106.
[12] S.R.P. Vellenki, A. Sahu, A.K. Katukuri, S. Kothari, V.S. Ponnekanti, D. Datta, An Improved Process for the Preparation of Darunavir, Indian Patent IN 2009CH02257 A (issued on March 9, 2012), 2012.
[13] A.E. Stappers, Y.L. Lang, S.B. Robinson, A Simplified Procedure for the Preparation of Darunavir, PCT Int. Appl. WO 2018091559 A1 (issued on May 5, 2018), 2018.
[14] X. Jiang, J. Xu, R. Liu, Method for Preparing Darunavir Containing Butyl Lithium and N-Hydroxy Compound, Faming Zhuanli Shenqing CN 108822123 A (issued on November 16, 2018), 2018.
[15] J. Xu, X. Jiang, Y. Zhou, M. Zhang, Method for Preparing Darunavir, Faming Zhuanli Shenqing CN 111039951 A (issued on April 21, 2020), 2020.
[16] J. Xu, X. Jiang, M. Zhang, Y. Zhou, Method for Preparing Darunavir, Faming Zhuanli Shenqing CN 110698492 A (issued on January 17, 2020), 2020.
[17] Website: https://www.rsc.org/Merck-Index/monograph/m4099/darunavir?q=unauthorize (Accessed April 1, 2020).
[18] Website: https://pubchem.ncbi.nlm.nih.gov/source/hsdb/7788#section=Solubility (Accessed April 1, 2020).
[19] A.C. Kogawa, H.R.N. Salgado, Development and validation of infrared spectroscopy method for the determination of darunavir in tablets, Phys. Chem. 3 (2013) 1–6.
[20] M. Abdulsaleem, B. Jadhav, B. Rajender, S. Abdul Raheem, M.A. Bashir, M.A. Sulthana, Optimization of UV spectrophotometric method for estimation of darunavir in bulk drug and tablet formulations, Int. J. Pharm. Sci. Nanotechnol. 9 (2016) 3345–3348.

[21] S.K. Mastanamma, T.A. Kumar, B.S. Sankarnaik, D.S. Priya, UV-spectrophotometric determination of daurnavir in bulk and pharmaceutical dosage form using hydrotropic solubilization technique (8m urea), Am. J. PharmTech. Res. 5 (2015) 574–583.
[22] N. Fathima, M. Brahmaiah, G. Abdul Huq, P.R. Reddy, Development and validation of UV spectroscopic methods for simultaneous estimation of darunavir and rilpivirine HCl in tablet formulation, Int. J. Pharm. Pharm. Res. 9 (2017) 282–292.
[23] M.M. Eswarudu, B.P. Srinivasa, K.P. Siva, L.G. Sneha, N. Anusha, M.S.K. Mounika, R.N. Punna, Method development and validation of UV spectrophotometric methods for the estimation of darunavir in bulk and its pharmaceutical dosage form, Indo Am. J. Pharm. Sci. 5 (2018) 2230–2237.
[24] K.V.V.K.K. Rao, B. Phanindra, K. Rajesh, Spectrophotometric method for estimation of darunavir ethanolate by using MBTH reagent in bulk and pharmaceutical dosage form, Pharm. Anal. Qual. Assur. 2013 (2013) 1–3.
[25] D. Mrinalini, D. Ajinkya, UV spectrophotometric absorbance correction method and absorbance ratio method for the simultaneous estimation of ritonavir and darunavir ethanolate, World J. Pharm. Pharm. Sci. 5 (2016) 1179–1187.
[26] M.R. Ghante, R.S. Shelar, M.M. Kadam, Development and validation of UV spectrophotometric method for estimation of darunavir ethanolate in bulk and tablet dosage form, Int. J. Pharm. Pharm. Sci. 6 (2014) 240–242.
[27] S.S. Vanukuri, S.K. Mastanamma, G. Alekhya, Validated UV-spectrophotometric methods for the estimation of darunavir by absorption maxima, first order derivative and area under curve in bulk and its tablet dosage form, Int. J. Pharm. Pharm. Sci. 6 (2014) 568–571.
[28] J.C.R. Corrêa, C.H.d.R. Serra, H.R.N. Salgado, Development and validation of first derivative spectrophotometric method for quantification of darunavir in tablets, Br. J. Pharm. Res. 4 (2014) 722–730.
[29] H. Nimje, P. Kore, S. Wankhede, A. Gadekar, Simultaneous estimation of darunavir ethanolate and ritonavir in combined dosage form, Int. J. Acad. Res. Dev. 2 (2017) 2455–4197.
[30] K.P. Rao, Determination of darunavir in pharmaceutical dosage form, Der Pharm. Lett. 8 (2016) 64–69.
[31] K.P. Rao, Validation of visible spectrophotometric methods of darunavir in pure and dosage forms, Der Pharm. Chem. 8 (2016) 54–61.
[32] R. Vijayalakshmi, Y.N.S. Ramya, A.D. Mani, M.D. Dhanaraju, Spectrophotometric determination of darunavir ethanolate by condensation technique, Int. J. PharmTech Res. 9 (2016) 301–306.
[33] M.P. Reddy, N. Ramireddy, Spectrophotometric estimation of darunavir in bulk and pharmaceutical formulations, Int. J. Chem. Sci. 11 (2013) 614–618.
[34] R. Vijayalakshmi, D. Anjani, M. Sriramamurthy, M.D. Dhanaraju, Analytical method development for the estimation of darunavir by diazotization and coupling by visible spectrophotometry, Asian J. Pharm. Clin. Res. 11 (2018) 149–152.
[35] R. Vijayalakshmi, D. Anjani, M.D. Dhanaraju, Analytical method development for the estimation of Darunavir by ion-pair complex using visible spectrophotometry, Int. J. Pharm. Pharm. Res. 12 (2018) 182–191.
[36] R.D. Godambe, J.I. Disouza, C.M. Jamkhandi, P.S. Kumbhar, Development of spectrophotometric and fluorometric methods for estimation of darunavir using QBD approach, Int. J. Curr. Pharm. Res. 10 (2018) 13–19.
[37] N.R. Das, G. Vidyasagar, Analytical method development of darunavir and ropivacain as bulk drug and formulation using chromogenic reagents, BioMedRx 1 (2013) 529–535.
[38] A.C. Kogawa, F.A. Aguiar, C.M. Gaitani, Validation of a stability indicating capillary electrophoresis method for the determination of darunavir in tablets and comparison with the of infrared absorption spectroscopic method, World J. Pharm. Pharm. Sci. 3 (2014) 283–297.

[39] S. Leonard, A.V. Schepdael, T. Iványi, I. Lázár, J. Rosier, M. Vanstockem, H. Vermeersch, J. Hoogmartens, Development of a capillary electrophoretic method for the separation of diastereoisomers of a new human immunodeficiency virus protease inhibitor, Electrophoresis 26 (2005) 627–632.
[40] A.C. Kogawa, H.R.N. Salgado, Development of analytical methods for the qualitative analysis of darunavir tablet, Rev. Ciênc. Farm. Básica Apl. 34 (2013) 207–213.
[41] A.C. Kogawa, J.N. Mendonca, N.P. Lopes, H.R.N. Salgado, Stability-indicating thin-layer chromatographic method for determination of darunavir in complex darunavir-β-cyclodextrin in the presence of its degradation products, Anal. Methods 6 (2014) 3689–3693.
[42] B.N. Patel, B.N. Suhagia, C.N. Patel, H.J. Panchal, A simple and sensitive HPTLC method for quantitative analysis of darunavir ethanolate tablets, J. Planar Chromatogr.-Mod. TLC 24 (2011) 232–235.
[43] B. Ramesh, S. Ramakrishna, R.K.K. Reddy, K.H. Babu, HPTLC method for determination of darunavir in rat plasma and its application in pharmacokinetic studies, J. Liq. Chromatogr. Related Technol. 36 (2013) 167–179.
[44] R.T. Heine, C.G. Alderden-Los, H. Rosing, M.J.X. Hillebrand, E.C.M. Gorp, A.D.R. Huitema, J.H. Beijnen, Fast and simultaneous determination of darunavir and eleven other antiretroviral drugs for therapeutic drug monitoring: Method development and validation for the determination of all currently approved HIV protease inhibitors and non-nucleoside reverse transcriptase inhibitors in human plasma by liquid chromatography coupled with electrospray ionization tandem mass spectrometry, Rapid Commun. Mass Spectrom. 21 (2007) 2505–2514.
[45] M. Takahashi, Y. Kudaka, N. Okumura, A. Hirano, K. Banno, T. Kaneda, The validation of plasma darunavir concentrations determined by the HPLC method for protease inhibitors, Biol. Pharm. Bull. 30 (2007) 1947–1949.
[46] S.P. García, D.G. Tunica, M.B. Serra, Development and validation of a method for determination of darunavir in plasma by LC-MS/MS, Rev. Clin. Lab. 4 (2011) 127–133.
[47] L. Goldwirt, S. Chhuna, E. Rey, O. Launay, J. Viard, G. Pons, V. Jullien, Quantification of darunavir (TMC114) in human plasma by high-performance liquid chromatography with ultra-violet detection, J. Chromatogr. B 857 (2007) 327–331.
[48] A. Gupta, P. Singhal, P.S. Shrivastav, M. Sanyal, Application of a validated ultra performance liquid chromatography–tandem mass spectrometry method for the quantification of darunavir in human plasma for a bioequivalence study in Indian subjects, J. Chromatogr. B 879 (2011) 2443–2453.
[49] A. Fayet, A. Béguin, B. Zanolari, S. Cruchon, N. Guignard, A. Telenti, M. Cavassini, H.F. Günthard, T. Buclin, J. Biollaz, B. Rochat, L.A. Decosterd, A LC-tandem MS assay for the simultaneous measurement of new antiretroviral agents: raltegravir, maraviroc, darunavir, and etravirine, J. Chromatogr. B 877 (2009) 1057–1069.
[50] N.L. Rezk, N.R. White, S.H. Jennings, A.D.M. Kashuba, A novel LC-ESI-MS method for the simultaneous determination of etravirine, darunavir and ritonavir in human blood plasma, Talanta 79 (2009) 1372–1378.
[51] J. Martin, G. Deslandes, E. Dailly, C. Renaud, V. Relequet, Raffi, P. Jolliet, A liquid chromatography–tandem mass spectrometry assay for quantification of nevirapine, indinavir, atazanavir, amprenavir, saquinavir, ritonavir, lopinavir, efavirenz, tipranavir, darunavir and maraviroc in the plasma of patients infected with HIV, J. Chromatogr. B 877 (2009) 3072–3082.
[52] A. D'avolio, M. Simiele, M. Siccardi, L. Baietto, M. Sciandra, S. Bonora, G.D. Perri, HPLC-MS method for the quantification of nine anti-HIV drugs from dry plasma spot on glass filter and their long term stability in different conditions, J. Pharm. Biomed. Anal. 52 (2010) 774–780.

[53] L. Else, V. Watson, J. Tjia, A. Hughes, M. Siccardi, S. Khoo, D. Back, Validation of a rapid and sensitive high-performance liquid chromatography–tandem mass spectrometry (HPLC-MS/MS) assay for the simultaneous determination of existing and new antiretroviral compounds, J. Chromatogr. B 878 (2010) 1455–1465.
[54] R. Heine, H. Rosing, E.C.M. Gorp, J.W. Mulder, W.A. Steeg, J.H. Beijnen, A.D.R. Huitema, Quantification of protease inhibitors and non-nucleoside reverse transcriptase inhibitors in dried blood spots by liquid chromatography–triple quadrupole mass spectrometry, J. Chromatogr. B 867 (2008) 205–212.
[55] R.B. Ganduri, R.A. Lanka, S. Pamidi, J.R. Peddareddigari, J.V.L.N.S. Rao, New RP-HPLC method for the determination of darunavir in tablet dosage form, Asian J. Pharm. Res. 1 (2011) 10–14.
[56] L. Satyanarayana, S.V. Naidu, M.N. Rao, A. Kumar, K. Suresh, The estimation of darunavir in tablet dosage form by RP-HPLC, Asian J. Res. Pharm. Sci. 1 (2011) 2231–5640.
[57] B.V.R. Reddy, G. Jyothi, B.S. Reddy, N.V.V.S.S. Raman, K.S.C. Reddy, C. Rambabu, Stability indicating HPLC method for the determination of darunavir ethanolate, J. Chromatogr. Sci. 51 (2013) 471–476.
[58] J.C.R. Correa, C.H.D.R. Serra, H.R.N. Salgado, Stability study of darunavir ethanolate tablets applying a new stability-indicating HPLC method, Chromatogr. Res. Int. 2013 (834173) (2013) 1–7.
[59] R.N. Rao, K.N. Kumar, B.S. Kumar, S. Bompelli Sravan, Liquid chromatographic separation and thermodynamic investigation of stereoisomers of darunavir on Chiralpak AD-H column, J. Sep. Sci. 35 (2012) 2671–2677.
[60] R.N. Rao, K.N. Kumar, C.G. Naidu, Liquid chromatographic separation of darunavir enantiomers on coated and immobilized amylose tris(3,5-dimethylphenylcarbamate) chiral stationary phases, Chirality 24 (2012) 652–660.
[61] R. Kanneti, K.I. Jaswanth, K.R. Neeraja, P.A. Bhatt, Development and validation of LC-MS/MS method for determination of Darunavir in human plasma for application of clinical pharmacokinetics, Int. J. Pharm. Pharm. Sci. 3 (2011) 491–496.
[62] S. Quaranta, C. Woloch, A. Paccou, M. Giocanti, C. Solas, B. Lacarelle, Validation of an electrospray ionization LC-MS/MS method for quantitative analysis of raltegravir, etravirine, and 9 other antiretroviral agents in human plasma samples, Ther. Drug Monit. 31 (2009) 695–702.
[63] A. D'Avolio, L. Baietto, M. Siccardi, M. Sciandra, M. Simiele, V. Oddone, S.G.D. Perri, An HPLC-PDA method for the simultaneous quantification of the HIV integrase inhibitor raltegravir, the new nonnucleoside reverse transcriptase inhibitor etravirine, and 11 other antiretroviral agents in the plasma of HIV-infected patients, Ther. Drug Monit. 30 (2008) 662–669.
[64] A. D'Avolio, M. Siccardi, M. Sciandra, B. Lorena, S. Bonora, L. Trentini, G.D. Perri, Giovanni, HPLC-MS method for the simultaneous quantification of the new HIV protease inhibitor darunavir, and 11 other antiretroviral agents in plasma of HIV-infected patients, J. Chromatogr. B 859 (2007) 234–240.
[65] A. Patel, A. Patel, A. Makwana, An ESI-LC-MS/MS method for simultaneous estimation of darunavir and ritonavir in human plasma, Int. J. Res. Pharm. Biomed. Sci. 4 (2013) 1138–1147.
[66] G.R. Babu, A.L. Rao, J.V. Rao, Development and validation of novel HPLC method for estimation of darunavir in pharmaceutical formulations, Int. J. Res. Pharm. Biomed. Sci. 3 (2013) 438–443.
[67] A.C. Kogawa, H.R.N. Salgado, Characteristics, complexation and analytical methods of darunavir, Br. J. Pharm. Res. 4 (2014) 1276–1286.
[68] M.M. Eswarudu, P.S. Krishna, P.S. Babu, P.N. Reddy, Darunavir: a review on its analytical methods, Int. J. Pharm. Pharm. Res. 12 (2018) 119–137.

[69] G. Kashish, P. Nisha, K. Hiren, Analytical techniques in the analysis of darunavir and ritonavir: a review, Curr. Pharm. Anal. 16 (2020) 447–455.
[70] R.N. Herqash, Development of New Enzyme Immunoassay for Therapeutic Monitoring of Darunavir, Master Thesis King Saud University (Saudi Arabia), 2019 Approved on Nov. 24.
[71] A. Karthika, K. Rama, A. Chenthilnathan, RP-HPLC method development and validation for quantification of darunavir ethanolate in tablet dosage form, Der Pharm. Lett. 8 (2016) 291–296.
[72] K.V. Bhuse, P.B. Jadhav, RP-HPLC method development and validation for determination of darunavir in bulk and dosage form, World J. Pharm. Pharm. Sci. 6 (2017) 1752–1766.
[73] S.M. Ramdharane, R. Vays, K.S. Nimava, A novel RP-HPLC method with core shell tecnology column for assay analysis of darunavir ethanolate in tablet dosage form, Int. J. Sci. Res. Sci. Eng. Technol. 3 (2017) 896–899.
[74] H.K. Jain, U.S. Jadhav, Development and validation of RP-HPLC method for estimation of darunavir ethanolate in bulk and tablets, Int. J. Pharm. Pharm. Sci. 7 (2015) 386–389.
[75] S.K. Mastanamma, V.S. Sirisha, G. Alekhya, K. Haritha, V.A. Babu, New validated RP-HPLC method for the estimation of darunavir in bulk and its dosage form, Int. Res. J. Pharm. 5 (2014) 13–16.
[76] G.A.S. Rao, M.M. Eswarudu, C. Devadasu, P.S. Babu, Novel RP-HPLC method for the determination of darunavir in pure and tablet dosage form, Am. J. PharmTech Res. 7 (2017) 255–266.
[77] A.V.D. Nagendrakumar, S.R. Battula, M.V.B. Rao, Validated RP-HPLC method for the determination of darunavir in bulk and pharmaceutical formulation, Res. J. Pharm. Biol. Chem. Sci. 5 (2014) 63–72.
[78] M.V. Nalini, P.R. Veni, B. Haribabu, Determination of darunavir and cobicistat simultaneously using stability indicating RP-HPLC method, Marmara Pharm. J. 20 (2016) 293–302.
[79] G. Estan-Cerezo, A. García-Monsalve, L. Soriano-Irigaray, F.J. Rodríguez-Lucena, A. Navarro-Ruiz, A rapid validated UV-HPLC method for the simultaneous determination of the antiretroviral compounds darunavir and raltegravir in their dosage form, Rev. Esp. Quimioter. 30 (2017) 195–200.
[80] N.M. Rao, D.G. Sankar, Simultaneous quantification of novel antiretroviral drug combination by stability-indicating high performance liquid chromatography method, Indian J. Pharm. 78 (2016) 755–762.
[81] S. Rachha, V.J. Rao, Tejaswi, A. Deevika, B. Swetha, Method development and validation of RP-HPLC for simultaneous estimation of darunavir and elvitegravir in API, J. Global Trends Pharm. Sci. 8 (2017) 4024–4029.
[82] K. Devi, N. Kannappan, Development of an RP-HPLC method for multicomponent tablet formulation containing cobicistat and darunavir, Int. J. Res. Pharm. Sci. 9 (2018) 691–695.
[83] K.P. Rao, B.V. Ramesh, C.S. Prasad, G.V. Ramana, M.C. Rao, Validated isocratic reversed phase liquid chromatographic method for the determination of darunavir in pure and formulations, Der Pharm. Lett. 8 (2016) 222–228.
[84] R.M. Paul, G.V. Chowdary, A.S.K. Reddy, RP-HPLC method development and validation for simultaneous estimation of cobicistat and darunavir in tablet dosage form, Eur. J. Biomed. Pharm. Sci. 3 (2016) 639–644.
[85] P.K. Bichal, R. Sharma, N. Kumar, A. Lawal, A. Ibrahim, M. Umar, Analytical method development and validation for the simultaneous estimation of darunavir and cobicistat by RP-HPLC method, Int. J. Res. Pharm. Chem. 10 (2020) 131–136.

[86] J. Peris-Vicente, M. Villarreal-Traver, I. Casas-Breva, S. Carda-Broch, J. Esteve-Romero, Use of micellar liquid chromatography to analyze darunavir, ritonavir, emtricitabine, and tenofovir in plasma, J. Sep. Sci. 37 (2014) 2825–2832.
[87] Z. Ahmad, S. Sonawane, A.S. Bhalerao, S.J. Kshirsagar, Development of a validated RP-HPLC method for estimation of darunavir in spiked human plasma with UV detection, Int. Res. J. Pharm. 1 (2016) 24–28.
[88] T. Mishra, P.S. Shrivastav, Validation of simultaneous quantitative method of HIV protease inhibitors atazanavir, darunavir and ritonavir in human plasma by UPLC-MS/MS, Sci. World J. 2014 (2014) 48269312 pages.
[89] A.V. Reddy, J. Jaafar, A.B. Aris, Z.A. Majid, K. Umar, J. Talib, G. Madhavi, Development and validation of a rapid ultra high performance liquid chromatography with tandem mass spectrometry method for the simultaneous determination of darunavir, ritonavir, and tenofovir in human plasma: application to human pharmacokinetics, J. Sep. Sci. 38 (2015) 2580–2587.
[90] E. Yamada, R. Takagi, K. Sudo, S. Kato, Determination of abacavir, tenofovir, darunavir, and raltegravir in human plasma and saliva using liquid chromatography coupled with tandem mass spectrometry, J. Pharm. Biomed. Anal. 114 (2015) 390–397.
[91] B. Ramesh, N. Manjula, S. Ramakrishna, P.S. Devi, Direct injection HILIC-MS/MS analysis of darunavir in rat plasma applying supported liquid extraction, J. Pharm. Anal. 5 (2015) 43–50.
[92] R.N. Rao, K.G. Prasad, LC-Q-TOF-MS/MS determination of darunavir and its metabolites in rat serum and urine: application to pharmacokinetics, J. Pharm. Biomed. Anal. 94 (2014) 92–98.
[93] A.T. Podany, L.C. Winchester, B.L. Robbins, C.V. Fletcher, Quantification of cell-associated atazanavir, darunavir, lopinavir, ritonavir, and efavirenz concentrations in human mononuclear cell extracts, Antimicrob. Agents Chemother. 58 (2014) 2866–2870.
[94] L. Belkhir, D.M. Laveleye, B. Vandercam, F. Zech, K.A. Delongie, A. Capron, J. Yombi, A. Vincent, L. Elens, V. Haufroid, Quantification of darunavir and etravirine in human peripheral blood mononuclear cells using high performance liquid chromatography tandem mass spectrometry (LC-MS/MS), clinical application in a cohort of 110 HIV-1 infected patients and evidence of a potential drug-drug interaction, Clin. Biochem. 49 (2016) 580–586.
[95] V. Adepu, K.E.V. Nagoji, G. Vedula, Method development and validation for the estimation of Darunavir in rat plasma by RP-HPLC, World J. Pharm. Sci. 4 (2016) 490–494.
[96] Y. Zheng, R. Aboura, S. Boujaafar, G. Lui, D. Hirt, N. Bouazza, F. Foissac, J.M. Treluyer, S. Benaboud, I. Gana, HPLC-MS/MS method for the simultaneous quantification of dolutegravir, elvitegravir, rilpivirine, darunavir, ritonavir, raltegravir and raltegravir-β-D-glucuronide in human plasma, J. Pharm. Biomed. Anal. 182 (2020) 113119.
[97] H. Gouget, G. Noe, A. Barrail-Tran, V. Furlan, UPLC-MS/MS method for the simultaneous quantification of bictegravir and 13 others antiretroviral drugs plus cobicistat and ritonavir boosters in human plasma, J. Pharm. Biomed. Anal. 181 (2020) 113057.
[98] M. Vanaja, J. Sreeramulu, Stability indicating determination of darunavir with HPLC in blood plasma samples, Rasayan J. Chem. 12 (2019) 839–848.
[99] V.M. Suvarna, P.C. Sangave, HPLC estimation, Ex vivo everted sac permeability and In Vivo pharmacokinetic studies of darunavir, J. Chromatogr. Sci. 56 (2018) 307–316.
[100] D. Nagano, T. Araki, T. Nakamura, K. Yamamoto, Determination of intracellular darunavir by liquid chromatography coupled with fluorescence detection, J. Chromatogr. Sci. 52 (2014) 1021–1025.

[101] K.C. Janjanam, B.P.K. Bimireddy, S.P. Kovvasu, V.J. Ratna, Bioanalysis of darunavir in human plasma using liquid chromatography coupled with tandem mass spectrometry, Int. J. PharmTech Res. 11 (2018) 28–34.
[102] R.N. Rao, B. Ramachandra, B. Sravan, S. Khalid, LC-MS/MS structural characterization of stress degradation products including the development of a stability indicating assay of darunavir: an anti-HIV drug, J. Pharm. Biomed. Anal. 89 (2014) 28–33.
[103] N.K. Sahoo, R. Rajitha, K.S. Muralikrishna, A newer RP–HPLC stability indicating method development and validation of darunavir in its bulk and formulation, Austin J. Anal. Pharm. Chem. 6 (1123) (2019) 1–5.
[104] H.R.K. Mahesh, K.S. Babu, A novel rapid stability indicating UPLC method for simultaneous estimation of darunavir and ritonavir in tablet dosage form, Pharm. Anal. Qual. Assur. 2015 (2015) 33–37.
[105] M. Ramakrishna, M. Subbarao, D. Mogilireddy, A selective and stability indicating high performance liquid chromatographic method for estimation of isomers content in darunavir drug substance, Indo Am. J. Pharm. Res. 6 (2016) 5267–5276.
[106] M. Ghante, S.D. Sawant, Development and validation of stability indicating method for darunavir with forced degradation studies using LC-ESI-MS/MS, Asian J. Chem. 28 (2016) 1729–1733.
[107] B. Prathap, V. Haribaskar, B. Kumar, S. Harika, M. Kavitha, C. Bhavya, Stability indicating RP-HPLC method for simultaneous estimation of ritonavir and darunavir in bulk and its synthetic mixture, J. Global Trends Pharm. Sci. 9 (2018) 5549–5560.
[108] P.D. Sindu, S.D. Gowri, S.K. Masthanamma, Stability-indicating reversed-phase high performance liquid chromatography method for the simultaneous estimation of darunavir and ritonavir, Asian J. Pharm. Clin. Res. 9 (2016) 71–76.
[109] V.B.A. Reddy, Z. Yusop, J. Jaafar, A.B. Aris, Z.A. Majid, K. Umar, J. Talib, Development and validation of a selective, sensitive and stability indicating UPLC-MS/MS method for rapid, simultaneous determination of six process related impurities in darunavir drug substance, J. Pharm. Biomed. Anal. 128 (2016) 141–148.
[110] B.P.V. Mantena, S.V. Rao, K.M.C.A. Rao, K. Ramakrishna, R.S. Reddy, Method development and validation for the determination of four potential impurities present in darunavir tablets by reverse phase-ultra-performance liquid chromatography coupled with diode-array detector, J. Liq. Chromatogr. Related Technol. 38 (2015) 1236–1246.
[111] A.A. Almehizia, R.N. Herqash, I.A. Darwish, Development of a highly sensitive ELISA for determination of darunavir in plasma samples using a polyclonal antibody with high affinity and specificity, Bioanalysis 12 (March 27, 2020) 355–366. Epub ahead of print. https://doi.org/10.4155/bio-2020-0034.
[112] K.H. Busse, S.R. Penzak, Darunavir: a second-generation protease inhibitor, Am. J. Health Syst. Pharm. 64 (2007) 1593–1602.
[113] C. Fenton, C.M. Perry, Darunavir: in the treatment of HIV-1 infection, Drugs 67 (2007) 2791–2801.
[114] S. De Meyer, H. Azijn, D. Surleraux, D. Jochmans, A. Tahri, R. Pauwels, P. Wigerinck, M.P. de Béthune, TMC114, a novel human immunodeficiency virus type 1 protease inhibitor active against protease inhibitor-resistant viruses, including a broad range of clinical isolates, Antimicrob. Agents Chemother. 49 (2005) 2314–2321.
[115] K. McKeage, C.M. Perry, S.J. Keam, Darunavir: a review of its use in the management of HIV infection in adults, Drugs 69 (2009) 477–503.
[116] Y. Koh, H. Nakata, K. Maeda, H. Ogata, G. Bilcer, T. Devasamudram, J.F. Kincaid, P. Boross, Y.F. Wang, Y. Tie, P. Volarath, L. Gaddis, R.W. Harrison, I.T. Weber, A.K. Ghosh, H. Mitsuya, Novel bis-tetrahydrofuranylurethane-containing nonpeptidic protease inhibitor (PI) UIC-94017 (TMC114) with potent activity against multi-PI-resistant human immunodeficiency virus in vitro, Antimicrob. Agents Chemother. 47 (2003) 3123–3129.

[117] K. Nivesanond, A. Peeters, D. Lamoen, C. Van Alsenoy, Conformational analysis of TMC114, a novel HIV-1 protease inhibitor, J. Chem. Inf. Model. 48 (2008) 99–108.
[118] Website: https://www.poz.com/basics/hiv-basics/hiv-drug-resistance (Accessed April 1, 2020).
[119] E. Lathouwers, E.Y. Wong, D. Luo, S. Seyedkazemi, S. De Meyer, K. Brown, HIV-1 resistance rarely observed in patients using darunavir once-daily regimens across clinical studies, HIV Clin. Trials 18 (2017) 196–204.
[120] S. De Meyer, T. Vangeneugden, B. van Baelen, E. de Paepe, H. van Marck, G. Picchio, E. Lefebvre, M.P. de Béthune, Resistance profile of darunavir: combined 24-week results from the power trials, AIDS Res. Hum. Retroviruses 24 (2008) 379–388.
[121] I. Dierynck, M. De Wit, E. Gustin, I. Keuleers, J. Vandersmissen, S. Hallenberger, K. Hertogs, Binding kinetics of darunavir to human immunodeficiency virus type 1 protease explain the potent antiviral activity and high genetic barrier, J. Virol. 81 (2007) 13845–13851.
[122] K. Brown, L. Stewart, J.M. Whitcomb, D. Yang, R.E. Nettles, E. Lathouwers, Prevalence of darunavir resistance in the United States from 2010 to 2017, AIDS Res. Hum. Retroviruses 34 (2018) 1036–1043.
[123] E.J. Arts, D.J. Hazuda, HIV-1 antiretroviral drug therapy, Cold Spring Harb. Perspect. Med. 2 (2012) a007161.
[124] M. Rittweger, K. Arasteh, Clinical pharmacokinetics of darunavir, Clin. Pharmacokinet. 46 (2007) 739–756.
[125] J.C.R. Corrêa, D.M. D'Arcy, C.H. Serra, H.R.N. Salgado, Darunavir: a critical review of its properties, use and drug interactions, Pharmacology 90 (2012) 102–109.
[126] A. Calcagno, A. Yilmaz, J. Cusato, M. Simiele, R. Bertucci, M. Siccardi, L. Marinaro, A. D'Avolio, G.D. Perri, S. Bonora, Determinants of darunavir cerebrospinal fluid concentrations: impact of once-daily dosing and pharmacogenetics, AIDS 26 (2012) 1529–1533.
[127] K. Patterson, S. Jennings, R. Falcon, J. Mrus, A. Kashuba, Darunavir, ritonavir, and etravirine pharmacokinetics in the cervicovaginal fluid and blood plasma of HIV-infected women, Antimicrob. Agents Chemother. 55 (2011) 1120–1122.
[128] Website: http://www.ema.europa.eu/docs/en_GB/document_library/EPAR_-_Product_Information/human/000707/WC500041756.pdf (Accessed April 1, 2020).
[129] Website: https://www.accessdata.fda.gov/drugsatfda_docs/label/2012/021976s021lbl.pdf (Accessed April 1, 2020).
[130] Website: https://www.pediatriconcall.com/drugs/darunavir/464 (Accessed April 1, 2020).
[131] Website: https://reference.medscape.com/drug/prezista-darunavir-342647#4 (Accessed April 1, 2020).
[132] Website: https://aidsinfo.nih.gov/drugs/397/darunavir/0/patient (Accessed April 1, 2020).

CHAPTER TWO

Bisoprolol: A comprehensive profile

Ahmed H. Bakheit[a,c,*], Raisuddin Ali[b,*], Ali D. Alshahrani[a], and Adel S. El-Azab[a]
[a]Department of Pharmaceutical Chemistry, College of Pharmacy, King Saud University, Riyadh, Kingdom of Saudi Arabia
[b]College of Pharmacy, King Saud University, Riyadh, Kingdom of Saudi Arabia
[c]Department of Chemistry, Faculty of Science and Technology, Al-Neelain University, Khartoum, Sudan
*Corresponding authors: e-mail address: abujazz76@gmail.com; aliraisuddin786@gmail.com

Contents

Profiles of Drug Substances, Excipients, and Related Methodology, Volume 46
ISSN 1871-5125
https://doi.org/10.1016/bs.podrm.2020.07.006

1. Description

1.1 Nomenclature

1.1.1 IUPAC names

(a) Bisoprolol
1-(propan-2-ylamino)-3-[4-(2-propan-2-yloxyethoxymethyl)phenoxy] propan-2-ol [1].
(b) Bisoprolol Fumerate
(*E*)-but-2enedioic acid;1-(propan-2-ylamino)-3-[4-(2-propan-2-yloxyethoxy methyl) phenoxy] propan-2-ol [2].
(c) Bisoprolol Hemifumarate
but-2-enedioic acid;1-(propan-2-ylamino)-3-[4-(2-propan-2-yloxyethoxy methyl)phenoxy] propan-2-ol [3].

1.1.2 Nonproprietary name

Generic: Bisoprolol.

Synonyms: Bisoprolo, Bisoprololum, (+-)-1-((alpha-(2-Isopropoxyethoxy)-*P*-tolyl)oxy)-3-(isopropylamino)-2-propanol, (*RS*)-1-(4-(2-isopropoxy ethoxymethyl)phenoxy)-3-(isopropyl amino)-2-propanol [4].

1.1.3 Proprietary names

Brand Names:

Cardicor, Concor,Detensiel; Emconcor; Emcor; Euradal; Isoten, Monocor, Soprol, Zebeta, Ziac, Apo-Bisoprolol, Novo-Bisoprolol, PMS-Bisoprolol, Sandoz-Bisoprolol, Monocor, Zebeta [5,6].

1.2 Formulae

1.2.1 Empirical formula, molecular weight, and CAS no

See Table 1.

Table 1 Empirical formula, molecular weight, and CAS number of bisoprolol, and its salts.

Name	Empirical formula	Molecular weight (g/mol)	CAS number	References
Bisoprolol	$C_{18}H_{31}NO_4$	325.45	[66722-44-9]	[1]
Bisoprolol fumarate	$C_{22}H_{35}NO_8$	441.52	[104344-23-2]	[2]
Bisoprolol hemifumarate	$C_{40}H_{66}N_2O_{12}$	766.96	[104344-23-2]	[3]

1.2.2 Structural formula

See Table 2.

Table 2 structure formula of bisoprolol, bisoprolol fumarate and bisoprolol hemifumarate [7].

Bisoprolol	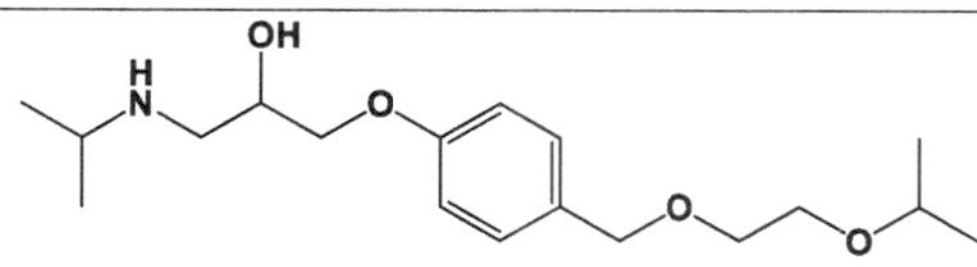
Bisoprolol Fumarate	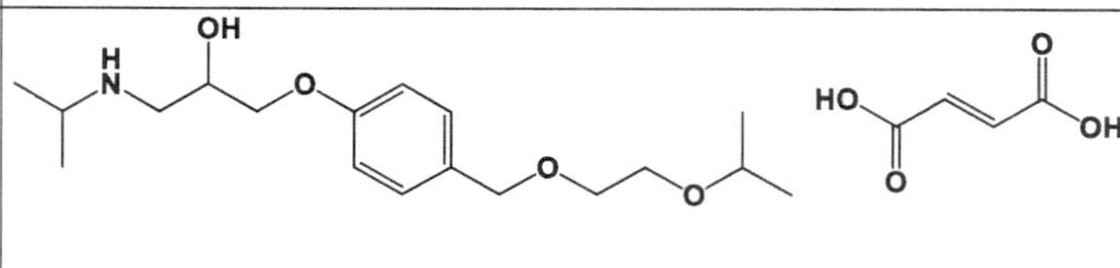
Bisoprolol Hemifumarate	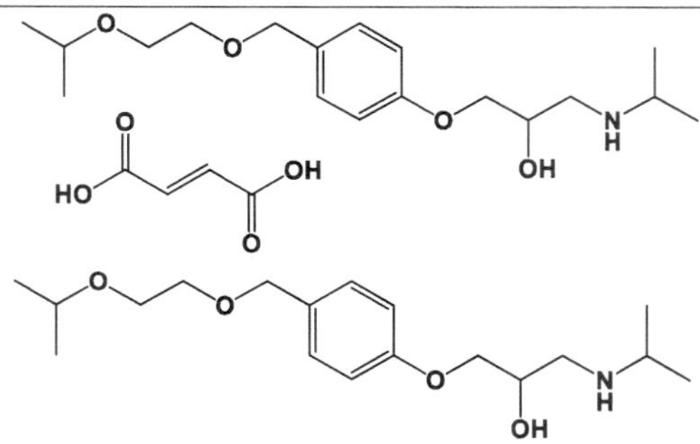

1.3 Elemantal analysis

- Bisoprolol: C: 66.43%, H: 9.60%, N: 4.30%, O: 19.66% [7]
- BisoprololFumarate: C: 59.58%, H: 8.41%, N: 3.16%, O: 28.86% [7]

1.4 Appearance

White or almost white, slightly hygroscopic powder [8].

2. Methods of preparation of bisoprolol

US patent No. US4258062A [9] describes a process for the synthesis of Bisoprolol using intermediate compound 4-(2-isopropoxyethoxy methyl) phenol (**2**), wherein the preparation of compound (**2**) comprises reaction of 4-hydroxy benzyl alcohol (**3**) with 2-isopropoxy ethanol (**4**). This reaction requires a high temperature of around 150 °C. However, the process yields the product with a higher concentration of impurities because of the high reaction temperature. The process described in said patent is illustrated below in Scheme 1.

Dmitri V. Soloviev et al. [10] synthesized (*R*,*S*)-, (*R*)- and (*S*)-des-isopropyl bisoprolol to be used for radiolabeling in three steps starting from commercially available hydroxy- benzyl alcohol (Scheme 2). The [4-(2-isopropoxyethoxy)-methyl]phenol**1** was obtained by mixing

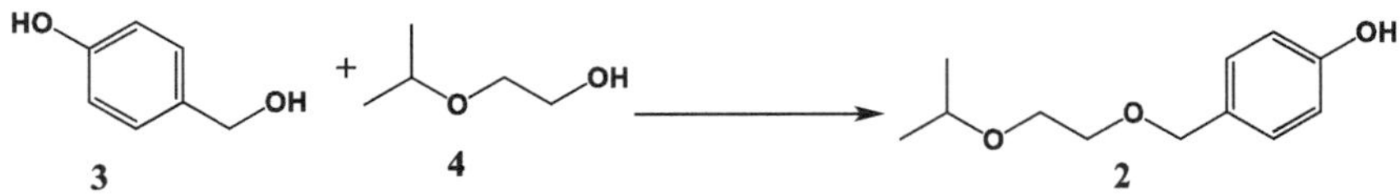

Scheme 1 Synthesis of intermediate compound 4-(2-isopropoxyethoxy methyl) phenol (2).

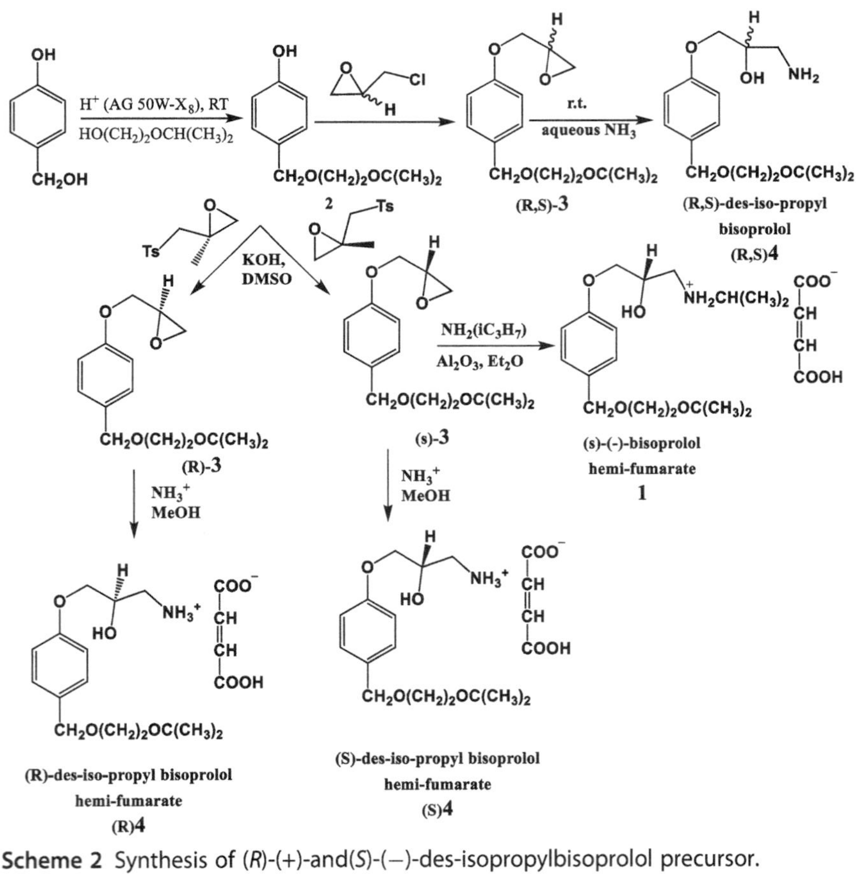

Scheme 2 Synthesis of (*R*)-(+)-and(*S*)-(−)-des-isopropylbisoprolol precursor.

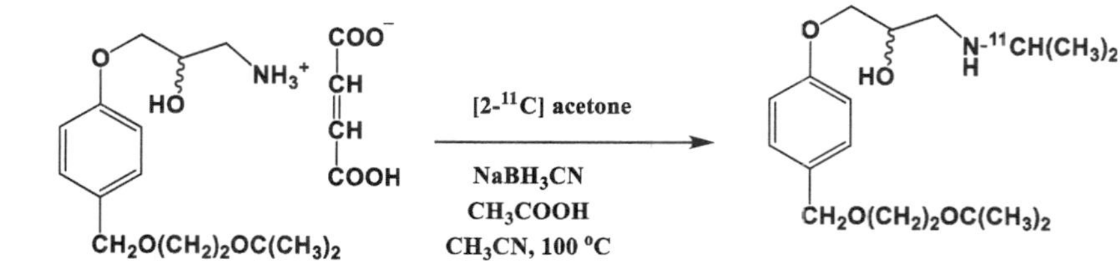

(R)- or (S)-des-iso-propyl precursor hemi-fumarate **(R)- or (S)-[11C] bisoprolol**

Scheme 3 Synthesis of (*R*)-(+)-and(*S*)-(−)-des-isopropylbisoprolol precursor.

2-isopropoxyethanol with hydroxybenzyl alcohol in the presence of strong cation-exchange resin. The resulting phenol 1 was reacted with the tosyl- or chloro-methyloxirane to give (*R*)-, (*S*)- or (*R*,*S*)-[4-(2-isopropoxyethoxy)-methyl]phenoxymethyloxirane 2. When the oxirane ring was opened with ammonia, the des-isopropyl pre-cursors (*R*,*S*)-3,$\alpha._D=0$, (*S*)-3,$\alpha._D=0$ and (*R*)-3 $\alpha_D=+11.0°$ were obtained. When the (*S*)-epoxide 2 was reacted with isopropylamine in the presence of aluminum oxide the reference standard of (*S*)-bisoprolol was obtained with enantiomeric purity of 92%, as measured by chiral HPLC. The α_D of (*S*)-bisoprolol (−12.2°, 1% in methanol) corresponded well to that of authentic (*S*)-bisoprolol $\alpha_D=-12.4°$, −13.0°, 1% in water.

Radiolabeling of bisoprolol was achieved by reductive alkylation of an amino moiety on des-isopropyl precursor by acetone in the presence of sodium cianoborohydride (Scheme 3) as described with minor modifications [11].

3. Physical characteristics

3.1 Color/form

Bisoprolol: Colorless and white to off-white [12]
Bisoprolol Fumarate: white to off-white powder [12]

3.2 Specific rotation

Specific rotation of (10 mg/mL) Bisoprolol Fumarate in methanol is found between −2° and +2° [12].

3.3 Melting point [5]

Bisoprolol: 100 °C.
Bisoprolol Fumarate: 100 °C.

3.4 Dissociation constant [13]

pK_a (strongest Acid) = 14.09.
pK_a (strongest Basic) = 9.27.

3.5 Solubility and partition coefficient

The free base form of Bisoprolol is very less water-soluble. At 25 °C, its water solubility was reported at 2.24 mg/mL [14]. Bisoprolol fumarate is very soluble in water [12] and methyl alcohol [8], freely soluble in alcohol, chloroform, and glacial acetic acid; slightly soluble in acetone and ethyl acetate [8].

The partition coefficient of Bisoprolol and Bisoprolol fumarate was reported as 0.93 and −1.10 [15]. A log *P* value for Bisoprolol also was reported as 2.15 [16].

3.6 Crystallographic properties

Bisoprolol fumarate is available as two anhydrous (form I and form II) and a hydrate form. A various solid state analytical methods were used to study crystalline forms. X-ray powder diffraction (XRPD), fourier transform infrared (FT-IR) spectroscopy, thermoanalytical methods (differential scanning calorimetry and thermogravimetry) and dynamic vapor sorption (DVS). The stability of thermodynamic and solubility of the Bisoprolol fumarate polymorphs were also investigated. Thermoanalytical experiments presented that Form I and Form II obviously different in melting points. Further, form II and hydrate forms too were converted into Form I at higher temperature. The DVS experiment results was proved that both Form I and Form II were became metastable and converted into hydrate form under extremely humid conditions (>80% RH). studies of thermodynamic stability was shown that polymorphs of Form I and Form II were in enantiotropic relationship with an enantiotropic point at about 40–45 °C. Studies of solubility shown that all of the prepared forms were highly soluble, and no difference was found between them [17].

Skotnicki et al. collected X-ray diffraction (XRPD) patterns of crystalline Bisoprolol fumarate as well as quench cooled Bisoprolol fumarate at different temperatures. Comparison of crystalline Bisoprolol fumarate XRPD at 60 and 80 °C shows insignificant changes in peak positioning and very minimal changes in the peak intensities. This indicates that there are non-significant changes in crystal structure upon heating. When bisoprolol quench cooled below the recrystallization temperature the XRPD showed characteristic halo pattern due to conversion of crystalline form into amorphous form. When quench cooled bisoprolol reheated to 60 °C, it regained crstalline structure but the XRPD showed comparatively diminished peak intensities suggesting reduction in the crystallinity [18].

3.7 Thermal analysis

3.7.1 Differential scanning calorimetry

Skotnicki et al. [18] subjected the Bisoprolol fumarate to a cycle of heating, cooling and then again heating within a temperature range from −50 °C to 140 °C. The heating or cooling rate was same and used as 10 °C/min. The first heating curve showed a sharp endothermic peak at onset temperature 102.3 °C. The sharpness of peak was the representation of crystalline nature of Bisoprolol fumarate. After heating sample was cooled to −50 °C. During reheating three events occurred. First was glass transition at −3.6 °C, second was recrystallization of Bisoprolol at 24.2 °C and the third was an endothermic melting peak with an onset temperature 95.4 °C.

Lavanya et al. [19] collected DSC data ofbisoprolol fumarate within a temperature range from 35 to 225 °C at a heating rate of 5 °C/min. Nitrogen gas purging at a flow rate of 50 mL/min was used during temperature scan. The melting point of pure bisoprolol was observed to be 102.4 °C.

3.7.2 Thermogravimetric analysis (TGA)

Detrich et al. [17] conducted thermogravimetric analysis of form I, II and hydrate form of bisoprolol fumarate up to 120 °C from room temperature at a temperature scanning rate of 10°C/min. Form I and II showed no significant weight loss up to 120 °C while the hydrate form lost approximately 2.0% weight. It confirms the form I and II are anhydrous while hydrate form contains water as volatile content. Similarly Skotnicki et al. [18] reported insignificant mass loss up to 170 °C, the sample was started weight loss after 17 °C and up to 99% weight loss of sample weight was observed at 600 °C.

3.8 Spectroscopy

3.8.1 Ultraviolet spectroscopy

UV absorption spectra of 20 ppm Bisoprolol in acetonitrile is shown in Fig. 1.

The UV spectrum of bisoprolol fumarate in acetonitrile was recorded using Shimadzu UV-1800 UV spectrophotometer with a pair of matched, 1 cm quartz cuvette. The absorption spectra of blank as well as the drug were

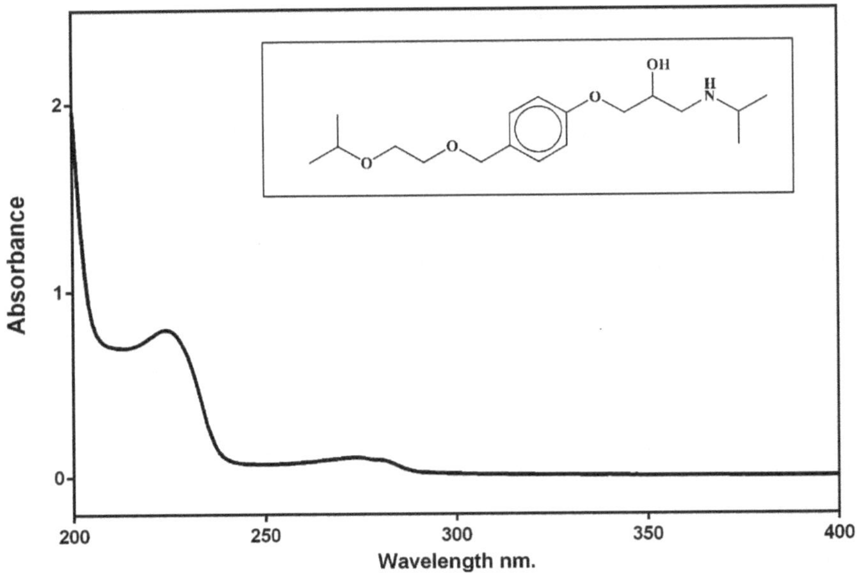

Fig. 1 Ultraviolet absorption spectrum of bisoprolol fumarate dissolved in acetonitrile.

recorded over a wavelength range from 200 to 400 nm. The values of the molar absorptivity of bisoprolol fumarate at 224.20 and 273.80 nm were 17,572.5 and 2362.13 $L\,mol^{-1}\,cm^{-1}$, respectively.

3.8.2 Infrared spectroscopy

The infrared absorption spectrum of bisoprolol fumarate was recorded by the KBr disk method using the Perkin Elmer Spectrum BX FT-IR instrument. Fig. 2 shows the detailed IR spectra of Bisoprolol fumarate. The IR absorption band of aliphatic –CH stretching appeared at 2858 cm^{-1}; CH stretching appeared at 1654 cm^{-1}; C—O stretching appeared at 1654 and 1610 cm^{-1}; C—C appeared at 1243. (Please check all values).

3.8.3 Mass spectroscopy

The mass spectrum of Bisoprolol fumarate was acquired using a mass spectrometer an Agilent Triple Quadrupole 6410 QQQ LC/MS equipped with an ESI (electrospray ionization) source. A connector was used instead of a column. The Mobile phase composed a mixture of HPLC grade water and acetonitrile (40:60). A 10 mg of Bisoprolol fumarate powder was transferred to 10 mL of volumetric flask and it was dissolved in acetonitrile and

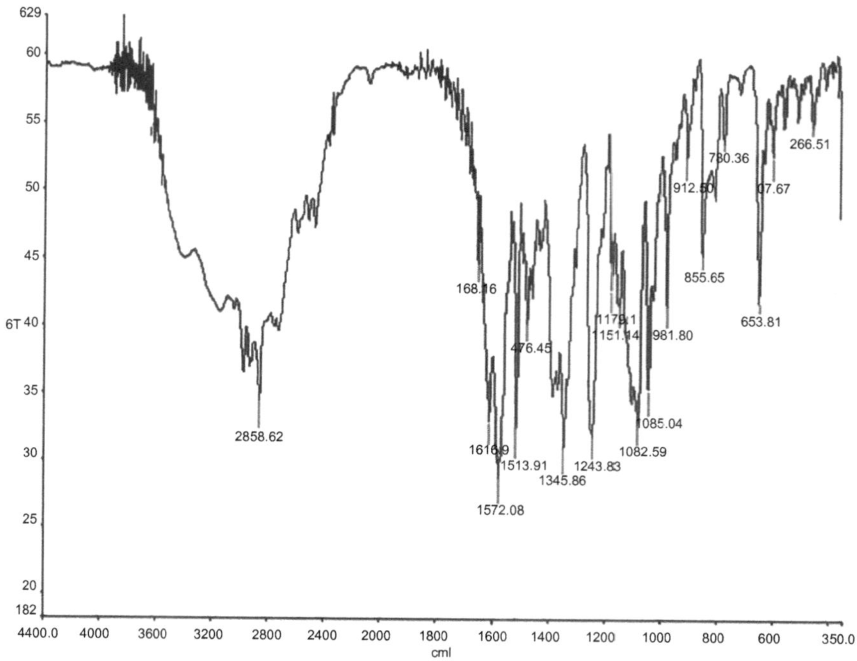

Fig. 2 IR spectrum of bisoprolol fumarate.

diluted to mark with a mobile phase. Based on the intensities of the ions the test solution was diluted to 10^{-3} μgmL^{-1} with the mobile phase. The flow rate was 0.4 mL min^{-1} and the runtimes was 5 min. The mass spectrometry parameters were optimized. The scan was in ultra-scan mode. In the mass range 50–1000 *m/z*, MS2 scans were performed. The ESI was in a positive mode. The source temperature was set to 350 °Cdegrees Celsius. The mass spectrum of Bisoprolol fumarate was shown in Fig. 3. the detailed mass fragmentation pattern at $m/z = 326$; $[M\text{-}115]^+$, 116; $[M\text{-}325]^+$, 74; $[M\text{-}367]^+$ were observed with a nebulizer gas pressure of 55.00 psi, dry gas flow rate of 12.00 L min.

3.8.4 NMR spectroscopy

3.8.4.1 ^{1}H NMR spectroscopy

^{1}H NMR Spectrum of Bisoprolol fumarate was recorded in CDCl3 on a Bruker NMR spectrometer operating at 500 MHz. Chemical shifts are expressed in δ-values (ppm) relative to TMS as an internal standard. Coupling constants (*J*) are expressed in Hz (Table 3 and Fig. 4).

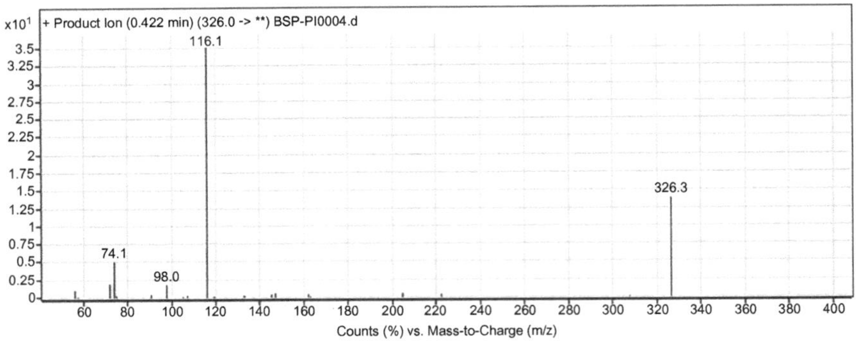

Fig. 3 Mass spectrum of bisoprolol fumarate.

Table 3 ^{1}H NMR of bisoprolol fumarate ($CDCl_3$).

	Location (δ)	Shape	Integration	Correspondences
1	1.18	d, J=6.0 Hz	6H	2CH3
2	1.27	s	1H	–CH
3	1.33	d, J=6.0 Hz	6H	2CH3
4	3.09	t, J=12.0, 11.5	2H	CH2
5	3.29	t, J=6.0 Hz	1H	CH
6	3.64–3.58	m	5H	Aliphatic H
7	3.96	d, J=6.0 Hz	1H	CH
8	4.04	d, J=5.0 Hz	1H	CH
9	4.49	s	3H	CH2 and NH
10	6.66	s	1H	OH
11	6.86	d, J=8.5 Hz	2H	Ar CH
12	7.24	d, J=8.0 Hz	2H	Ar CH
13	7.95	br s	3H	COOH and 2C=H

3.8.4.2 ^{13}C NMR spectroscopy

^{13}C NMR Spectrum of Bisoprolol fumarate was recorded in $CDCl_3$ on a Brucker NMR spectrometer operating at 125 MHz. Chemical shifts are expressed in δ-values (ppm) relative to TMS as an internal standard (Table 4).

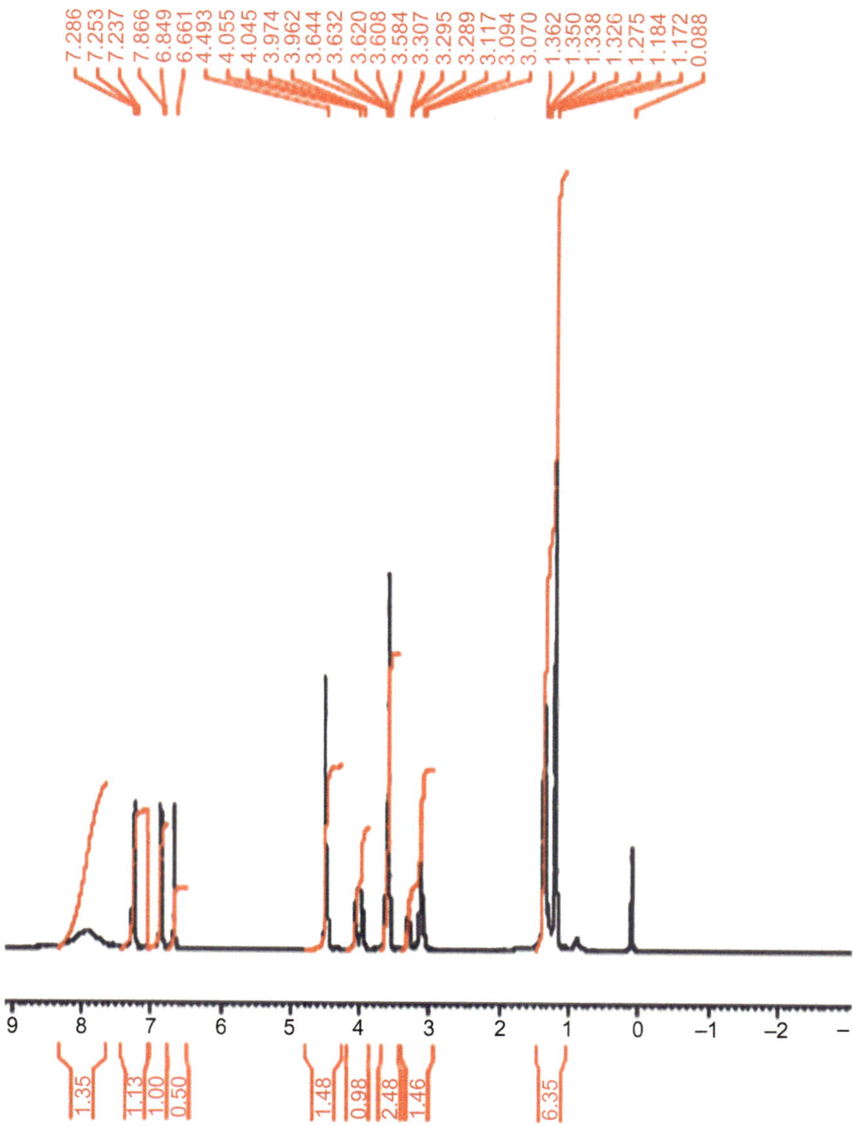

Fig. 4 ^{1}H NMR Spectrum of bisoprolol fumarate.

3.8.4.3 ^{15}N NMR spectrometry

^{15}N NMR Spectrum of Bisoprolol Fumarate was recorded in CDCl3 on a Brucker NMR spectrometer operating at 50 MHz. Chemical shifts are expressed in δ-values (ppm) relative to TMS as an internal standard. ^{15}N NMR (CDCl3) was δ 199.97 (Fig. 5).

Table 4 ^{13}C NMR of bisoprolol fumarate (DMSO-d_6).

Signal	Location (δ)	Correspondences
1	19.19	–CH3
2	19.54	–CH3
3	22.09	2CH3
4	48.22	CH2
5	50.46	CH
6	65.59	CH2
7	67.45	CH
8	69.49	CH2
9	70.08	CH2
10	71.89	CH2
11	72.79	CH
12	114.38, 129.34	Ar-C
13	131.05, 136.23	C=H
14	157.96	Ar-C
15	173.46	C=O

4. Method of analysis

4.1 Compendial method

4.1.1 United states pharmacopeia methods [12]

4.1.1.1 Bisoprolol fumarate $(C_{18}H_{31}NO_4)_2 \bullet C_4H_4O_4$ [12]

The quantity of Bisoprolol Fumarate $(C_{18}H_{31}NO_4)_2 \bullet C_4H_4O_4$ calculated on anhydrous basis is not less than (NLT) 97.5% and not more than (NMT) 102.0%.

4.1.1.1.1 Identification [12]

- *Test A*: Infrared Absorption (197%).
- *Tet B*: The retention time of the main peak in the chromatogram of test preparation is consistent with that of the standard preparation chromatogram, which is obtained in the assay.

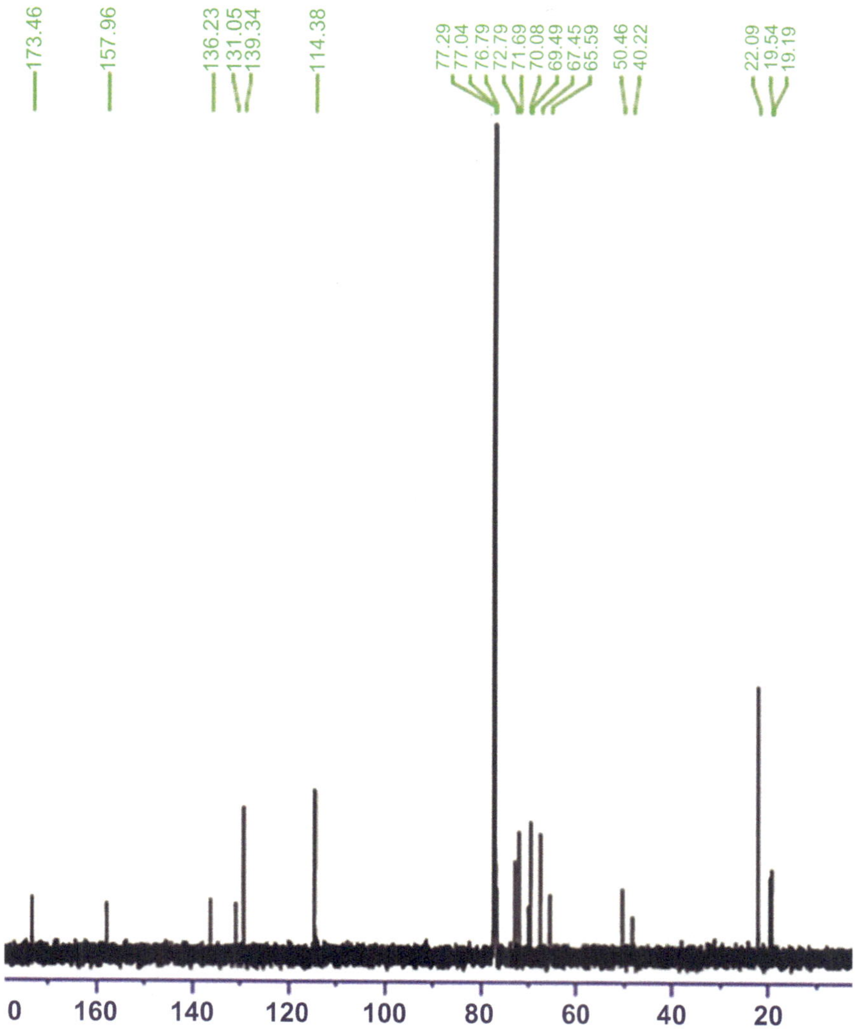

Fig. 5 ^{13}C NMR spectra of bisoprolol fumarate.

- *Specific rotation* The test should be performed according to the Method (781): Specific rotation in a test solution of (10 mg/mL) Bisoprolol Fumarate in methanol is measured and it is found between −2° and + 2.
- *Water Determination*, The test should be performed according to the Method I (921): The result of water contain on sample must be not more than 0.5%.

- *Residue on ignition* (281): The test should be performed according to the general procedure (281), the result of residue must be not more than 0.1.
- *Heavy metal*, The test should be performed according to the "Method I" (231), the result of heavy metal must be not more than 0.002%.

4.1.1.1.2 Related compounds [12] A standard preparation (1 mg/mL) is prepared using a quantity of USP Bisoprolol Fumarate RS in diluent which is a mixture of a water/acetonitrile (65:35). While the Assay solution of Bisoprolol Fumarate sample is prepared using 50 mg of sample powder afterward it is transferred into a 50-mL volumetric flask, then it is dissolved and diluted with diluent to the mark. When, System suitability solution is consisted a mixture of 0.5 mg of propranolol hydrochloride and 1 mg of Bisoprolol Fumarate in diluent per mL.

Chromatographic system is employed to determination of related compounds according to the general method (621). The liquid chromatograph is equipped with a 4.6-mm × 12.5-cm column that contains packing L7, the mobile phase is combined a mixture of 5 mL heptafluorobutyric acid, 5 mL diethylamine and 2.5 mL formic acid into a 1-L portion of diluent. After that, the mixture is filtered, degassed, and adjusted if necessary, the flow rate is 1 mL/min with detection wavelength is 273 nm. the injection volume of test solution into the chromatograph is 10 μL. The resulting chromatogram is recorded and the peak areas are measured. The total impurities percentage are calculated in the portion of Bisoprolol Fumarate according to the formula:

$$100(r_i/r_s)$$

in which r_i is the areas sum for all the peaks, except the fumaric acid and bisoprolol peaks; and r_s is the areas sum of all the peaks in the chromatogram. The resulting percentage of total impurities should be not more than 0.5%.

Chromatograph: the System suitability solution, and the peak responses areas are recorded as directed for procedure (System Suitability under Chromatography (621)). the resolution (Rs) between bisoprolol and propranolol should be not less than 7.0. the tailing factor should be not more than 2.0; and the relative standard deviation for replicate injections should be not more than 2.0%.

The assay of content of fumaric acid in sample substance is performed according to the general method of titrimetry (541).

About 500 mg of Bisoprolol Fumarate is transfered into a beaker, then it is dissolved in 70 mL of dehydrated alcohol followed by addition of 8.0 mL of 0.1 N tetrabutylammonium hydroxide (a Volumetric Solution that is standardized (VS)), this solution is stirred for 2 min, then it is titrated with 0.1 N tetrabutylammonium hydroxide VS, the endpoint potentiometrically is determined using a glasscalomel electrode system. The blank sample is performed in the determination, from the results of titration, each 1 mL of 0.1 N tetrabutylammonium hydroxide is equivalent to 5.804 mg of fumaric acid. Finally, the furmaric acid percentage which is estimated on the anhydrous basis should be in the range 14.8–15.4%.

4.1.1.1.3 Assay [20] About 10 μL of the standard solution and assay solution are Separately Injected into the chromatograph, the chromatograms is recorded, and the main peak areas are measured. The quantity is calculated in mg, of $(C_{18}H_{31}NO_4)_2 \bullet C_4H_4O_4$ in the portion of Bisoprolol Fumarate according to the formula:

$$50C(r_u/r_s)$$

in which C is the concentration, in mg/mL, of USP Bisoprolol Fumarate RS in the Standard solution; and r_u and r_s are the peak areas achieved from the Assay solution and the Standard solution, respectively [12].

4.1.1.2 Bisoprolol fumarate tablets [20]

4.1.1.2.1 Definition The tablets of Bisoprolol Fumarate contain NLT 90.0% and NMT 105.0% of the labeled quantity of bisoprolol fumarate $[(C_{18}H_{31}NO_4)_2 \bullet C_4H_4O_4]$.

4.1.1.2.2 Identification *Thin-Layer Chromatographic* is used for identification of bisoprolol fumarate in bulk and standard according to the general method (201). Equivalent to 40 mg of bisoprolol fumarate is weighed from powdered Tablets (NLT 5) thereafter, it is transferred into a 50-mL volumetric flask. About 20 mL of a mixture of dichloromethane and methanol (7:3) are added to solution then the mixture is shaked for 30 min, and centrifuged. The clear solution of a mixture is used as sample solution. The chromatographic separation is achieved using solvent system is consisted dichloromethane, methanol, and ammonia Test Solution (TS), stronger (70:10:0.8). the procedure is proceed as directed in the bisoprolol fumarate general methodology in previous section, chromatogram development is excepted until the front of solvent is moved about 2/3 of the plate length and the plate is dried in a current of cold air.

4.1.1.2.3 Assay Chromatographic system is used to assay Bisoprolol Fumarate RS in tablets according to Chromatography in the general method (621), System Suitability. An equivalent of 25 mg of bisoprolol fumarate from powdered Tablets (NLT 20) is Transferred into a 25-mL volumetric flask. Followed by Addition of 10 mL of diluent which is a solution of acetonitrile and water in ratio (7:13), and the mixture is sonicated for 10 min, Cooled, and diluted by Diluent to volume, then it is Centrifuged for 20 min. Finally, the clear supernatant solution form the mixture is used as assay solution.

A system suitability solution is prepared by mixing a solution containing 0.5 mg/mL of propranolol hydrochloride and 1 mg/mL of bisoproiol fumarate into diluent.

The concentration of standard solution of USP Bisoprolol Fumarate RS is 1 mg/mL which is prepared in Diluent solution.

The chromatographic separation is performed on 4.6-mm × 12.5-cm; packing L7 column, the mobile phase is consisted of 5 mL of heptafluorobutyric acid, 5 mL of diethylamine, and 2.5 mL of formic acid into the 1 L portion of diluent, the flow rate is 1 mL/min. and detection wavelength And the injection volume are 273 nm and 0 μL, respectively.

4.1.1.2.4 System suitability The requirement of System suitability: The resolution between bisoprolol peak and propranolol peaks should not less than (NLT) 7.0, Tailing factor for each peak should not more than (NMT) 2.0, the concentration relative standard deviation (RSD) of the standard solution should not more than (NMT)2.0%,

Analysis: The percentage of $(C_{18}H_{31}NO_4)_2 \bullet C_4H_4O_4$ is calculated in the portion of Tablets taken:

$$Result = (r_u/r_s) \times (C_s/C_u) \times 100$$

r_u = response of peak from the solution of Sample.r_s = response of peak from the solution of Standard.Cs = USP Bisoprolol Fumarate RS concentration in the solution of Standard (mg/mL).C_u = bisoprolol furnarate nominal concentration in the solution of Sample (mg/mL).

The acceptable range of percentage of Bisoprolol Fumarate in assay is 90.0–105.0%.

4.1.1.2.5 Dissolution test

➢ Test 1

Dissolution test is performed according to the general method **(711)**, the test of dissolution is performed in Apparatus 2 with 75 rpm for 20 min. and medium of dissolution is 900 mL of water.

The sample solution is prepared using a portion of the solution under dissolution test that is Withdrawn, filtered, and diluted with an equal volume of diluent (which is prepared by mixing Methanol, triethylamine, phosphoric acid, and water in ratio (160: 5: 2.5: 35)). While Standard stock solution of USP Bisoprolol Fumarate RS is prepared in water to obtain a solution having a known concentration of about twice the concentration of bisoprolol fumarate in the Sample solution. A sample soltution of $(C_{18}H_{31}NO_4)_2 \cdot C_4H_4O_4$ is measured by using Chromatographic method accoding to the general method (621), System Suitability. The separation is achieved on 4.6-mm × 33-mm; packing L7, the mobile phase is composed methanol, triethylamine, and water in ratio (34:1:50). Which is adjusted with phosphoric acid to give a pH of 4.0 ± 0.1. the flow rate is 1 mL/min with UV detection at 225 nm and injection volume is 50 μL.

The criteria of system Suitability should be required the concentration relative standard deviation (RSD) of the standard solution should not more than (NMT) 2.0%. the tolerances should not less than 80% of the labeled amount of Bisoprolol Fumarate that is dissolved.

➢ Test 2

The assay method for determination of Bisoprolol Fumarate (that is dissolved in the dissolution test) is used the same LC method for assay of Bisoprolol Fumarate in dissolution test 1. The exception is the solution of medium which is 900 mL of 0.5 M sodium chloride, and the diluent solution is a mixture of methanol, 0.1 N hydrochloric acid, triethylamine, and phosphoric acid in ration (160:35:5:2.5), The dimensions of the column are 4.6 mm × 25 cm.

4.1.2 British pharmacopeia methods [21]

The Bisoprolol Fumarate, 99.0–101.0% (anhydrous substance).

4.1.2.1 Identification [21]

- ➢ *Test A*: Infrared absorption spectrophotometry is used to identify of Bisoprolol fumarate using general method (2.2.24).
- ➢ *Water* is estimated according to general method **(2.5.12)** Maximum amount of water in sample is 0.5%, the water contain is determined on 1.000 g.
- ➢ *Sulfated ash* is estimated according to general method **(2.4.14)** Maximum amount of Sulfated ash in substance 0.1%, determined on 1.0 g.

4.1.2.2 Related substances [21]

- ➢ Liquid chromatography (2.2.29).

A diluent solution is consisted a mixture of acetonitrile R1 and water (20:80, v/v). this diluent Test solution is prepared by transferred 25 mg of the sample of Bisoprolol fumarate into 25.0 mL of volumetric flask then it is dissolved and diluted with the diluent to mark. From this test solution about 1.0 mL is diluted with diluent to 100.0 mL, after that, 2.0 mL of final solution is diluted with diluent to give Reference solution (a). while, Reference solution (b) is prepared by dissolved a vial contents of bisoprolol for peak identification CRS (containing impurities A and E) in 1.0 mL of the diluent. Whereas, Reference solution (c) is prepared by dissolved a vial contents of bisoprolol for system suitability CRS (containing impurity G) in 1.0 mL of the diluent. The separation and quantitation are achieved on a C_{18} column (4.6 × 250 mm, 5 μm), and a mobile phase composed of 10 g/L solution of phosphoric acid R (solvent A) and 10 g/L solution of phosphoric acid R in acetomtrile R1 (solvent B, v/v). For the elution of combinative, elution is performed using the following conditions: (i) isocratic elution with 95% A from the injection time until 4 min, (ii) a gradient from 95% A to 80% A over the next 4 min, (iii) isocratic elution with 80% A over the next 7 min, (iv) a further linear gradient from 80% to 20% A for 19 min, and finally (v) isocratic elution with 20% A for 2 min (total run time 36 min). The detection wavelength was set at 225 nm, and the flow rate was 1.0 mL/min. The column temperature was set at 20 ± 2 °C, and the injection volume was 10 μL. according to this conditions, and each peak area is determined by this method: the each peak area of the impurities which appear after the peak of bisoprolol from the sample solution is not larger than the peak area of bisoprolol from the standard solution, and the total peak area of these impurities is not larger than 2.5 times of the peak area from the standard solution.

4.1.2.3 Assay [21]

Dissolve 0.300 g of sample substance is dissolved in 50 mL of anhydrous acetic acid R. this solution is titrated with 0.1 M perchloric acid, the endpoint is determined using potentiometrically according to general method (2.2.20). the resulting 1 mL of 0.1 M perchloric acid is equivalent to 38.35 mg of $C_{40}H_{66}N_2O_{12}$.

4.1.2.4 Storage [21]

The sample substance is preserved in an airtight container, and it must be protected from light.

4.1.2.5 Impurities [21]

The specified impurities A, E, G are detected using one or other of the tests in this reference. Other detectable impurities (if present at a sufficient level, or unspecified impurities), therefore, it is not necessary to identify these impurities. B, C,D, F, K, L, N, Q,R, S, T, and U.

A. (2*RS*)-1-(4-hydroxymethyl-phenoxy)-3-isopropylaminopropan-2-ol,

B. (2*RS*)-1-isopropylamino-3-[4-(2-propoxyethoxymethyl)phenoxy] propan-2-ol,

C. 1-[4-[4-(2-hydroxy-3-isopropylaminopropoxy)benzyl]phenoxy]-3-isopropylaminopropan-2-ol,

D. 1-[4-[4-(2-hydroxy-3-isopropylaminopropoxy)benzyloxylmethyl] phenoxy]-3-isopropylamino propan-2-ol,

E. (EZ)-[3-[4-(2-isopropoxyethoxymethyl)phenoxy]allyl] isopropylamine,

F. (2*RS*)-2-[4-(2-isopropoxy-ethoxymethyl) phenoxy]-3-isopropylaminopropan-2-ol,

G. [[(2-isopropoxyethoxy)methoxy]methy]] phenoxy]-3-isopropylaminopropan-2-ol,

K. 2-isopropoxyethyl 4-[[(2*RS*)-2-hydroxy-3-(isopropylamino) propyl]oxy]benzoate,

L. 4-[[(2.*RS*)-2-hydroxy-3-(isopropylamino)propyl]oxy] benzaldehyde,

N. (2*RS*)-1-[4-[(2-ethoxyethoxy)methyl]phenoxy]-3-isopropylaminopropan-2-ol.

Q. (2*RS*)-1-(isopropylamino)-3-[4-(2-methoxyethoxy)methyl] phenoxypropan-2-ol,

R. (2*RS*)-1-(Gsopropylamino)-3-(4-methylphenoxy)propan-2-ol,

S. 4-hydroxybenzaldehyde,

T. 4-[(3-isopropyl-2-oxo-1,3-oxazolidin-5-yl)methoxy] benzaldehyde,

U. 5-[[4-(hydroxymethyl)phenoxy]methyl]-3-isopropyl-1,3-oxazolidin-2-one.

4.2 Reported methods of analysis

4.2.1 Spectroscopic methods

4.2.1.1 Ultraviolet–visible spectroscopy

Mohammed et al. [22] developed and validated a UV spectrometric method for the estimation of Bisoprolol fumarate in aqueous media. The λ_{max} at 271 nm was used. The method followed Beer Lambert's law in the concentration range 5–25 μg/mL with correlation coefficient value, not less than 0.9986. The limit of detection and limit of quantification of the validated method was 0.22 and 0.66 μg/mL, respectively.

Sahu et al. [23] reported specific ratio spectra derivative spectrophotometry and a simple UV spectrophotometry using simultaneous equation methods for the simultaneous determination of bisoprolol fumarate and hydrochlorothiazide from combined pharmaceutical dosage forms. For ratio spectra derivative spectrophotometry, the amplitudes were measured at 202.6 nm for bisoprolol fumarate and the difference in amplitudes at 212.6 and 230 nm were measured for the determination of hydrochlorothiazide. In the simultaneous equation method, the signals were measured at 223 and 271.6 nm. The concentration of each drug was obtained by using the absorptivity values calculated for the drugs at two wavelengths, 223 and 271.6 nm. Commercial tablet formulations were successfully analyzed using the developed methods.

Kumbhar et al. [24] demonstrated a colorimetric method for the estimation of bisoprolol. The method is based on the formation of a blue-colored chromogen when reacted with ferric chloride and potassium ferricyanide. The method showed linearity over a range of 1–13 μg/mL. The blue-colored complex has absorption maxima at 770 nm with molar absorptivity and Sandell's 2.2387×10^4 L mol^{-1} cm^{-1} and 0.0154 μg/cm^2/0.001 absorbance units, respectively.

A first-order derivative UV spectroscopic method was reported for the estimation of Bisoprolol in bulk as well as tablet formulations. Bhusnure et al. prepared different dilutions of Bisoprolol fumarate in water then scanned the samples from 200 to 400 nm range and the first-order derivative of the spectrum was recorded. The method was linear from 2 to 12 μg/mL. The limit of detection and limit of quantification was calculated as 0.04753 and 0.1584 μg/mL, respectively [25].

A colorimetric method for the estimation of Bisoprolol in pharmaceutical formulations was reported by Panainte et al. [26] Bisoprolol forms a blue-colored ion-pair complex with bromocresol green in an acidic environment. The blue-colored complex was extracted in chloroform and recorded the absorbance at 402 nm (λ_{max}). The method was linear from 7 to 80 μg/mL with a correlation coefficient of 0.9998. The method was robust and successfully applied for the assay of Bisoprolol in the pharmaceutical formulation such as Concor®, Bisotens®, and Bisoblock®.

Panainte et al. [27] developed a colorimetric method for to estimate bisoprolol fumarate in tablets. This method was based on the reaction of Bisoprolol with tropaeolin 00. Bisoprolol forms a yellow colored complex when reacted with tropaeolin 00. this complex was extracted in dichloromethane with maximum absorbance at 412 nm. The method was validated and linear between 5 and 30 μg/mL.

Panainte et al. [28] developed a spectrophotometric method to estimate bisoprolol in drug preparations using green bromocresol in hydrochloric acid medium. The maximum absorbance of the product of reaction was found at 402 nm. this method was shown good linearity in the concentration range of 7–80 μg/mL with correlation coefficient ($r = 0.9998$), while the limit of detection (LOD) was found about 1.78 μg/mL and the limit of quantification (LOQ) was found about 5.41 μg/mL.

Atul et al. [29] developed two spectrophotometer methods for simultaneous determination of bisoprolol fumarate and hydrochlorothiazide in combined pharmaceutical dosage. The "method I" was depended on "simultaneous equations" while the "method II" was depended on the instrument's "multi-component method of analysis." 0.1 N NaOH was used as a solvent for both these methods. The maximum absorbance of bisoprolol fumarate and hydrochlorothiazide were observed at 224 and 273 nm, respectively. Linearity of bisprolol fumarate and hydrochlorothiazide lies between 3–21 μg/mL and 3–18 μg/mL, respectively.

Kakde et al. [30] developed a spectrophotometric method for simultaneous determination of amlodipine besylate and bisoprolol fumarate in combined dosage form. A simultaneous equation method was employed for analysis with 10% methanol as a solvent. For the measurement of bisoprolol fumarate and amlodipine besylate, two wavelengths 222 and 365 nm were selected, respectively. The linearity range of both drugs were observed at concentration range of 5–100 μg/mL.

4.2.1.2 Spectrofluorimetric method

Abdelwahab et al. [31] reported a spectrofluorimetric method for simultaneous estimation of Bisoprolol fumarate and rosuvastatin calcium in tablet formulation as well as in the human spiked plasma. The fluorescence intensity of Bisoprolol fumarate and Rosuvastatin calcium were measured in methanol at an emission wavelength of 297 and 485 nm upon excitation at 227 and 242 nm, respectively. The emission values of both drugs showed zero-emission at the emission wavelength of another drug thus making it suitable for simultaneous estimation. The method reproduced excellent linearity over the range of 10–500 ng/mL and 20–1000 ng/mL for Bisoprolol and rosuvastatin respectively.

4.2.2 Voltammetric method

Goyal et al. [32] investigated an electrochemical method using differential pulse voltammetry at single-wall carbon nanotubes (SWNTs) modified glassy carbon electrode (GCE). The electrode was prepared with excellent electrocatalytic activity against BF oxidation, which led to significant sensitivity improvement compared with the glass carbon electrode, where the study did not detect electrochemical activity. A sharp anodic peak with a 950 mV potential for BF oxidation was presented in the SWNT's adjusted GCE. The linear calibration curve was obtained in optimal conditions, in the BF 0.01–0.1 mM concentration range in 0.5 M solution of phosphate buffer (pH 7.2), thru a correlation coefficient of 0.9789 and detection limit of 8.27×10^{-7} M. The adapted electrode was administered in human urine and commercial tablets to evaluate the drug without any prior extraction.

Bozal et al. [33] developed Differential pulse voltammetry (DPV) and square wave voltammetry (SWV) methods for the simultaneous assay of bisoprolol fumarate and hydrochlorothiazide in a commercial tablet dosage form using a glassy carbon electrode. This methods were performed without any pretreatments of the samples for analyses. The method of simultaneous assay for bisoprolol fumarate and hydrochlorothiazide were achieved by measuring of its peaks potentials at about 1100 and 1400 mV for DPV and SWV, respectively.

The best single peak shape, reproducibility, and resolution between the peaks of two compounds were obtained with Robinson (BR) (pH 3.0). The linear calibration curve was obtained in optimal conditions, the concentration range of bisoprolol fumarate and hydrochlorothiazide was 5.00–80.00 μg/mL and 2.50–70.00 μg/mL for both DPV and SWV methods, respectively.

4.2.3 Capillary electrophoresis

Duan et al. [34] developed a new method for separation and estimation of metoprolol tartrate and bisoprolol fumarate by capillary electrophoresis coupled with tris(2,2′-bipyridyl)-ruthenium (II) electrochemiluminescence detection. The pH, running buffer concentration, separation voltage, and detection potential were optimized for electrochemiluminescence detection. The limit of detection was 1.9×10^{-8} mol/L and 9.4×10^{-8} mol/L for metoprolol tartrate and Bisoprolol fumarate respectively. The method was used for the determination of metoprolol tartrate and bisoprolol fumarate in human urine samples. The method was also used for the interaction studies between metoprolol tartrate or bisoprolol fumarate and human serum albumin.

4.2.4 *Chromatography*

4.2.4.1 Thin-layer chromatography

Witek et al. [35] reported a thin layer chromatographic method for the identification and determination of Bisoprolol in pharmaceutical preparations. They derivatized the Bisoprolol using dabsyl chloride and then chromatographed on silica-based TLC plates with a mobile phase composed of n-hexane-chloroform-acetone in ratio of 2:5:1, respectively. The densitometric quantitation of dabsylated Bisoprolol was done at 490 nm. TheRSD of quantitation was between 1.6 and 2.3%.

4.2.4.2 High-performance liquid chromatography methods

Zaki et al. [36] developed and validated a reversed-phase high-performance liquid chromatographic method for the simultaneous estimation of Bisoprolol fumarate and rosuvastatin calcium in a combination tablet dosage form. The chromatographic separation was performed on a C18 column (150 × 4.6 mm, 0.5 μm) with a mobile phase made up of acetonitrile and 0.05% aqueous solution of orthophosphoric acid (65:35 v/v). The mobile phase flow rate was 1 mL/min and UV detection was performed at 230 nm. The column temperature and the injection volume was 25 °C and 20 μL, respectively. The method was linear over 0.2–50 μg/mL concentration range.

Another HPLC method for the simultaneous estimation of Bisoprolol and enalapril in combined dosage form was developed by Logoyda et al. [37]. The separation and quantification of two compounds was achieved on C18 column (250 × 4.6 mm, 5 μm) with mobile phase containing a mixture of methanol: phosphate buffer (65:35, v/v) and the flow rate was 1 mL/min. The column temperature and the injection volume were 25 °C and 300 μL, respectively. The ultraviolet scanning of the analytes was performed at 225 nm. The reported retention time for Bisoprolol was 4.75 min. The method followed linearity between 0.625 and 5.0 mg/mL concentration range.

Panainte et al. [38] developed and validated a HPLC method for quantification of Bisoprolol in the bulk and pharmaceutical dosage form. Chromatographic elution was achieved using a Zorbax SB-C18 column (100 × 3 mm, 5 μm) at 25 °C and UV detection was performed at 225 nm. The mobile phase was run in isocratic mode, using a combination of phosphate buffer (pH 3.5) and acetonitrile at a ratio of 7:3, respectively. The temperature of the column and the volume of injection were 25 °C and 20 μL, respectively. The reported retention time for Bisoprolol was 1.158 min. The method followed beer's law over a concentration range of 5–90 μg/mL.

The limits of detection and quantification for Bisoprolol were 1.63 and 4.94 μg/mL, respectively.

Caudron et al. [39] described a method for the simultaneous determination of three cardiovascular drugs: celiprolol, bisoprolol, and irbesartan in human plasma. A solid-phase extraction was utilized for sample preparation. The separations of three analytes were performed using a C18 column (5 μm, 4.6 mm × 150 mm). The mobile phase was composed of two components(A and B). Component A was phosphate buffer (pH 3.4) and component B was pure acetonitrile. The flow rate was 1 mL/min. The starting mobile phase was linear with composition A:B (76:24) which changed to gradient run over 15 min in such a manner so that final composition changed to A:B (40:60). Then held constant for 6 min. The diode array detector was used for chromatographic detection within a range of 200 to 280 nm. The retention times for Celiprolol, Bisoprolol, and irbesartan were 5.273, 7.445, and 16.502 min, respectively.

Shaikh et al. [40] reported an HPLC method for the simultaneous estimation of Bisoprolol fumarate and hydrochlorthiazide using 4.6 mm × 250 mm, 5 μm cyano analytical column at 25 °C temperature. The mobile phase was composed of 0.1 M aqueous phosphate buffer, acetonitrile, and tetrahydrofuran (85:10:5, v/v/v). The chromatogram was recorded at a flow rate of 1.0 mL/min in an isocratic manner. Both analytes were detected at 225 nm using a UV detector in less than 10 min.

Braza et al. [41] developed HPLC method coupled with fluorimetric detection for the estimation of Bisoprolol fumarate in human plasma. The method showed linearity over a range of 6.25–200 ng/mL. The method involved liquid-liquid extraction of Bisoprolol fumarate using diethyl ether from human plasma. The method consisted of an isocratic mixture of acetonitrile: HPLC water containing 1.2% (w/v) of triethylamine and the pH adjusted to 3 with 85% orthophosphoric acid (18:82 v/v) as mobile phase. Reversed-phase C18 column (125 × 4 mm i.d.) was used as the separation phase, and the mobile phase flow was 1 mL/min. The injection volume was 100 μL. The excitation wavelength and the emission wavelength were 232 and 300 nm, respectively.

Another simple HPLC coupled with fluorescence detection method for estimation of Bisoprolol in plasma or serum was reported by Eastwood et al. [42]. Methyl t-butyl ether was used as an extraction solvent. A Spherisorb S5W silica packed column (125 × 5 mm i.d., 5 μm particle size) was used as separation phase. Camphor sulfonic acid (1 mM) in methanol was used as a mobile phase. The mobile phase flow rate was 2.0 mL/min.

The column eluent was subjected to fluorescence detection at an excitation wavelength 215 nm.

Patel et al. [43] developed a reverse phase high-performance liquid chromatographic method for the simultaneous estimation of Bisoprolol fumarate and hydrochlorothiazide in tablet dosage forms using a UV detector. Lichrospher 100 C18 column (5 μm, 20 cm × 4.6 mm) was used. The mobile phase was water, acetonitrile, and tetrahydrofuran in a proportion of 80:20:5 (v/v/v). The flow rate was 1 mL/min and the wavelength of the UV detector 225 nm, The retention time of Bisoprolol fumarate and hydrochlorothiazide was 1.48 ± 0.02 and 4.72 ± 0.03 min, respectively. The resolution factor was 9.0. The calibration curve for bisoprolol fumarate was found to be linear from 10 to 150 μg/mL. The limit of detection and limit of quantification for bisoprolol fumarate was found to be 3.5 and 8.5 μg/mL, respectively. The proposed was used for simultaneous estimation of bisoprolol fumarate and hydrochlorothiazide in the tablet dosage forms.

Hefnawy et al. [44] developed a chiral chromatographic method for separation and quantification of bisoprolol enantiomers in pharmaceutical formulations. For the better resolution of enantiomers, the teicoplanin macrocyclic antibiotic chiral stationary phase was used with fluorescence detection at 275 and 305 nm as excitation and emission wavelength respectively. The mobile phase was made up of methanol–glacial acetic acid-triethylamine at a ratio of100:0.020:0.025 (v/v/v) and the mobile phase flow rate was 1.5 mL/min. The reported method was linear over the range of 5–250 ng/mL for each enantiomer. The overall recoveries of S-(−)- and R-(+)-bisoprolol from pharmaceutical products ranged from 97.6 to 100.5% with %RSD ranging from 0.7 to 2.6%. The limit of quantification (LOQ) and limit of detection (LOD) for each enantiomer were 5 and 2 ng/mL, respectively. The method proved to be of chiral quality control for bisoprolol formulations by HPLC.

For the estimation of bisoprolol in the plasma, a derivatization HPLC method was reported by Ulu et al. [45]. A fluorogenic reagent, namely, 4-chloro-7-nitro-2,1,3-benzoxadiazole was used in borate buffer at pH 9.5. The chromatographic elution was achieved at 40 °C temperature using a reversed-phase C18 column (4 μm, 150 × 4.6 mm). The mobile phase was composed of methanol: water (70:30, v/v) at a flow rate of 1.2 mL/min. Excitation and emission wavelength were 458 and 525 nm respectively for fluorescent detection. The method followed the beer lambert law for a concentration range of 10–2000 ng/mL. The method was used for pharmacokinetic estimation of Bisoprolol formulations in healthy human volunteers.

Kondratova et al. [46] developed a RP-HPLC method for evaluating bisoprolol fumarate in bulk and pharmaceutical dosage form. A chromatographic separation of the ACQUITY Arc System was performed. An isocratic a mobile phase consisted of acetonitrile: phosphate buffer (25:75, v/v) was used at a flow rate of 1.4 mL/min. A Waters Symmetry C18 column (3.9 mm i.d. × 150 μm particle size) was used. Orthophosphoric acid was used to adjusted the pH of the mobile phase to 7.0 and the UV detector was 226 nm. It was found that the retention time of bisoprolol fumarate was 2.09 min. In linearity, precision, accuracy and stability, the proposed procedure has been validated according to ICH guidelines. The detection limit and the quantification limit are estimated at 0.4825 and 1.4621 μg/mL, respectively.

A HPLC method with fluorescence detection for estimating bisoprolol in human plasma was reported by Tatar Ulu & Aydoğmuş [47]. Bisoprolol with 4-chloro-7-nitro-2,1,3-benzoxadiazole as a fluorogenic reagent was used to generate a fluorescent product in the borate buffer at pH 9.5. The chromatographic separation was performed an isocratic mobile phase of methanol–water (70:30, % v/v). The flow rate was 1.2 mL/min at 40 °C using a reversed-phase C18 column (4 μm, 150 × 4.6 mm). A fluorescence detector was used at the excitation and emission wavelengths of 458 and 525 nm, respectively. The reported method validated for linearity, detection limit, quantification limit, precision, accuracy, recovery and suitability of the method. The analysis was linear over the 10–2000 ng/mL range. This approach was used for pharmacokinetic studies in healthy volunteers.

Joshi et al. [48] established and validated a stability- indicating HPLC process for simultaneous evaluation of bisoprolol fumarate and hydrochlorothiazides in pharmaceutical dosage form. The chromatographic separation was perform on an Inertsil ODS 3V (5 μm, 525 cm × 4.6 mm) column was used with isocratic flow. The mobile phase consisted of 0.1 M potassium dihydrogen phosphate buffer and acetonitrile (70:30, v/v) at a flow rate of 1.0 mL/min. The UV detection was carried out at 228 nm. A linear response was detected over the range 2,5–50 μg/mL and 6,25–125 μg/mL for bisoprolol fumarate and hydrochlorothiazide concentration range, respectively. Bisoprolol fumarate and hydrochlorothiazide detection limit and quantitative limit were 0.01 and 0.03 μg/mL and 0.01 and 0.05 μg/mL, respectively.

Panainte et al. [49] developed a RP-HPLC method for estimation of bisoprolol fumarate in tablets. The mobile phase was methanol: acetonitrile: 45 mM potassium dihydrogen phosphate buffer (30:25:45) at pH 3.0 and at flow rate of 0.3 mL/min. The chromatographic separation was perform on a SB-C18 Nucleosil column (125 × 4 mm). The UV detector was 225 nm. The method was validated in terms of linearity, detection limit, quantification limit, precision, accuracy, recovery and system suitability. The retention time of bisoprolol fumarate was 2.32 min. A linear response was detected over the range 0.3–10 μg/mL.

Mahu et al. [50] developed and validated a method of HPLC for the determination of bisoprolol fumarate in tablets. The method was based on an Eclipse XDB C18 type column (150 mm × 4.6 mm, 5 m) and mobile phase contained a mixture of water/methanol/acetonitrile in a ratio of 50:30:20 (v/v/v) with flaw rate 1 mL/min. A UV detection wavelength was performed at 225 nm. The volume of injection and the The temperature of column were 20 mL and at 25 °C, respectively. The method showed two linearity areas in the range of 0.8–80 g/mL and 80–1000 g/mL, respectively. The detection limit was 1.3 g/mL and the quantification limit was 3.98 g/mL.

Arjun et al. [51] developed the reverse phase isocratic method for determination of Bisoprolol fumarate in tablet dosing. The method was achieved on Gracesmart RP18, 5 mcm (150 mm × 4.6 mm) column and mobile phase was consisted of phosphate buffer solution pH 4.5 and acetonitrile in ratio (82:18) with flow rate and UV detection wavelength were 2.0 mL/min and 225 nm, respectively. Linear response was detected over the range 50–500 μg/mL, with correlation coefficient of 0.99921, and recovery percent was 99.0%.

Dobričić et al. [52] developed RP-HPLC method for simultaneous determination of bisoprolol fumarate, hydrochlorothiazide and impurities (A, L, K and B). Separation was carried out using BDS Hypersil C8 column (250 × 4.6 mm, 5 μm particle size), the temperature of column was 50 °C. and an optimized mobile phase contained of phosphate buffer solution (pH 3.35) (solvent A) and acetonitrile (solvent B, v/v). For the combinative elution, elution was executed using the following conditions: (i) isocratic elution with 85% A from the injection time until 2.5 min, (ii) a gradient from 85% A to 65.6% A over the next 0.10 min, (iii) a further linear gradient from 65.5% to 62% A for 4.4 min, (iv) an additional linear gradient from 62% to 50% A for 3 min, (vi) a more linear gradient from 50% to 85% A for 0.1 min, and finally (vi) isocratic elution with 85% A for 5 min (total run time 15 min). The flow rate and detection wavelength were 1.5 mL/min and 220 nm, respectively. The critical factors pairs separation for experiments and their influence were selected using Box-Behnken plan. Desirability function of Derringer was applied for choice of optimal chromatographic conditions.

The analysis was linear with acceptable correlation coefficient ($r \geq 0.9945$) for (impurities) and $r = 0.9998$ for (active substances). Recoveries percent of Accuracy was (93.5–107.1%) for (impurities) and 98.1–101.9% for (active substances).

Reverse phase ultra-performance liquid chromatography (RP-UPLC) method was developed for determination of bisoprolol hemifumarate and hydrochlorothiazide in tablets by Dinç et al. [53]. The optimization of the method was carried out using experimental design which was depended on a 3^3 full factorial design for identify the factors effects (column temperature, flow rate, and 0.1 M H_3PO_4% in mobile phase), the column temperature, flow rate, and (0.1 M H_3PO_4) (which were selected as the optimum conditions of chromatographic using experimental design) were 58.2 °C, 0.37 mL/min, and 23.6%, respectively. Linearity response of bisoprolol hemifumarate and hydrochlorothiazide were detected at 8.0–40.0 μg/mL and 2.0–26.0 μg/mL, respectively.

Madhusudhanareddy et al. [54] developed and validated a liquid chromatographic method for the quantification of bisoprolol fumarate in tablets. The separation was performed on a C18 column using mobile phase containing a buffer / acetonitrile (75:25, v/v, pH 5.6), the flow rate was 1.0 mL/min. The separation was carried out at room temperature. Detection that was used a diode array detector was carried out at 226 nm. The mean recovery was 99.87–100.43%. linearity response was detected over range 25–100 μg/mL with Correlation coefficient was (r) 0.9998, the detection Limit and the quantification Limit were 0.02 and 0.05 μg/mL, respectively.

Kurbanoglu et al. [55] developed and validated a UPLC method for the simultaneous quantification of bisoprolol fumarate and hydrochlorothiazide in combined dosage forms and in spiked samples of human urine. The chromatographic separation was carried out on an Acquity UPLC BEH C18 1.7 μm (2.1 × 50 mm) column, column temperature was 40 °C. The mobile phase was composed of acetonitrile: 20 mM phosphate buffer at pH 3.0 with a gradient elution at detection wavelength 225 nm. The gradient elution was performed by simultaneously changing the flow rate and mobile phase composition. Initially the mobile phase composition was acetonitrile: phosphate buffer in ratio 15:85 with a flow rate of 0.7 min mL^{-1}. Then it was changed to 20:80 in 0.6 min analysis time, furthers it was changed to 60:40 within 1.4 min analysis time with the flow rate reached 0.9 min mL^{-1}. At the end of 2 min, it was returned to the initial mobile phase composition with flow rate 0.7 min mL^{-1}. The linearity range of bisoprolol fumarate and hydrochlorothiazide were found 0.5–250 μg/mL and 0.5–150 μg/mL, respectively. The detection limit and quantitation limit for hydrochlorothiazide were found of 0.01 and 0.03 μg/mL, respectively, and for bisoprolol fumarate were 0.07 and 0.21 μg/mL, respectively.

Vora and Kadav [56] developed a stability indicating RP-HPLC method for simultaneous quantification of bisoprolol fumarate and amlodipine besylate in combination dosage form. The separation was achieved on was Luna C18-2 column (3 μ, 50 × 4.6 mm ID) with mobile phase containing 25 mM ammonium acetate (pH 5.0) and methanol (65:35), the flow rate was 0.8 mL/min, the detection wavelength at 230 nm. The retention times for both bisoprolol and amlodipine were 1.45, 3.91 min, respectively. Linearity response was detected over range 8–33 μg/mL. the mean recovery of bisoprolol fumarate and amlodipine besylate were 99.1% and 98.6%, respectively.

Wang et al. [57] developed HPLC method for determination of bisoprolol fumarate in tablets. The chromatographic separation was achieved on YWG-C18 column, with mobile phase was composed methanol: 0.1 M sodium acetate buffer (pH 5) in ratio (6:4). The detection wavelength was 270 nm. *p*-hydroxybenzoyl propylate was used as internal standards. Linearity of the peak area ratio was calculated in range of concentration from 2 to 10 μg/mL and the internal standard was 0–5 μg/mL, respectively with ($r = 0.999$).

A enantioselective high-performance chromatographic liquid (HPLC) technique for the quantitation of S-(−)- and R-(+)-bisoprolol in human plasma was developed and validated by Hefnawy et al. [58]. A solid-phase extraction procedure was used to extract analytes and IS from human plasma samples prior to HPLC analysis. Resolution of baseline was obtained through the teicoplanin macrocyclic antibiotic chiral stationary phase (CSP) that was called Chirobiotic T, a polar ionic mobile phase (PIM) was composed a mixture of methanol–glacial acetic acid-triethylamine (100: 0.02: 0.025, v/v/v) with flow rate of 1.5 mL/min. Fluorescence detector (excitation and emission) were 275 and 305 nm, respectively. The linearity of each enantiomer concentration was performed over the range of 20–200 ng/mL. mean recovery for *S*-(−)- and *R*-(+)-bisoprolol enantiomers were within the range of 95–102%.

4.2.4.3 Liquid chromatography-tandem mass spectrometry methods

Logoyda et al. [59] developed a liquid chromatography-tandem mass spectrometric method for simultaneous identification and quantification of amlodipine, Bisoprolol, and enalapril maleate in human plasma. Chromatographic separation was performed on the Eclipse C18 column

(4.6 × 100 mm, 5 μm) by maintaining column temperature at 30 °C. The mobile phase was composed of acetonitrile: 0.01% formic acid (70:30, v/v) and pumped in isocratic mode at a flow rate of 0.7 mL/min. Injection volume was 5 μL for each injection. Acetonitrile was used for protein precipitation and extraction of analytes from human plasma. The method showed linearity from 0.5–50 ng/mL for Bisoprolol and successfully employed for pharmacokinetic study in healthy human volunteers.

A liquid chromatography-mass spectrometric method for the quantitative estimation of amlodipine and bisoprolol in rat plasma was reported by Chang et al. [60]. Clenbuterol was used as internal standard was extracted by liquid-liquid extraction from plasma. Analyte separation was achieved using a reversed-phase C18 column (50 mm × 4.6 mm, 5 μm). A methanol: water: formic acid (75:25:0.01, v/v/v) was used as a mobile phase at a flow rate of 0.3 mL/min. The method was set to multiple reaction monitoring modes with the transition of m/z 409.1 → 237.9, 326.2 → 116.0, and 277.0 → 237.9 for amlodipine, bisoprolol, and clenbuterol, respectively. The method followed the linearity in concentration range 0.2–50 ng/mL with a lower limit of quantification 0.2 ng/mL.

Turner et al. [61] reported an HPLC MS/MS method for the estimation of atorvastatin, bisoprolol, and clopidogrel in human plasma. Analyte extraction from plasma was achieved by protein precipitation using acetonitrile with acetic acid (0.1%, v/v). A chromatographic separation was achieved by a 2.7 μm Halo C18 (50 × 2.1 mm) column. The mobile phase consisted of acetonitrile (X) and water (Y) each containing 0.1% formic acid v/v at flow rate 0.5 mL/min. The gradient elution was employed for the separation of analytes for a total run time of 6 min. Gradient parameters were 0.00–2.50 min 10% X, 2.51–3.75 min 50% X, 3.76–4.25 95% X, and from 4.26 to 6.0 min returned to 10% X. Analytes and internal standards were ionized in positive ionization mode and monitored using multiple ionization monitoring modes. For bisoprolol m/z transition 326.2 → 116.2 was taken into consideration.

Liu et al. [62] developed an HPLC MS/MS method for the quantification of Bisoprolol in human plasma. A sample plasma spiked with internal standard diphenhydramine was cleaned with protein precipitation using acetonitrile. Then centrifuged and the supernatant was transferred to another tube and dried then reconstituted with a fixed volume of the mobile phase before injection. Mobile phase was composed of methanol: 10 mM ammonium acetate: formic acid (70:30:0.1 v/v/v). The method showed linearity over the range of 0.5–100 ng/mL and the limit of quantification was 0.5 ng/mL.

Peste et al. [63] developed and validated a liquid chromatography-tandem mass spectrometry technique (LC/MSD) for estimating bisoprolol in human plasma samples and metoprolol as an internal standard. Separation was achieved on Zorbax SB-C18 Solvent Saver Plus, 3 × 100 mm, 3.5 μm as a reverse phase column at a temperature of 40 °C, the injection volume was 5 μL. The mobile phase was consisted a mixture of 0.1% formic acid solution—acetonitrile in ration (50:50, v/v) the flow rate was 0.3 mL/. The linearity response was covered the concentration range for bisoprolol 1 and 100 ng/mL with a correlation coefficient of 0.998599.

Hemavathi and Hipparagi [64] developed and validated a liquid chromatography-tandem mass spectrometric method for the simultaneous determination of bisoprolol and triamterene in human plasma with an internal standard (IS) was metoprolol, a protein precipitation extraction method was used to extract both the analytes and IS. The separation was carried out on Welchrom XB C18, 50 × 4.6 mm0, 5 μm column, the mobile phase was consisted a mixture of 2 mM ammonium formate: acetonitrile in ratio (70:30, v/v) with flow rate of 0.6 mL/min.

The interface between the chromatograph and a triple quadrupole mass detector was an electrospray ionization source operated in the positive ion mode. Highly selective identification of the analytes and IS were confirmed by multiple reaction monitoring (MRM) using the most representative transitions from the precursor ion to the different product ions. The transition the precursor ion to product ions were selected 326 → 116 *m/z* for bisoprolol, 254 → 237 *m/z* for triamterene; and 268 → 116 *m/z* for IS (metoprolol) for quantitative response. The parameters of compounds such as declustering potential, entrance potential, collision energy, and collision cell exit potential were 70, 10, 50, and 15 V for bisoprolol; 50, 10, 35, and 18 V for triamterene; and 75, 10, 30 and 8 V for IS, respectively. The run time was 3.5 min, the retention time of bisoprolol, triamterene, and IS were occurred at ~2.57, 1.30, and 1.57 min, respectively. The linear response of both analytes in human plasma were detected over range of 2.04–210 ng/mL. The accuracy of intra- and interday were about 1.12% and 7.87%, respectively for bisoprolol 1.26% and 6.36%, respectively for triamterene. Moreover, precisions of intra- and interday were in the range of 1.46–6.13% and 1.65–7.34% for bisoprolol and triamterene, respectively.

Ding et al. [65] developed a chromatography–electrospray ionization-mass spectrometry technique (LC–ESI-MS) for quantitation of bisoprolol in human plasma and metoprolol as internal standard (IS). A liquid-liquid extraction with ethyl acetate was used to extract analyte from samples after addition of sodium hydroxide. Chromatographic separation was achieved on a ZORBAX SB-C18 column, the mobile phase was consisted 10 mM ammonium acetate buffer including 0.1% formic acid–methanol (32:68, v/v), the flow rate was 1 mL/min. and the run time was less than 5 min. The linearity of bisoprolol in human plasma was coved the concentration range of 0.05–120 ng/mL. The intra- and standard deviation of inter-run was less than 3.8% and 7.5%, respectively.

Li et al. [66] developed and validated a LC–MS/MS method for simultaneous estimation of 10 antiarrhythmic drugs (diltiazem, amiodarone, mexiletine, propranolol, sotalol, verapamil, bisoprolol, metoprolol, atenolol, carvedilol) and a metabolite (norverapamil) in human plasma. a deproteinization method was performed using acetonitrile for the precipitation of protein, therefore extract the analytes and IS. The separation was achieved on a Capcell C18 column (50 mm × 2.0 mm, 5 μm) using a gradient mixture of acetonitrile and 0.02% formic acid solution as a mobile phase at rate of flow of 0.3 mL/min. In the positive electrospray ionization (ESI) interface, the analytes were protonated and detected in multiple reaction monitoring (MRM) mode. For all analytes, calibration curves were linear across wide ranges, from sub- to over-therapeutic plasma concentration, and the linearity of bisoprolol between 2–400 ng/mL, with correlation coefficient was 0.9977, intra- and interday analysis precision of was less than 12.0%, range of accuracy was between 90% and 110%, mean recovery between 85.0% and 99.7%.

4.2.4.4 High-performance thin-layer chromatography (HPTLC) methods

A densitometric method for simultaneous estimation of Bisoprolol fumarate and hydrochlorthiazide in bulk, and its formulations, was reported by Yadav et al. [67]. Ethyl acetate: methanol: ammonia (10:0.5:0.5 v/v) was used as a solvent system. The densitometric wavelength detection was performed at 225 nm. The RD values for Bisoprolol fumarate and hydrochlorthiazide were found to be 0.60 and 0.38 respectively. The method followed the linearity equation between concentration range from 150 ng/spot to 900 ng/spot for Bisoprolol fumarate.

A HPTLC method for simultaneous determination of Bisoprolol fumarate and amlodipine besylate in tablet formulation was developed and validated by Nilam et al. [68]. The separation was performed on precoated silica gel HPTLC aluminum plate 60F254 with mobile phase consisting chloroform: ethanol: glacial acetic acid 2: 8: 0.1 (v/v/v). The quantitative detection was carried out using densitometric scanning at 231 nm.

The Retention Factor (R_f) of Bisoprolol fumarate and Amlodipine besylate were observed at 0.72 and 0.53, respectively. Linearity of the amlodipine besylate and Bisoprolol fumarate were detected over the range of 200–1200 ng/spot. The detection limits and quantitation limits were 50 and 100 ng/spot, respectively for Bisoprolol fumarate and 40 and 120 ng/spot, respectively for Amlodipine besylate. The mean recoveries was acquired for Bisoprolol fumarate and Amlodipine besylate at 99.11 ± 0.13% and 99.20 ± 0.41%, respectively.

5. Stability

Dulin et al. [69] reported accelerated degradation of Bisoprolol fumarate in tablet dosage forms when granular dicalcium phosphate anhydrous USP was used in place of milled dicalcium phosphate anhydrous USP. The degradation of Bisoprolol fumarate was mainly due to the generation of the acidic microenvironment in the presence of moisture. The granular dicalcium phosphate anhydrous USP produced a low pH environment (4.7–5.0) while milled dicalcium phosphate anhydrous USP was relatively neutral when come in contact with moisture. Bisoprolol was found to be more prone to degradation at acidic pH as compared to neutral.

Vujanovic et al. [70] demonstrated the order of degradation of Bisoprolol fumarate in different environments. A maximum degradation was found with oxidative stress agents, then acid and finally with a basic environment. Water at neutral pH and light had no significant effect on the stability of Bisoprolol fumarate.

6. Clinical applications

6.1 Pharmacology

Bisoprolol is a synthetic β_1-selective (cardioselective) adrenoceptor blocking agent without intrinsic sympathomimetic activity in therapeutic doses. It has a low affinity for the β_2-receptor present in blood vessels, bronchial smooth muscles, and fat cells. At lower doses, it does not produce any β_2 receptor-mediated effects. Bisoprolol reduces heart rate, reduces cardiac output, and inhibits renin release from the kidneys [71].

6.2 Mechanism of action

The exact mechanism of action and the antihypertensive effect of Bisoprolol has not been established completely. Its antihypertensive effects are supposed to be the result of an antagonistic effect on β_1 adrenoceptor to reduce cardiac output, inhibition of renin release from kidneys, or diminution of sympathetic flow from the vasomotor centers in the brain. Electrophysiological evidences are available that bisoprolol significantly reduces tachycardia and prolongs AV node refractory period. β_1 receptors are mainly found in heart muscles and heart conductive tissue as well as juxtaglomerular cells in the kidney [72].

6.3 Pharmacokinetics

6.3.1 Absorption

Bisoprolol is quickly and almost completely absorbed (>90%), only a small portion (<10%) goes hepatic metabolism. Thus the bisoprolol has higher bioavailability 90% after oral dosing. The bioavailability is not affected by food. Following oral administration of bisoprolol fumarate 10 mg tablet to healthy subjects, a mean peak plasma concentration (Cmax) of Bisoprolol of approximately 36 ± 7 ng/mL was achieved within approximately 3 h (Tmax) [16].

6.3.2 Distribution

Bisoprolol is extensively distributed. The volume of distribution is 3.5 L/kg. Binding to plasma proteins is approximately 35%; uptake into human blood cells was not observed [15]. The widespread distribution includes the highest concentration in the heart, liver, lungs, and saliva. It crosses the blood-brain barrier and enters breast milk [6].

6.3.3 Metabolism

Bisoprolol is a racemic mixture consisting of the R and S enantiomers. Oxidative metabolic pathways are responsible for the metabolism of Bisoprolol in human being. Major metabolites of Bisoprolol are pharmacologically inactive. Bisoprolol primarily metabolized by CYP3A4 pathway and CYP2D6 very less involved. All metabolites of Bisoprolol are very polar and eliminated through urine. The intrinsic clearance by human recombinant CYP3A4 appears to be non-stereoselective while the metabolism by CYP2D6 is stereoselective (R/S = 1.50) [73].

6.3.4 Elimination

The clearance of bisoprolol is "balanced" between renal elimination of the unchanged drug (~50%) and hepatic metabolism (~50%) to metabolites which are also excreted through the renal system. The total clearance of the drug is 15.6 ± 3.2 L/h with renal clearance being 9.6 ± 1.6 L/h. In a study with ^{14}C-labeled bisoprolol, the total urinary and fecal excretion was 90 ± 2.7% and 1.4 ± 0.1% of the dose, respectively (mean ± SEM recoveries of the total dose within 168 h). Bisoprolol has an elimination half-life of 10–12 h [74].

6.3.5 Elimination half-life

Bisoprolol is removed from the plasma with a half-life of 10–12 h. The plasma elimination half-life of bisoprolol was investigated both in volunteers and in younger (mean age 49, age range 26–63 years) and elderly hypertensive patients (age range 69–80 years). Under steady-state conditions, the plasma elimination half-life was always in the range of 10–12 h. This means that Bisoprolol may be administered once daily, and this has a beneficial effect on the compliance of the patient. Maximum plasma levels are reached 1–3 h after administration. Normal renal function: 9–12 h; Cl_{cr} less than 40 mL/min: 27–36 h; hepatic cirrhosis: 8–22 h. Time to peak: 2–4 h [6].

References

[1] https://pubchem.ncbi.nlm.nih.gov/compound/Bisoprolol, *National Center for Biotechnology Information, PubChem Database, Bisoprolol, CID=2405.* Apr. 26, 2020.

[2] https://pubchem.ncbi.nlm.nih.gov/compound/Bisoprolol-fumarate, *National Center for Biotechnology Information, PubChem Database, Bisoprolol fumarate, CID=6917733.* Apr. 26, 2020.

[3] https://pubchem.ncbi.nlm.nih.gov/compound/Bisoprolol, *National Center for Biotechnology Information, PubChem Database, Bisoprolol hemifumarate, CID=53394791,* Apr. 26, 2020.

[4] D.S. Wishart, Y.D. Feunang, A.C. Guo, E.J. Lo, A. Marcu, J.R. Grant, T. Sajed, D. Johnson, C. Li, Z. Sayeeda, N. Assempour, I. Iynkkaran, Y. Liu, A. Maciejewski, N. Gale, A. Wilson, L. Chin, R. Cummings, D. Le, A. Pon, C. Knox, M. Wilson, DrugBank 5.0: a major update to the DrugBank database for 2018, Nucleic Acids Res. 46 (2018) D1074.

[5] D. Zhong, H. Blume, HPLC-determination of loratadine and its active metabolite descarboethoxyloratadine in human plasma, Pharmazie 49 (10) (1994) 736–739.

[6] Lacy, C.F., Drug Information Handbook with International Trade Names Index 2010–2011: Lexi-Comp.

[7] Bisoprolol, 2020; from https://www.rsc.org/Merck-Index/monograph/m2565/bisoprolol?q=unauthorize.

[8] C.S. Sean, Martindale, The complete drug Reference, Thirty-six edition, Pharmaceutical Press, London, 2005.

[9] R. Jonas, K.-H. Becker, H.-J. Enenkel, K. Minck, H.-J. Schliep, Phenoxy-amino-propanols, U.S. Patent and Trademark Office, Washington, DC, 1981. U.S. Patent No. 4,258,062.

[10] D.V. Soloviev, M. Matarrese, R.M. Moresco, S. Todde, T.A. Bonasera, F. Sudati, P. Simonelli, F. Magni, D. Colombo, A. Carpinelli, Asymmetric synthesis and preliminary evaluation of (R)-and (S)-[11C] bisoprolol, a putative Î²1-selective adrenoceptor radioligand, Neurochem. Int. 38 (2) (2001) 169–180.

[11] M.S. Berridge, E.H. Cassidy, A.H. Terris, J.M. Vesselle, Preparation and in vivo binding of [11C]carazolol, a radiotracer for the beta-adrenergic receptor, Int. J. Rad. Appl. Instrum. 19 (5) (1992) 563–569.
[12] M. Rockville, United States pharmacopeia and national formulary (USP 41-NF 36), in: United States Pharmacopeial Convention, in Bisoprolol Fumarate, 2018, pp. 542–543.
[13] D.S. Wishart, Y.D. Feunang, A.C. Guo, E.J. Lo, A. Marcu, J.R. Grant, T. Sajed, D. Johnson, C. Li, Z. Sayeeda, N. Assempour, I. Iynkkaran, Y. Liu, A. Maciejewski, N. Gale, A. Wilson, L. Chin, R. Cummings, D. Le, A. Pon, C. Knox, M. Wilson, DrugBank 5.0: a major update to the DrugBank database for 2018, Nucleic Acids Res. 46 (D1) (2018) D1074–D1082.
[14] W.M. Meylan, P.H. Howard, R.S. Boethling, Improved method for estimating water solubility from octanol/water partition coefficient, Environ. Toxicol. Chem. 15 (2) (1996) 100–106.
[15] W. Song, D. Cun, H. Xi, L. Fang, The control of skin-permeating rate of bisoprolol by ion-pair strategy for long-acting transdermal patches, AAPS PharmSciTech 13 (3) (2012) 811–815.
[16] G. Leopold, W. Ungethum, J. Pabst, Z. Simane, K.U. Buhring, H. Wiemann, Pharmacodynamic profile of bisoprolol, a new beta 1-selective adrenoceptor antagonist, Br. J. Clin. Pharmacol. 22 (3) (1986) 293–300.
[17] A. Detrich, K.J. Domotor, M.T. Katona, I. Markovits, J. Vargane Lang, Polymorphic forms of bisoprolol fumarate, J. Therm. Anal. Calorim. 135 (6) (2019) 3043–3055.
[18] M. Skotnicki, J.A. Aguilar, M. Pyda, P. Hodgkinson, Bisoprolol and bisoprolol-valsartan compatibility studied by differential scanning calorimetry, nuclear magnetic resonance and X-ray powder diffractometry, Pharm. Res. 32 (2) (2015) 414–429.
[19] B. Lavanya, V. Shanmugam, Formulation and evaluation of bisoprolol fumarate optiyorb dispersible tablet to improve tablet disintegration, World J. Parm. Pharm. Sci. 4 (1) (2015) 561–576.
[20] M. Rockville, United States pharmacopeia and national formulary (USP 41-NF 36), in: United States Pharmacopeial Convention, in Bisoprolol Fumarate Tablets, 2018, pp. 544–545.
[21] Commission, B.P, G.B.M. Commission, G.M. Council, British Pharmacopoeia 2016, Vol. 1, Bernan Press, PA, 2016. I-(303–305).
[22] S.A. Mohammed, M.E. Adam, S.W. Shantier, Development and validation of UV spectrophotometric method for determination of bisoprolol fumarate in bulk and pharmaceutical dosage forms, Mediterr. J. Chem. 6 (5) (2017) 196–199.
[23] R. Sahu, Simultaneous spectrophotometric estimation of hydrochlorothiazide and bisoprolol fumarate in combined dosage forms, Indian J. Pharm. Sci. 68 (6) (2006) 764.
[24] S.T. Kumbhar, P.P. Shinde, D.B. Shinde, P.B. Solankar, Visible spectrophotometric method for estimation of bisoprolol from its bulk and tablet formulation, Asian J. Pharm. Clin. Res. 6 (4) (2013) 103–105.
[25] O.G. Bhusnure, R.B. Dongare, S.B. Gholve, T.M. Rajmane, A.B. Munde, P.S. Giram, Development and validation of uv spectroscopic method for the determination of bisoprolol fumarate tablets, Int. J. Pharm. Biol. Sci. 8 (2) (2018) 338–344.
[26] A.D. Panainte, N. Bibire, G. Tantaru, M. Apostu, M. Vieriu, V. Dorneanu, A new method for the assay of bisoprolol using bromocresol green, Rev. Chim-Bucharest 65 (8) (2014) 916–920.
[27] A.D. Panainte, N. Bibire, G. Tantaru, M. Apostu, M. Vieriu, V. Dorneanu, Spectrophotometric method for estimation of bisoprolol fumarate in tablets, Med. Surg. J. 118 (2) (2014) 558–563.
[28] A. Panainte, A. Murăra¸u, V. Dorneanu, Spectrophotometric determination of bisoprolol using methyl orange as reagent, Farmacia 60 (2012) 634–641.

[29] A.S. Atul, R.T. Rajesh, J.S. Sanjay, Simultaneous spectrophotometric estimation of Bisoprolol fumarate and hydrochlorothiazide in tablet dosage form, Pak. J. Pharm. Sci. (2008) 21(4).
[30] R. Kakde, V. Kotak, A. Barsagade, N. Chaudhary, D. Kale, Spectrophotometric method for simultaneous estimation of amlodipine besylate and bisoprolol fumarate in pharmaceutical preparations, Res. J. Pharm. Technol. 1 (4) (2008) 513–515.
[31] N.S. Abdelwahab, N.W. Ali, M.M. Zaki, A.A. Ali, M.M. Abdelkawy, Spectrofluorimetric determination of Bisoprolol fumarate and Rosuvastatin calcium in a novel combined formulation and in human spiked plasma, Eur. J. Chem. 9 (4) (2018) 331–337.
[32] R.N. Goyal, A. Tyagi, N. Bachheti, S. Bishnoi, Voltammetric determination of bisoprolol fumarate in pharmaceutical formulations and urine using single-wall carbon nanotubes modified glassy carbon electrode, Electrochim. Acta 53 (6) (2008) 2802–2808.
[33] B. Bozal, M. Gumustas, B. Dogan-Topal, B. Uslu, S.A. Ozkan, Fully validated simultaneous determination of bisoprolol fumarate and hydrochlorothiazide in their dosage forms using different voltammetric, chromatographic, and spectrophotometric analytical methods, J. AOAC Int. 96 (1) (2013) 42–51.
[34] H.-B. Duan, J.-T. Cao, H. Wang, Y.-M. Liu, Determination of metoprolol tartrate and bisoprolol fumarate by capillary electrophoresis coupled with tris (2, 2â€^{2}-bipyridyl)-ruthenium (ii) electrochemiluminescence detection and study on the interaction between the drugs and human serum albumin, Anal. Methods 7 (9) (2015) 3946–3951.
[35] A. Witek, H. Hopkala, G. Matysik, TLC-densitometric determination of bisoprolol, labetalol and propafenone, as dabsyl derivatives, in pharmaceutical preparations, Chromatographia 50 (1–2) (1999) 41–44.
[36] M.M. Zaki, N.S. Abdelwahab, A.A. Ali, S.M.Z. Sharkawi, Simultaneous determination of bisoprolol fumarate and rosuvastatin calcium in a new combined formulation by validated RP-HPLC, Eur. J. Chem. 10 (1) (2019) 52–56.
[37] L. Logoyda, S. Kovalenko, A.M. AbdelMegied, I. Zhulkevych, I. Drapak, I. Demchuk, O. Netsyuk, HPLC method development for the analysis of bisoprolol in combined dosage form containing Bisoprolol and Enalapril and in vitro dissolution studied, Int. J. Appl. Pharm. 11 (3) (2019) 186–194.
[38] A.D. Panainte, M. Vieriu, G. Tantaru, M. Apostu, N. Bibire, Fast RP-HPLC method for the determination of bisoprolol, Med. Surg. J. 120 (3) (2016) 720–726.
[39] E. Caudron, S. Laurent, E.M. Billaud, P. Prognon, Simultaneous determination of the acid/base antihypertensive drugs celiprolol, bisoprolol and irbesartan in human plasma by liquid chromatography, J. Chromatogr. 801 (2) (2004) 339–345.
[40] S. Shaikh, O.A. Thusleem, M.S. Muneera, J. Akmal, A.V. Kondaguli, K. Ruckmani, A simple and rapid high-performance liquid chromatographic method for the determination of bisoprolol fumarate and hydrochlorothiazide in a tablet dosage form, J. Pharm. Biomed. Anal. 48 (3) (2008) 1055–1057.
[41] A.J. Braza, P. Modamio, C.F. Lastra, E.L. Marino, Development, validation and analytical error function of two chromatographic methods with fluorimetric detection for the determination of bisoprolol and metoprolol in human plasma, Biomed. Chromatogr. 16 (8) (2002) 517–522.
[42] R.J. Eastwood, J.C. Jerman, R.K. Bhamra, D.W. Holt, High performance liquid chromatographic measurement of bisoprolol in plasma, Biomed. Chromatogr. 4 (4) (1990) 178–180.
[43] L.J. Patel, B.N. Suhagia, P.B. Shah, R.R. Shah, Simultaneous estimation of bisoprolol fumarate and hydrochlorothiazide in tablet dosage form by RP-HPLC method, Indian J. Pharm. Sci. 68 (5) (2006) 635–638.

[44] M.M. Hefnawy, M.A. Sultan, M.M. Al-Shehri, Development of an HPLC method for the quantitation of bisoprolol enantiomers in pharmaceutical products using a teicoplanin chiral stationary phase and fluorescence detection, J. Liq. Chromatogr. Relat. Technol. 29 (20) (2006) 2901–2914.
[45] S.T. Ulu, Z. Aydogmus, An HPLC method for the determination of bisoprolol in human plasma and its application to a pharmacokinetic study, J. Chromatogr. Sci. 50 (7) (2012) 615–619.
[46] Y. Kondratova, L. Logoyda, Y. Voloshko, A. Abdel-Megied, D. Korobko, Y. Soroka, Development and validation of HPLC-DAD method for the determination of bisoprolol in tablet dosage forms, Int. J. Appl. Pharm. 9 (2017) 54–59.
[47] S. Tatar Ulu, Z. Aydoğmuş, An HPLC method for the determination of bisoprolol in human plasma and its application to a pharmacokinetic study, J. Chromatogr. Sci. 50 (7) (2012) 615–619.
[48] S.J. Joshi, P.A. Karbhari, S.I. Bhoir, K. Bindu, C. Das, RP-HPLC method for simultaneous estimation of bisoprolol fumarate and hydrochlorothiazide in tablet formulation, J. Pharm. Biomed. Anal. 52 (3) (2010) 362–371.
[49] A.-D. Panainte, L. Agoroaei, N. Bibire, G. Tantaru, M. Apostu, M. Vieriu, F.A. Spac, A HPLC method for the determination of bisoprolol in tablets and its application to a bioequivalence study, Rev. Chim. 66 (11) (2015) 1791–1795.
[50] S.C. Mahu, A.F. Spac, C. Ciobanu, M. Hancianu, L. Agoroaei, E. Butnaru, Quantitative determination of bisoprolol fumarate by HPLC I. Method validation, Rev. Chim. 67 (3) (2016) 414–417.
[51] G. Arjun, S. Dinakaran, B. Madhavi, M.M. Naga, R. Ramalingam, R.N. Anisetti, A simple HPLC method for quantitation of bisoprolol fumarate in tablet dosage form, Indian Drugs 46 (7) (2009) 39–42.
[52] V. Dobričić, M. Vulović-Tadić, B. Jančić-Stojanović, S. Vladimirov, O. Čudina, Desirability based optimization and validation of new RP-HPLC method for simultaneous determination of bisoprolol fumarate, hydrochlorothiazide and their impurities, Chromatographia 79 (9–10) (2016) 571–579.
[53] E. Dinç, Z.C. Ertekin, G. Rouhani, A new RP-UPLC method for simultaneous quantitative estimation of bisoprolol hemifumarate and hydrochlorothiazide in tablets using experimental design and optimization, J. Liq. Chromatogr. Relat. Technol. 38 (9) (2015) 970–976.
[54] I. Madhusudhanareddy, R. Bhagavan, P. Rajendra, Validated and stability indicating liquid chromatography method for quantification of bisoprolol fumarate tablet dosage form, Int. J. Pharm. 2 (1) (2012) 64–70.
[55] S. Kurbanoglu, P.R. San Miguel, B. Uslu, S.A. Ozkan, Stability-indicating UPLC method for the determination of bisoprolol fumarate and hydrochlorothiazide: application to dosage forms and biological sample, Chromatographia 77 (3–4) (2014) 365–371.
[56] D. Vora, A. Kadav, Development and validation of a simultaneous HPLC method for estimation of bisoprolol fumarate and amlodipine besylate from tablets, Indian J. Pharm. Sci. 70 (4) (2008) 542.
[57] W. Wang, L. Gao, T. Shi, HPLC determination of content of bisoprolol fumarate tablets, Chin. J. Pharm. Anal. 19 (5) (1999) 308–310.
[58] M.M. Hefnawy, M.A.-A. Sultan, M.M. Al-Shehri, Enantioanalysis of bisoprolol in human plasma with a macrocyclic antibiotic HPLC chiral column using fluorescence detection and solid phase extraction, Chem. Pharm. Bull. 55 (2) (2007) 227–230.
[59] L. Logoyda, S. Kovalenko, M. Gaafar, A.M. Abdel-Megied, F.A. Elbarbry, A validated LC-MS/MS method for the quantification of amlodipine, bisoprolol, enalapril and enalaprilate-application to pharmacokinetic study in healthy volunteers, Microchem. J. 155 (2020) 104700.

[60] H. Chang, J. Li, J. Li, X. Guan, F. Sun, Z. Qian, K. Bi, G. Fan, Simultaneous determination of amlodipine and bisoprolol in rat plasma by a liquid chromatography/tandem mass spectrometry method and its application in pharmacokinetic study, J. Pharm. Biomed. Anal. 71 (2012) 104–110.
[61] R.M. Turner, V. Fontana, M. Bayliss, S. Whalley, A. Santoyo Castelazo, M. Pirmohamed, Development, validation and application of a novel HPLC-MS/MS method for the quantification of atorvastatin, bisoprolol and clopidogrel in a large cardiovascular patient cohort, J. Pharm. Biomed. Anal. 159 (2018) 272–281.
[62] M. Liu, D. Zhang, Y. Sun, Y. Wang, Z. Liu, J. Gu, Liquid chromatographic-electrospray tandem mass spectrometric determination of bisoprolol in human plasma, Biomed. Chromatogr. 21 (5) (2007) 508–513.
[63] G. Peste, C. Oniscu, A. Vlase, Experimental research for determination of bisoprolol fumarate in human plasma samples using liquid chromatography-tandem mass spectrometry (LC-MS/MS) technique, Rom. Biotechnol. Lett. 15 (2) (2010) 5141.
[64] G. Hemavathi, S. Hipparagi, Sensitive liquid chromatography tandem Mass spectrometric method for the simultaneous determination of bisoprolol and triamterene in human plasma, Asian J. Pharm. Clin. Res. 10 (4) (2017) 341–348.
[65] L. Ding, X. Zhou, X. Guo, Q. Song, J. He, G. Xu, LC–ESI-MS method for the determination of bisoprolol in human plasma, J. Pharm. Biomed. Anal. 44 (2) (2007) 520–525.
[66] S. Li, G. Liu, J. Jia, Y. Liu, C. Pan, C. Yu, Y. Cai, J. Ren, Simultaneous determination of ten antiarrhythic drugs and a metabolite in human plasma by liquid chromatography—tandem mass spectrometry, J. Chromatogr. B 847 (2) (2007) 174–181.
[67] S.S. Yadav, J.R. Rao, Simultaneous HPTLC analysis of bisoprolol fumarate and hydrochlorthiazide in pharmaceutical dosage form, Int. J. Pharm. Pharm. Sci. 5 (2) (2013) 286–290.
[68] N.B. Jadhav, S.S. Yadav, Analysis of bisoprolol fumarate and amlodipine besylate in tablet dosage form by using HPTLC, Int. J. Pharm. Sci. Rev. Res. 58 (1) (2019) 75–78.
[69] W.A. Dulin, Degradation of bisoprolol fumarate in tablets formulated with dicalcium phosphate, Drug Dev. Ind. Pharm. 21 (4) (1995) 393–409.
[70] I. Kasagic-Vujanovic, B.J. Stojanovic, D. Ivanovic, Monitoring of bisoprolol fumarate stability under different stress conditions, in: A. Badnjevic (Ed.), CMBEBIH 2017, Springer Singapore, Singapore, 2017, pp. 415–424.
[71] MYLAN-BISOPROLOL^{Pr}, Mylan Pharmaceuticals ULC, Etobicoke, ON, Canada, 2012, pp. 1–31.
[72] Papich, M.G. and M.G. Papich, Bisoprolol fumarate, in Saunders Handbook of Veterinary Drugs (fourth ed.), W.B. Saunders: St. Louis. p. 76–77.
[73] A.P.P. Eisele, E.R. Sartori, Simple and rapid determination of loratadine in pharmaceuticals using square-wave voltammetry and a cathodically pretreated boron-doped diamond electrode, Anal. Methods 7 (20) (2015) 8697–8703.
[74] G. Leopold, Balanced pharmacokinetics and metabolism of bisoprolol, J. Cardiovasc. Pharmacol. 8 (1986) S16–S20.

CHAPTER THREE

Betaxolol: A comprehensive profile

Majed J. Al-wadei[a], Ahmed H. Bakheit[a,b], Alaa A.-M. Abdel-Aziz[a], and Tanveer A. Wani[a,*]

[a]Department of Pharmaceutical Chemistry, College of Pharmacy, King Saud University, Riyadh, Kingdom of Saudi Arabia
[b]Department of Chemistry, Faculty of Science and Technology, Al-Neelain University, Khartoum, Sudan
*Corresponding author: e-mail address: twani@ksu.edu.sa

Contents

Profiles of Drug Substances, Excipients, and Related Methodology, Volume 46
ISSN 1871-5125
https://doi.org/10.1016/bs.podrm.2020.07.002

1. Description

1.1 Nomenclature

1.1.1 Chemical names

Betaxolol: [1,2]

(±)-1-(Isopropylamino)-3-[p-(cyclopropylmethoxyethyl)phenoxy]-2-propanol

1-[(4-{2-[(cyclopropylmethyl)oxy]ethyl}phenyl)oxy]-3-[(1-methylethyl)amino]propan-2-ol

1-[4-[2-(Cyclopropylmethoxy)ethyl]phenoxy]-3-[(1-methylethyl)amino]-2-propanol

1-{4-[2-(Cyclopropylmethoxy)ethyl]phenoxy}-3-(isopropylamino)-2-propanol

1-{4-[2-(Cyclopropylmethoxy)ethyl]phenoxy}-3-(isopropylamino)propan-2-ol

1-{4-[2-(cyclopropylmethoxy)ethyl]phenoxy}-3-(propan-2-ylamino)propan-2-ol

1-{4-[2-(cyclopropylmethoxy)ethyl]phenoxy}-3-[(1-methylethyl)amino]propan-2-ol

Betaxolol Hydrochloride: [3,4]

(±)-1-(p-(2-(Cyclopropylmethoxy)ethyl)phenoxy)-3-(isopropylamino)-2-propanol Hydrochloride

1-{4-[2-(Cyclopropylmethoxy)ethyl]phenoxy}-3-(isopropylamino)-2-propanol hydrochloride (1:1)

1-{4-[2-(Cyclopropylmethoxy)ethyl]phenoxy}-3-(isopropylamino)propan-2-ol hydrochloride (1:1)

1-{4-[2-(cyclopropylmethoxy)ethyl]phenoxy}-3-[(1-methylethyl)amino]propan-2-ol hydrochloride

1-{4-[2-(Cyclopropylmethoxy)ethyl]phenoxy}-3-[(1-methylethyl)amino]propan-2-olhydrochlorid2-propanol

1-[4-[2-(cyclopropylmethoxy)ethyl]phenoxy]-3-[(1-methylethyl)amino]-, hydrochloride

2-Propanol,1-[4-[2-(cyclopropylmethoxy)ethyl]phenoxy]-3-[(1-methylethyl)amino]-, hydrochloride (1:1)

Levobetaxolol: [5]

(-)-βxolol

(2S)-1-(4-(2-(Cyclopropylmethoxy)ethyl)phenoxy)-3-((1-methylethyl)amino)-2-propanol

(2S)-1-{4-[2-(Cyclopropylmethoxy)ethyl]phenoxy}-3-(isopropylamino)-2-propanol

(2S)-1-{4-[2-(Cyclopropylmethoxy)ethyl]phenoxy}-3-(isopropylamino)-2-propanol

(2s)-1-{4-[2-(cyclopropylmethoxy)ethyl]phenoxy}-3-(isopropylamino) propan-2-ol

(S)-(-)-βxolol

(S)-βxolol

2-Propanol, 1-[4-[2-(cyclopropylmethoxy)ethyl]phenoxy]-3-[(1-methylethyl)amino]

Levobetaxolol hydrochloride: [6]

(2S)-1-(4-(2-(cyclopropylmethoxy)ethyl)phenoxy)-3-((1-methylethyl) amino)-2-propanol hydrochloride

(2S)-1-{4-[2-(Cyclopropylmethoxy)ethyl]phenoxy}-3-(isopropylamino)-2-propanol hydrochloride (1:1)

2-Propanol, 1-[4-[2-(cyclopropylmethoxy)ethyl]phenoxy]-3-[(1-methylethyl)amino]-, (2S)-, hydrochloride (1:1)

(2S)-1-{4-[2-(cyclopropylmethoxy)ethyl]phenoxy}-3-(propan-2-ylamino)propan-2-ol hydrochloride

(2S)-3-{4-[2-(cyclopropylmethoxy)ethyl]phenoxy}-2-hydroxy-N-(propan-2-yl)propan-1-aminium chloride

(S)-βtaxolol hydrochloride

(S)-βxolol HCl

(S)-βxolol hydrochloride

1.1.2 Non-propriety name

Generic: betaxolol

Synonyms: 1-(4-(2-(Cyclopropylmethoxy)ethyl)phenoxy)-3-((1-methylethyl)amino)-2-propanol [1]

1-(Isopropylamino)-3-[P-(cyclopropylmethoxyethyl)phenoxy]-2-propanol, Betaxololum, (S)-betaxolol [7]

SLD-212; SL-75.212, NU-1779 [8]

1.1.3 Propriety name

Brand names:

Betoptic; Betoptima; Betoquin, Betaxololum, Kerledex, Oxodal, Kerlone [8–10]

1.1.4 Multi-ingredient

Betoptic S® (betaxolol hydrochloride Ophthalmic Suspension 0.25% as Base, Sterile Topical Ophthalmic Drops), Kerledex™ (betaxolol hydrochloride and chlorthalidone), Kerlone® (betaxolol hydrochloride Tablets) [10].

Betoptic/pilo (betaxolol/Pilocarpine), Kerledex (betaxolol hydrochloride; chlorthalidone) [9,10].

1.2 Formulae

1.2.1 Empirical formula, molecular weight, and CAS number

CAS name	Empirical formula	Molecular weight	CAS number	
Betaxolol	$C_{18}H_{29}NO_3$	307.4278 g/mol	[63659-18-7]	[8]
Betaxolol Hydrochloride	$C_{18}H_{30}ClNO_3$	343.8887 g/mol	[63659-19-8]	[8]
Levobetaxolol	$C_{18}H_{29}NO_3$	307.427 g/mol	[93221-48-8]	[7]
Levobetaxolol hydrochloride	$C_{18}H_{30}ClNO_3$	343.8887 g/mol		[6]

1.2.2 Structural formula

See Fig. 1.

1.3 Elemental analysis

Betaxolol: (C: 70.32%, H: 9.51%, N: 4.56%, O: 15.61%), Total percent composition = 99.999%

Betaxolol Hydrochloride: (C: 62.87%, H: 8.79%, N: 4.07%, O: 13.97%, Cl: 10.31%), Total percent composition = 100

Levobetaxolol: (C: 70.32%, H: 9.51%, N: 4.56%, O: 15.61%), Total percent composition = 99.999%

Levobetaxolol hydrochloride: (C: 62.87%, H: 8.79%, N: 4.07%, O: 13.97%, Cl: 10.31%), Total percent composition = 100 [12]

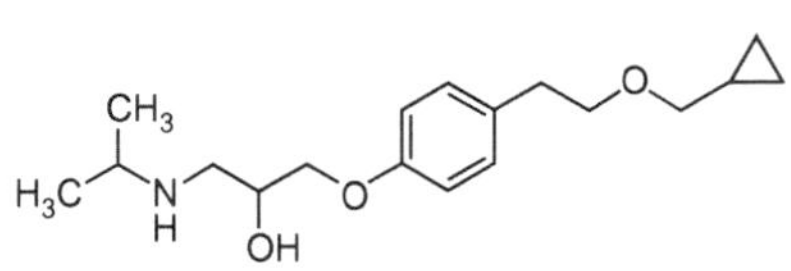

Fig. 1 Chemical structure of betaxolol [11].

1.4 Appearance

A white or almost white crystalline powder [11].

1.5 Uses and applications

Betaxolol is a cardioselective beta blocker. It is reported to lack intrinsic sympathomimetic activity and is known to have little membrane-stabilizing activity. Betaxolol hydrochloride is used in the management of hypertension, angina pectoris, and glaucoma [11,13].

2. Method of preparation

Betaxolol was synthesized by Manoury et al. [14] (Scheme 1). (S)-(-)-2-phenyl-3-isopropyl-5-(hydroxylmethyl)oxazolidinyl tosylate **(3)** was used for the alkylation of phenol **2**. The resulting product **4** after acid-catalyzed hydrolysis afforded the (S) enantiomer of **1**. Compound **2** was also allowed to react with (2R)-3-(tosyloxy)-1,2-propanediol acetonide **5** followed by hydrolysis. The resulting diol **6** was tosylated and converted to epoxide **7**. Treatment of epoxide 10 with isopropylamine furnished the (R) enantiomer of **1**.

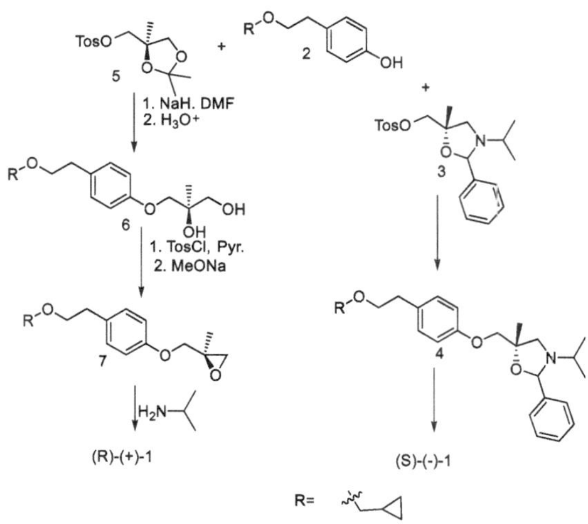

Scheme 1 Synthesis of betaxolol enantiomers.

Allen and Tizot [15] also synthesized betaxolol starting with bromination of 4-[2-(cyclopropylmethoxy)ethyl]phenol, an intermediate employed for the preparation of betaxolol (Scheme 2). The product 2,6-dibromo phenol **8** was alkylated using excess epichlorohydrin under basic conditions to afford

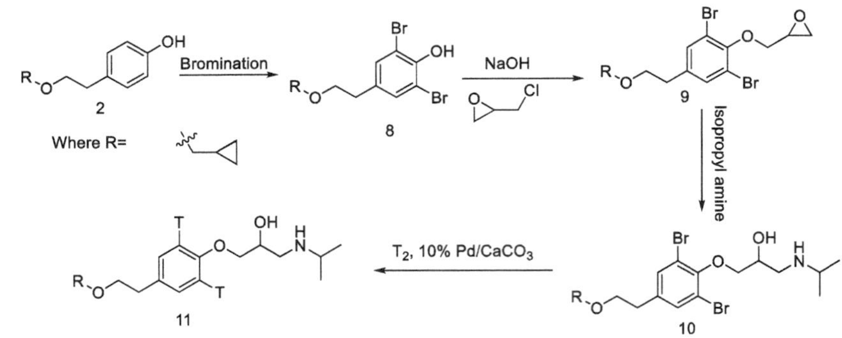

Scheme 2 Synthesis of tritium-labeled betaxolol.

epoxypropane **9**. Treatment of epoxypropane **9** with excess of isopropylamine at 100 °C in a bomb provided 2,6 dibromobetaxolol **10**. Debromination of **10** yielded betaxolol, which was purified and identified by instrumental techniques. Exercising the same procedure with deuterium and tritium in the presence of catalyst yielded [$^2H_2$]-betaxolol and [^{3}H]-betaxolol, respectively. The specific radioactivity of [^{3}H]-betaxolol was 49 C1/mmol and 98% radiochemical purity determined by thin layer radio chromatography.

Datta et al. [16] synthesized betaxolol via Heck arylation of vinyl ethers (Scheme 3). The staring material for the synthesis was *p*-chloronitrobenzene (**12a**). Palladium-catalyzed transvinylation of cyclopropyl methanol and ethyl vinyl ether using 2,2′-bipyridyl ligand resulted in the formation of cyclopropylmethylvinyl ether (**13c**). Compound **15** (60% yield) in highly regioselective coupling was prepared using Heck arylation in aqueous DMF (Method A). The hydrogenation of the olefinic bond and nitro group of **15** furnished compound **16** (79% yield). Aryl chloride **12a** was also

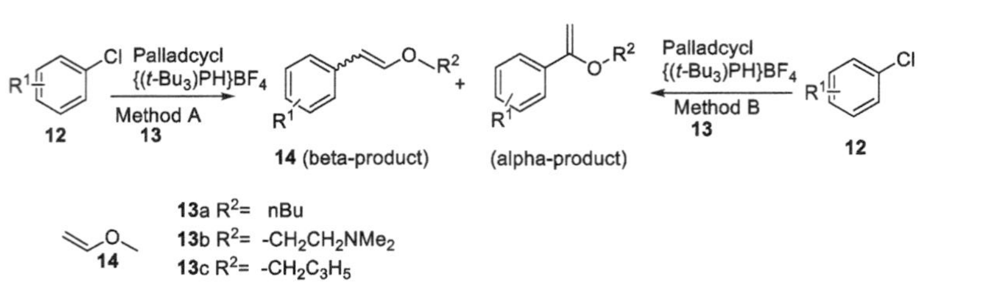

method A= 1.0mmol of **12**, 3.0 mmol of **13**, 0.05 mmol of palladacycle, 0.10 mmol of [(t-Bu$_3$)PH]BF$_4$, 3.0 mmol of Cy$_2$NMe,200 mL of H_2O, and 2mL of DMF in sealed vessels. Microwave heating, 160°C for 60 min.
Method B: 1.0mmol of **12**, 3.0 mmol of **13**, 0.05 mmol of palladacycle, 0.10 mmol of [(t-Bu$_3$)PH]BF$_4$, 5.0 mmol of PMP and 2mL of PEG-200 in sealed vessels. Microwave heating, 160°C for 60 min.

Scheme 3 Heck arylation of vinyl ethers **13**.

converted directly in one step to **16** in greater yield. Treatment of **16** with sodium nitrite (diazotization) and water resulted in phenol **5** (50% yield). Reaction of (R)-3-isopropylamine 1,2-epoxypropane and then refluxing in ethanol for 8 h provided active (S)-betaxolol. Treatment with dry HCl (gas) in dry ether and crystallization afforded (S)-**1**.HCl (Scheme 4).

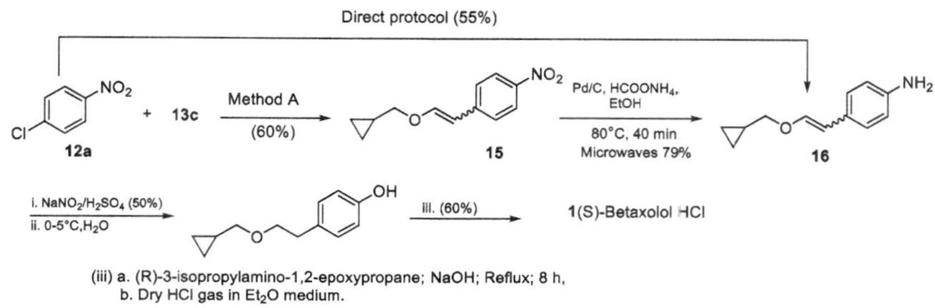

Scheme 4 Synthesis of betaxolol.HCl (1).

A concise method for synthesis of (S)-betaxolol was proposed by Muthukrishnan et al. [17]. Reaction of 2-(4-hydroxyphenyl) ethanol (17) with (±) epichlorohydrin in anhydrous 2-butanone, K_2CO_3 base, and a phase transfer catalyst for 15 h under refluxing temperature resulted in the racemate epoxide **18** (86% yield). HPLC-monitored HKR was performed for racemic epoxide **18** for 30 h at ambient temperature using Jacobsen catalyst and water in isopropanol. (S)-epoxide **19** (42% yield, 99% *ee*) and (R)-diol **17** (47% yield, 92% *ee*) were obtained over silica gel column chromatography. (S)-epoxide **20** (47% yield, 92% *ee*) was achieved when hydroxyl group of (S)-epoxide **18** was selectively O_alkylated using chloromethyl cyclopropane in the presence of KO*t*-Bu. Treatment of (S)-epoxide **20** with excess *N*–isopropyl amine and refluxing for 2{10 h provided (S)-betaxolol. Silica gel column chromatography was then performed to purify the crude (S)-betaxolol (1) (96% yield, 99% *ee*) (Scheme 5).

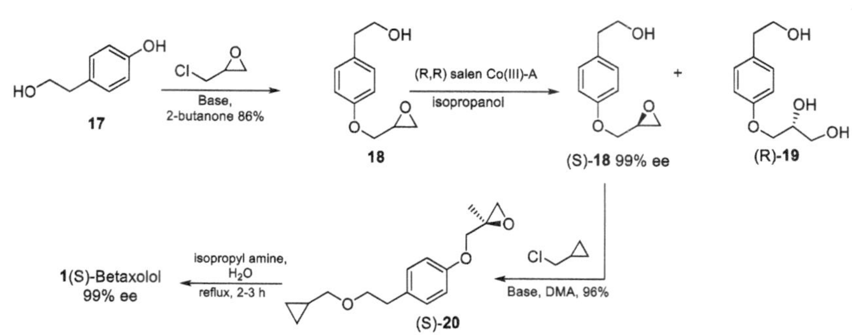

Scheme 5 Synthesis of (S)-betaxolol.

Joshi et al. [18] proposed a method for preparation of S-(-)-betaxolol and salts by condensation of 2-4-hydroxyphenyl-ethanol (5) with benzyl halide in the presence of a base, phase transfer catalyst and organic solvent was done to obtain 2-4-benzyloxyphenyl-ethanol (6). 2-4-benzyloxyphenyl-ethanol (6) was condensed with an allyl halide in the presence of a base and an organic Solvent to obtain 1-(2-allyloxy ethyl)-4-benzyloxybenzene, (7). Cyclopropanating 1-(2-allyloxyethyl)-4-benzyloxybenzene of formula (7) by conventional methods Such as Simmon Smith reaction and Furukawa modification of Simmon Smith reaction to obtain the 1-benzyloxy-4-(2-cyclopropyl methoxy-ethyl)-benzene (8). Deprotecting 1-benzyloxy-4-(2-cyclopropyl methoxy-ethyl)-benzene of formula (8) by hydrogenation to obtain 4-(2-cyclopropylmethoxyethyl)-phenol (9). O-Alkylation of 4-(2-cyclopropylmethoxyethyl)-phenol of formula, (9) by treating with R-(-)-epichlorohydrin in the presence of alkali to obtain the mixture of compounds of the formulae (10 + 11) treating the mixture of compounds of the formulae, (10 + 11) with isopropylamine to give S-(-)-betaxolol (12). Treating S-(-)-betaxolol of the formula, (3) with alcoholic hydrochloric acid in organic solvent to give S-(-)-betaxolol HCl, of formula (1) or maleic acid in organic solvent to give S-(-)-betaxolol maleate salt (Scheme 6).

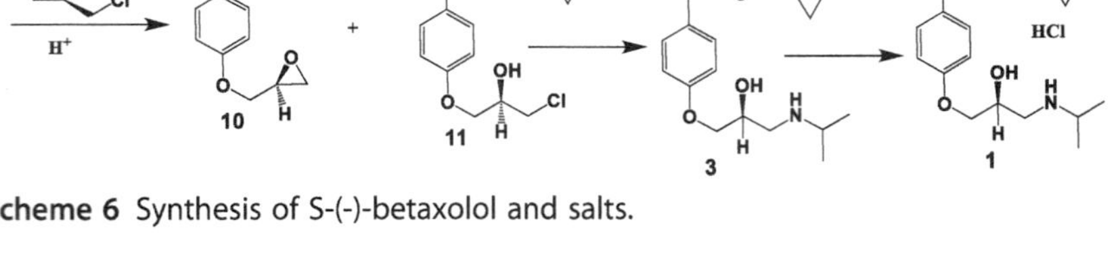

Scheme 6 Synthesis of S-(-)-betaxolol and salts.

3. Physical characteristics

3.1 Color/form

Betaxolol hydrochloride occurs as white or almost white crystalline powder [8].

3.2 Melting point

Betaxolol has a melting point of 70–72 °C and betaxolol hydrochloride has a melting point of 113–117 °C [8].

3.3 Dissociation constant

$pK_a = 9.4$ [8]

$pK_a = 9.38$ [19]

3.4 Octanol/water partition coefficient

log $P = 2.81$ (Experimental properties)

log $P = 3$ (Predicted properties)

log $P = 2.54$ (Predicted properties)

Clog $P = 2.17$ theoretical *n*-octanol/water partition coefficient for non-ionized form solute calculated by the fragmental method [20]

3.5 Solubility

Betaxolol is a hydrophilic drug and soluble in water, DMSO and in ethanol and is insoluble in ether [11]. It is freely soluble in chloroform, ethanol and methanol; soluble in dichloromethane [8]. In vitro solubility of betaxolol at 25 °C is 62 mg/mL (201.67 mM) in DMSO, 62 mg/mL (201.67 mM) in water and <1 mg/mL (<1 mM) in ethanol [21].

3.6 Vapor pressure

The vapor pressure for betaxolol at 25 °C was found to be 0.0 ± 1.1 mmHg [1].

3.7 X-ray powder diffraction pattern (XRPD) of betaxolol

The XRPD was obtained by Maria et al. [22] using glass capillaries of 0.5 mm diameter filled with the powdered betaxolol. The samples were analyzed using ENRAF–NONIUS powder diffractometer (equipped with a CPS120 detector by INEL) and data was collected for 30 min using Debye–Scherrer geometry. Monochromatized Cu $K_{\alpha 1}$ radiation was used ($\lambda = 1.5406$ Å). Silicon was chosen as an external standard. The polymorphic behavior, in melt cooling experiments, of racemic betaxolol, a low aqueous solubility selective b1-adrenergic antagonist drug with a flexible molecular structure. A multidisciplinary approach is employed, using thermal analysis (differential scanning calorimetry, polarized light thermomicroscopy), spectroscopic methods (infrared spectroscopy, magic angle spinning ^{1}H NMR) and X-ray powder diffraction. A glass phase is obtained, $T_g \sim -10\,°C$, on cooling the melt, unless the cooling rate is $0.5 \leq °C\,min^{-1}$, while a new metastable form, polymorph II, $T_{fus} = 33\,°C$, is generated in subsequent heating runs in a two-step process. Although either partial crystallization from the melt in the first step or the formation of an intermediate, metastable, low ordered phase may explain these observations, our results favor the second hypothesis. The stable polymorph I, $T_{fus} = 69\,°C$, which crystallizes on further heating after form II melting, has also been obtained either from polymorph II or from the molten phase, on standing at 25 °C. The racemic betaxolol crystalline phases are found to exhibit some degree of disorder.

The structure of betaxolol obtained from ethanol: water solution by evaporating at 25 °C was studied by X-ray diffraction by Canotilho et al. [23]. X-ray data show the existence of two conformers in the unit cell differing only in the conformation of the cyclopropylmethoxy fragment. Differences in the bond lengths angles and dihedral between both conformations are observed. The cyclopropyl groups lie in approximately perpendicular planes. The crystal is a mixture of 59% of conformation A and 41% of conformation B. The conformations differ only in the cyclopropylmethoxy moiety. Natural Bond Orbital analysis shows that the particular spatial orientation of cyclopropylmethoxy fragment arises from stabilizing interactions. In conformer A, this is primarily caused by the overlap of the O_3 lone pair with the greater p-character, $n(LP_2O_3)$, with the C15–H″ and C15–H′ antibonding orbitals. In conformer B, the same $n(LP_2O3)$ orbital interacts mostly with the C15–H′ and C15–C16 antibonding orbitals. The conformations of the cyclopropylmethoxy are such that the cyclopropyl groups of both conformations lie in perpendicular planes.

The bent bond orbitals typical of cyclopropyl were observed and the orbital interactions accountable for its remarkable stability were analyzed. Optimizing the structure given by X-ray at the B3LYP/6-31(d, p) level, the structure of both betaxolol conformations is slightly distended and the differences between bond lengths and angles in the cyclopropyl groups are dissipated with the values of geometrical parameters of both groups tending to those of cyclopropane.

3.8 Thermal analysis

3.8.1 Boiling point

Boiling and flash point of betaxolol are 760 mmHg at 448 °C and 224.7 °C, respectively [24].

3.8.2 Differential scanning calorimetry (DSC)

The DSC was recorded by Maria et al. [22] employing a Perkin Elmer Pyris 1 DSC, calorimeter. The instrument had helium purge (20 mL/min) and a liquid nitrogen Cryofill cooling unit. Preliminary DSC experiments were carried out in the temperature range between $T = -170$ and 80 °C at 10 °C min^{-1}. No thermal events were detected below −20 °C. Perkin Elmer 40 μL aluminum pans and temperature calibration using cyclohexane, $T_{fus} = (6.66 \pm 0.04)$ C, $T_{trs} = (-86.9 \pm 0.02)$ C; benzophenone (Mettler Toledo Calibration substance, $T_{fus} = (48.1 \pm 0.2)$ C) and biphenyl (Product Code: CRM, LGC 2610), $T_{fus} = 68.93$ C. Biphenyl, $\Delta_{trs}H = (18.574 \pm 4)$ $J\,mol^{-1}$, was used for enthalpy calibration.

Heating curve (DSC) of betaxolol polymorph I at a scanning rate $\beta = 10$ °C min^{-1}, showed a single endothermic transition corresponding to fusion process $T_{fus} = (69.0 \pm 0.3)$ C, $\Delta_{fus}H = (47.3 \pm 1.6)$ kJ mol^{-1}.

3.9 Spectroscopy

3.9.1 Ultraviolet spectroscopy

The UV spectra for betaxolol HCl was recorded with Shimadzu UV-spectrophotometer, model no. UV-1800 (Shimadzu, Japan) equipped with 1 cm quartz cells. The UV absorption spectra of 20 μg/mL betaxolol HCl in methanol is given in Fig. 2. The absorption spectra were recorded in the range of 200–400 nm. The values of the molar absorptivity of

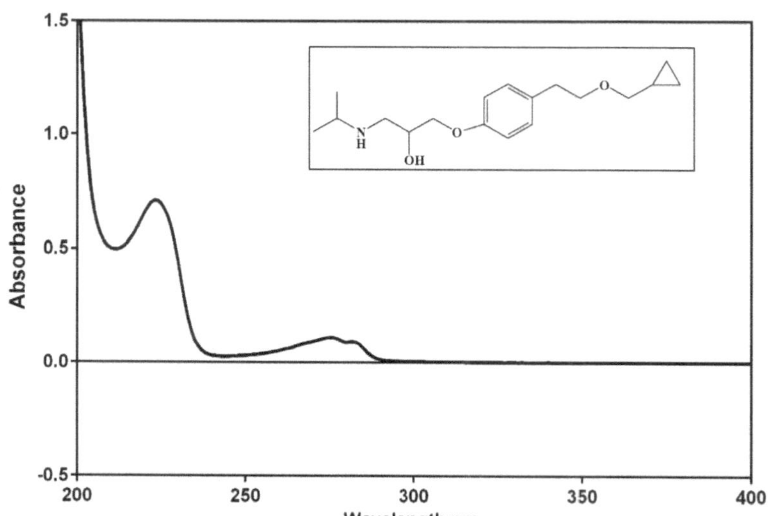

Fig. 2 Ultraviolet absorption spectrum of betaxolol HCl dissolved in acetonitrile.

betaxolol HCl at 282, 275.4 and 223.4 nm were 1564.69, 1874.19 and 12,225.24 L/mol cm, respectively.

3.9.2 Fluorescence spectroscopy

The fluorescence spectra for betaxolol HCl were recorded using Jasco FP-8200 Fluorescence Spectrometer (Jasco Corporation, Japan) equipped with a 150 W xenon lamp and 1 cm quartz cells. The spectra were recorded for betaxolol HCl (1 μg/mL) in acetonitrile Fig. 3. The slit widths for both the excitation and emission monochromators were set at 5.0 nm. Wavelength calibration was performed by measuring λ_{ex} at 225 and 277 nm and λ_{em} at 305 and 309.5 nm, respectively; no variation in the wavelength was observed. All recorded spectra converted to ASCII format by RFPC software.

3.9.3 Infrared spectroscopy

The infrared absorption spectrum ofbetaxolol HCl was recorded as KBr disk using the Perkin Elmer FT-IR Spectrum BX apparatus. The FTIR spectrum is given in Fig. 4 and shows the detailed IR (KBr) ν_{max}/cm; 3246 (NH), 2929, 2860, 2793 (Aliphatic-CH), 1611, 1560 (C=C), 1248 (C—O).

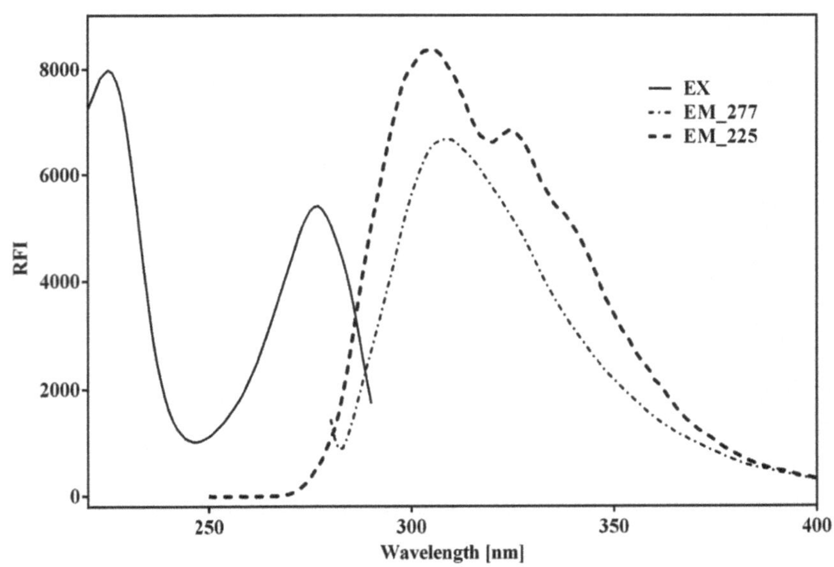

Fig. 3 Fluorescence spectrum of betaxolol HCl dissolved in acetonitrile.

(C=C),

1248

(C-O).

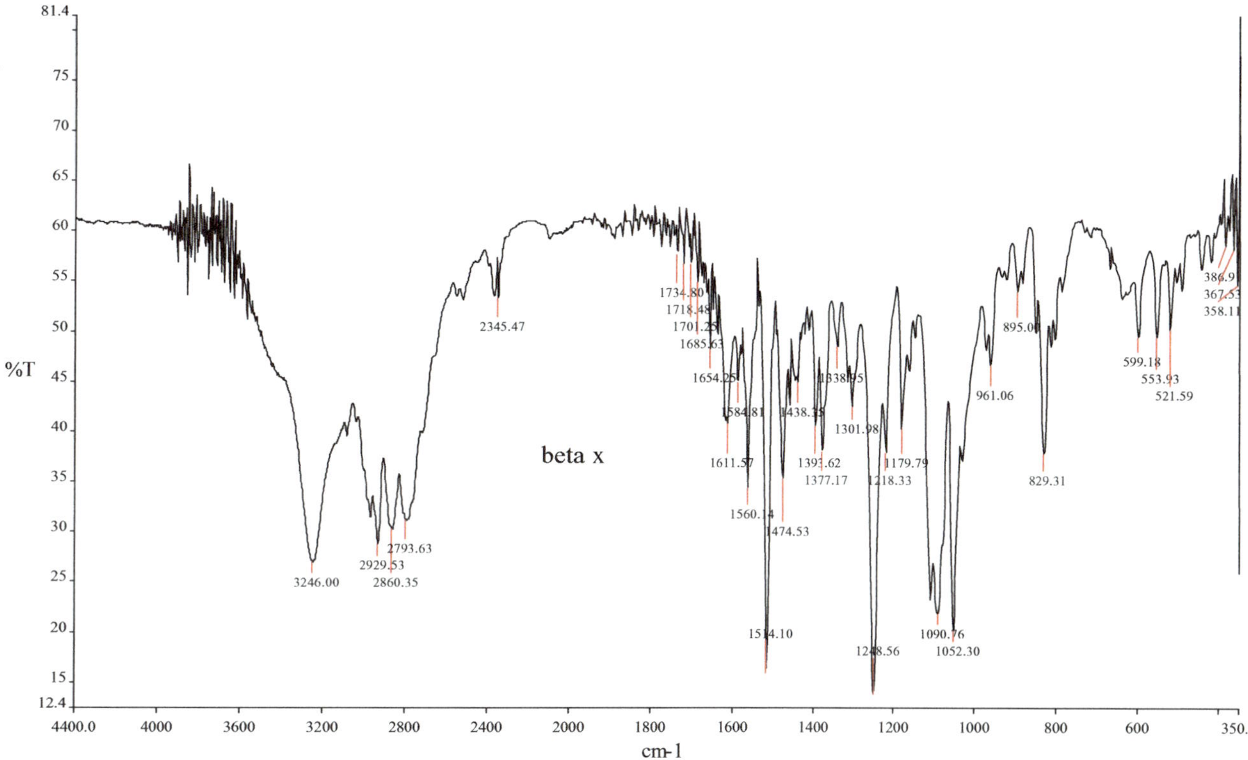

Fig. 4 IR spectrum of betaxolol HCl.

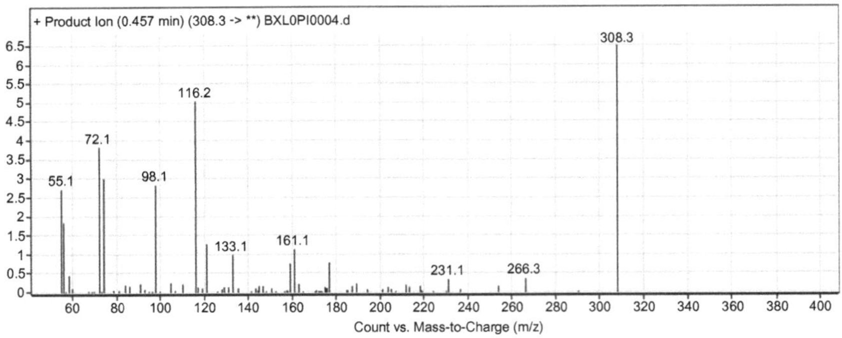

Fig. 5 Mass spectrum of betaxolol HCl.

3.9.4 Mass spectroscopy

The mass spectrum of betaxolol HCl was obtained using an Agilent Triple Quadrupole 6410 LC/MS equipped with an ESI (electrospray ionization) source. Fig. 5 shows the detailed mass fragmentation pattern $m/z = 308$; $[M-35]^+$, 266; $[M-77]^+$, 161; $[M-182]^+$, 133; $[M-210]^+$, 121; $[M-222]^+$, 116; $[M-227]^+$.

3.9.5 NMR spectrometry

3.9.5.1 ^1H NMR spectrometry

^1H NMR spectrum of betaxolol HCl was scanned in $CDCl_3$ on a Bruker NMR spectrometer operating at 500 MHz. Chemical shifts are expressed in δ-values (ppm) relative to TMS as an internal standard. Coupling constants (*J*) are expressed in Hz (Table 1 and Fig. 6A–C).

3.9.5.2 ^{13}C NMR spectrometry

^{13}C NMR spectrum of betaxolol HCl was scanned in $CDCl_3$ on a Bruker NMR spectrometer operating at 125 MHz. Chemical shifts are expressed in δ-values (ppm) relative to TMS as an internal standard (Table 2 and Fig. 7).

3.9.5.3 ^{15}N NMR spectrometry

^{15}N NMR spectrum of betaxolol HCl was scanned in $CDCl_3$ on a Brucker NMR spectrometer operating at 50 MHz. Chemical shifts are expressed in δ-values (ppm) relative to TMS as an internal standard. Fig. 8 showed ^{15} N NMR ($CDCl_3$): δ 199.97.

4. Method of analysis

4.1 Compendial methods

4.1.1 United States pharmacopeia compendial methods

4.1.1.1.1 Definition Betaxolol Hydrochloride contains NLT 98.5% and 101.5% of betaxolol hydrochloride ($C_{18}H_{29}NO_3 \cdot HCl$), calculated on the dried basis [25].

Table 1 ^1H NMR of betaxolol HCl ($CDCl_3$).

Signal	Location (δ)	Shape	Integration	Correspondences
1	0.19–0.20	d, J=3.5 Hz	2H	CH_2
2	0.52–0.54	t, J=5.0, 2.0 Hz	2H	CH_2
3	1.05–1.06	d, J=4.5 Hz	1H	CH
4	1.47–1.50	dd, J=7.0, 6.5 Hz	6H	$2CH_3$
5	2.81–2.84	t, J=7.5, 7.0 Hz	2H	CH_2
6	3.17–3.18	d, J=8.0 Hz	1H	CH
7	3.27–3.28	d, J=6.5 Hz	3H	CH + CH_2
8	3.47	S	1H	CH
9	3.58–3.61	t, J=7.5, 7.0 Hz	2H	CH_2
10	3.95–3.98	t, J=6.0, 3.0 Hz	1H	CH
11	4.05–4.08	dd, J=4.5, 5.0 Hz	1H	CH
12	4.65	S	1H	OH
13	5.42–5.43	d, J=5.0 Hz	1H	CH
14	6.80–6.81	d, J=8.0 Hz	2H	Ar-H
15	7.10–7.12	d, J=8.5 Hz	2H	Ar-H
16	8.54	S	1H	NH or HCl
17	9.55	S	1H	NH or HCl

4.1.1.1.2 Identification Infrared absorption ⟨197K⟩

Identification tests: it responds to test for chloride ⟨protocol specified in section 191⟩ when tested as directed for alkaloidal hydrochlorides.

4.1.1.1.3 Assay procedure *Buffer:* 0.025 M monobasic potassium phosphate containing 0.1% (w/v) of tetrabutyl ammonium bromide. Adjust with 0.025 M phosphoric acid to a pH of 3.0.

Mobile phase: Suitable buffer and acetonitrile and (85:15).

Resolution Solution: Prepare in mobile phase 2.0 mg/mL of USP Betaxolol Hydrochloride RS and 1.0 mg/mL of alprenolol hydrochloride.

Test preparation: 2.0 mg/mL of USP Betaxolol Hydrochloride in mobile phase.

Chromatographic system: ⟨Protocol specified in section 621⟩: Liquid chromatograph equipped with a 273 nm detector and a 4.6 mm × 15 cm column that contains 5 μM packing L1. The flow rate is about 1.5 mL per minute. Chromatograph the Resolution solution, and record the chromatogram

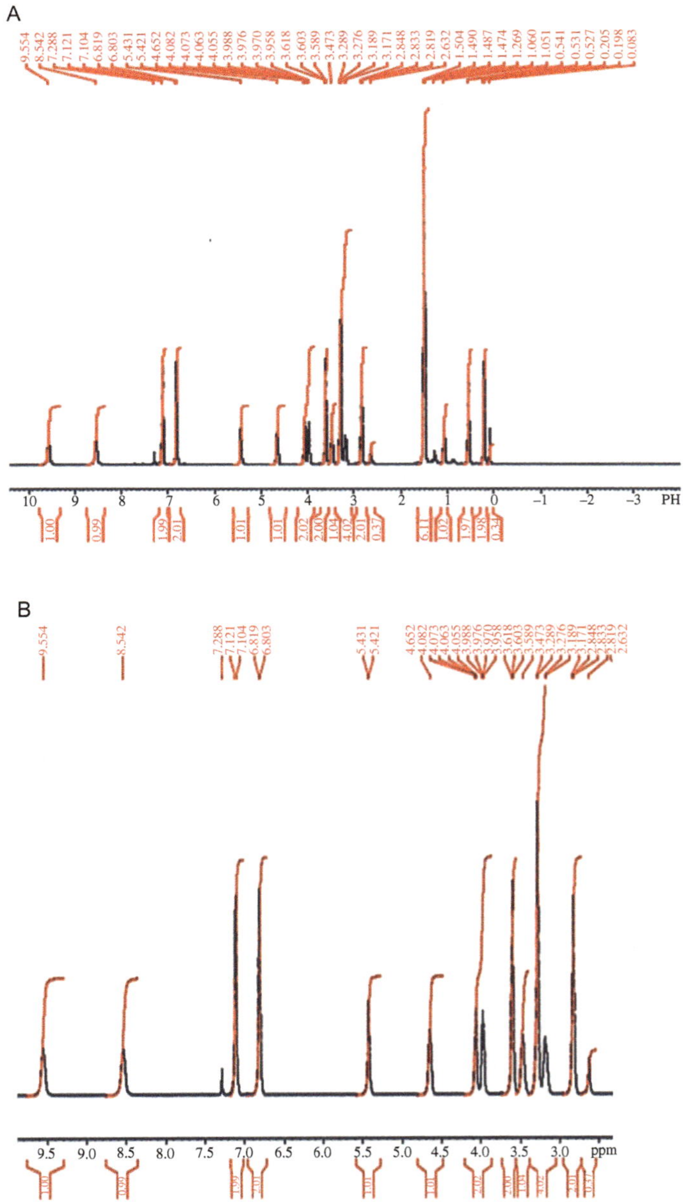

Fig. 6 (A) Full spectrum of ^{1}H NMR of betaxolol HCl. (B) Expansion of the ^{1}H NMR aromatic region of betaxolol HCl. (C) Expansion of the ^{1}H NMR aliphatic region.

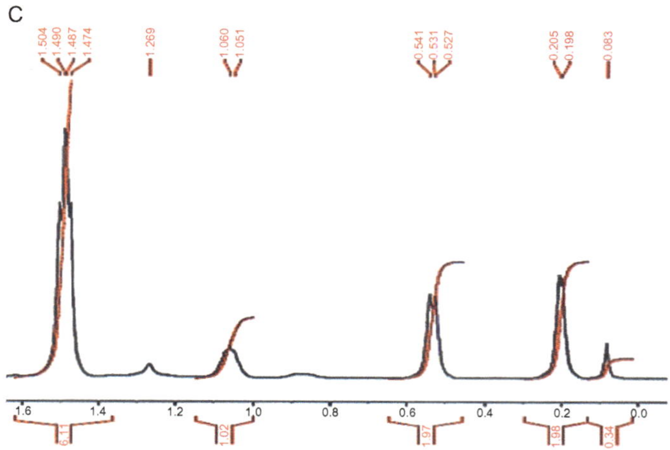

Fig. 6—Cont'd

Table 2 ^{13}C NMR of betaxolol HCl ($CDCl_3$).

Signal	**Location (δ)**	**Correspondences**
1	2.98	$\underline{C}H_2$
2	10.62	$\underline{C}H$
3	35.45, 48.12, 51.50, 65.70, 69.61, 71.73, 75.61	$\underline{C}H_2$ and $\underline{C}H$
4	114.46, 129.88, 131.82, 156.69	Aromatic-$\underline{C}_s$

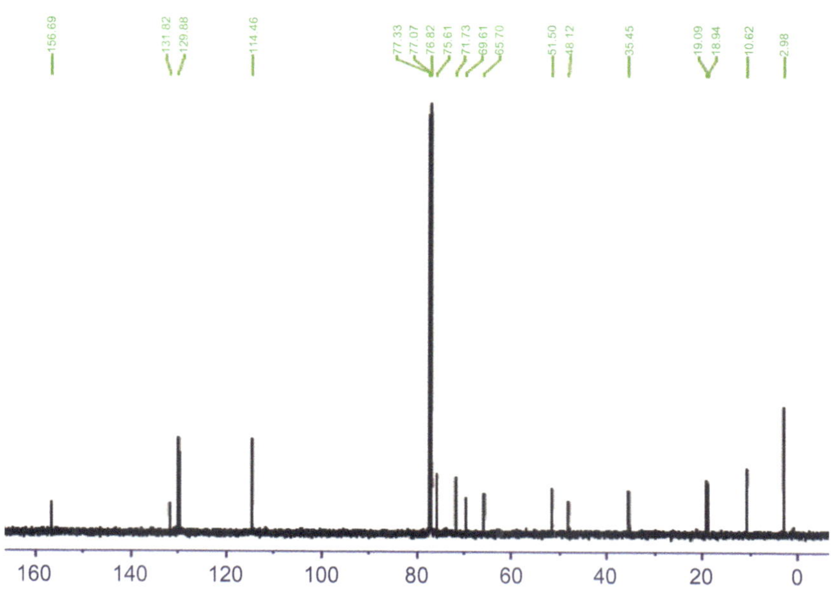

Fig. 7 ^{13}C NMR spectrum of betaxolol HCl.

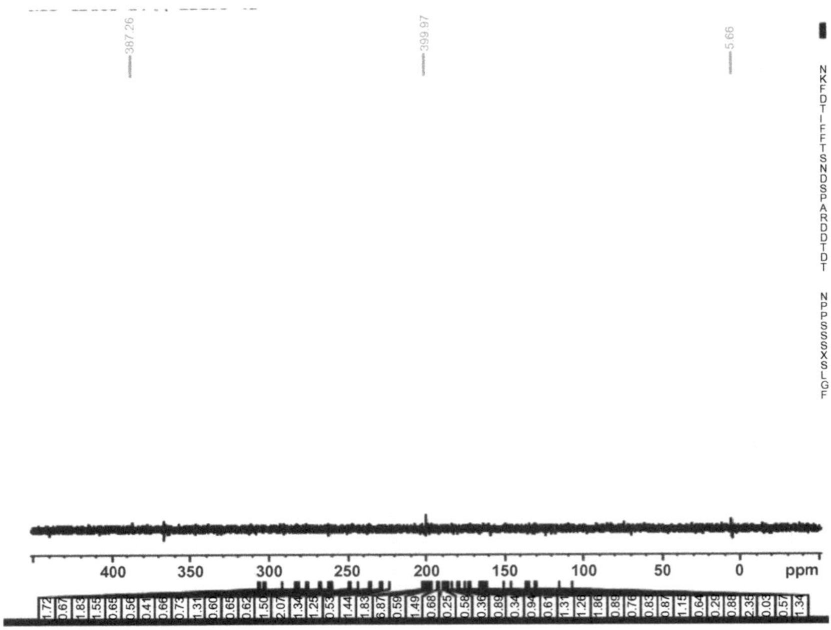

Fig. 8 ^{15}N NMR spectrum of betaxolol HCl.

and measure the peak response as directed. *Procedure*: The relative retention times are about 0.9 for alprenolol and 1.0 for betaxolol; the resolution R, between the two peaks is not less than 1.0; the tailing factor for the two peaks is not more than 2.0, and the relative standard deviation for replicate injections is not more than 2.0%.

Procedure: inject about (20 μL) of the test preparation into the chromatogram, record the chromatogram, and measure the areas of the peaks. Calculate the percentage of each impurity by the same formula:

$$100(r_i/r_s)$$

In which r_i is the response of each individual peak, other than the main betaxolol peak, in the chromatogram obtained from the test preparation, and r_s is the sum of the responses of all the peaks obtained in the chromatogram for the test preparation: the sum of all impurities found is not more than 1.0%.

Assay: Dissolve about 300 mg of betaxolol hydrochloride, accurately weighed in 50 mL of glacial acetic acid. Add 7 mL of mercuric acetate TS, and titrate with 0.1 N perchloric acid VS, determining the endpoint potentiometrically. Perform a blank determination and make any necessary correction. Each mL of 0.1 perchloric acid is equivalent to 34.39 mg of $C_{18}H_{29}NO_3 \cdot HCl$.

Loss on drying ⟨Protocol specified in section 731⟩: Dry under vacuum at 65° for 2 h: it loses not more than 1.0% of its weight.

pH ⟨Protocol specified in section 791⟩: between 4.5 and 6.5, in a solution (1 in 50).

Residue on ignition: not more than 0.1%.

Heavy Metals: Method II ⟨Protocol specified in section 731⟩: 0.002%.

4.2 Reported method of analysis

4.2.1 Spectrophotometric methods of analysis

A spectrophotometric method for a simultaneous determination of betaxolol hydrochloride and chlorthalidone in a binary mixture was developed by Patel et al. [26]. The absorbance was measured at 282.0 and 250.0 nm corresponding to the absorbance maxima of betaxolol and chlorthalidone in methanol, respectively. The linearity concentration range of 4–14 μg/mL for betaxolol HCl and 10–60 μg/mL for chlorthalidone. Concentration of each drug was obtained by the absorptivity values calculated for both drugs and solving the simultaneous equation. Developed method was applied successfully to the test mixture. The method was validated statistically and recovery study was performed to confirm the accuracy of the method.

A spectrophotometric method for the quantitative determination of betaxolol in pharmaceutical formulations was developed by Zhuk and Vasyuk [27]. This method was based on the formation of yellow complex as a result of the reaction between betaxolol and bromocresol green (3′,3″,5′,5″-tetrabromo-m-cresol-sulfophthalein) with maximum absorption at 408 nm in acetone. The proposed method validated as per Ukrainian Pharmacopeia. The proposed method was linear for betaxolol over the range of 1.65–2.75 mg/100 mL. The precision % RSD did not exceed 0.797 and the accuracy of the method was between 99.92% and 99.98%. According to the experimental data, the technique can be correctly reproduced and it is suitable for routine quality control.

A spectrophotometric method was developed for determination of betaxolol hydrochloride in bulk powder, and its pharmaceutical dosage forms by Suhagia et al. [28]. In the proposed method, betaxolol was oxidized with sodium periodate, to liberate formaldehyde, which is determined in situ, using acetyl acetone in the presence of ammonium acetate. A yellow colored chromogen was obtained having absorption maxima at

405 nm. The method is found to be linear for betaxolol in the concentration range of 5–30 μg/mL, with a regression co-efficient of 0.9997. The RSD was found to be 0.72%–4.48% and recovery was 99.10%–100.90% for betaxolol. This method was useful for routine in-process quality control for bulk drug, and its pharmaceutical preparations.

A spectrophotometric method for determination of atenolol, timolol maleate, propranolol hydrochloride, metoprolol tartarate, betaxolol hydrochloride, levobunolol hydrochloride and bisprolol fumarate in pure form as well as in their pharmaceutical formulations was developed by Salem [29]. The methods were based on the reaction of these drugs as n-electron donors with the sigma-acceptor iodine, and the pi-acceptors: 7,7,8,8-tetracyanoquinodimethane, 2,3-dichloro-5,6-dicyano-1,4-benzoquinone, tetracyanoethylene, 2,3,5,6-tetrabromo-1,4- benzoquinone (bromanil) and 2,3,5,6-tetrachloro-1,4-benzoquinone (chloranil). The obtained charge-transfer complexes were measured at 365 nm for iodine (in 1,2-dichloroethane), at 840, 420, and 470 nm for 7,7,8,8-tetracyanoquinodimethane, tetracyanoethylene and 2,3-di-chloro-5,6-dicyano-1,4-benzoquinone (in acetonitrile), respectively, and at 450 and 440 nm for bromanil and chloranil (in ethanol), respectively. Due to the rapid development of colors at ambient temperature, the obtained results were used on thin-layer chromatograms for the detection of the investigated drugs. Beer's plots were obeyed in a general concentration range of 4–120 μg/mL with correlation coefficients not less than 0.9991. The procedures could be applied successfully to the determination of the investigated drugs in tablets and ophthalmic solutions with percent ranged from 98.03 ± 0.98 to 100.30 ± 0.90. The association constants and standard free energy changes using Benesi-Hildebrand plots were studied.

A kinetic method for the determination of three cardiovascular drugs, namely betaxolol, clopidogrel, and imidapril in pharmaceutical formulations was developed by Yazbi et al. [30]. The method was based on the alkaline oxidation of the selected drugs with potassium manganate (VII) at room temperature (25 °C) to produce a bluish-green colored product. The reaction was followed spectrophotometrically by measuring the rate of change of absorbance at 610 nm as a function of time. The fixed-time method was adopted for constructing the calibration curves, which were found to be linear over the concentration ranges of 0.40–2.0 μg/mL, 1.0–10.0 μg/mL, and 6.0–16.0 μg/mL for betaxolol, clopidogrel, and imidapril, respectively. The retention times were 15 min for betaxolol and clopidogrel and 10 min for imidapril. The method was successfully applied to pharmaceutical formulations of each drug. The percentage relative error and the coefficient of variation were less than 0.6 and 1.2, respectively. The lower limits of quantitation (LOQ) were 9.75×10^{-7} M for betaxolol, 3.62×10^{-6} M for clopidogrel, and 4.22×10^{-6} M for imidapril.

4.2.2 Spectrofluorometric method

A spectrofluorometric method for estimation of betaxolol under different experimental conditions was developed by Sultan [31]. The highest fluorescence intensity was obtained in methanol at 305 nm using 275 nm for excitation. The fluorescence intensity-concentration plot was rectilinear over the range 0.1–1.0 μg/mL with lower detection limit of 0.025 $\mu g\,mL^{-1}$ (8.1×10^{-8} M). Interference likely to be introduced from co-formulated drugs such as benzalkonium chloride or co-administered drugs such as hydrochlorothiazide, was studied. The method was successfully applied to the determination of the drug in its tablets and ophthalmic solution. The mean percent recoveries were in agreement with those provided by the official methods. The method was further extended to the in-vitro determination of the drug in spiked human plasma using solid phase extraction, utilizing a C18 Bond-Elute column.

4.2.3 Electrochemical method

4.2.3.1 Potentiometric titration

A potentiometric titration method was developed for three β-blockers, namely, sotalol, carvedilol, and betaxolol by Wassil et al. [32]. In this method, the construction of PVC matrix-type β-blockers (sotalol, carvedilol, and betaxolol) ion selective electrodes was described. The proposed sensors were based on the complex ion associates of β-blockers with tungstophosphate (TP) and Ammonium Reineckate (Rein) ionophores in poly vinyl chloride membrane with Dioctylphthalate plasticizer. The four electrodes (Beta-TP), (Sota-TP), (Carve-TP), and (Cave-Rein) showed a stable potential response with a near Nernstian slope of 50.8, 33.7, 32.35, and 33 mv per decade. The range of concentration 10–2 to 10–7 M for β-blockers was reported. Selectivity coefficients data obtained for 11 different organic and inorganic ions were presented. The electrodes response time (30 and 40 s) were used over wide range of pH 4.5–8.5. Validation of the method according to the quality assurance studies was found to be suitable for the proposed sensors for use in the quality control assessment of these drugs. The average recoveries of the proposed method were 100.78% and the mean standard deviation was ±1.2. The data agree well with those obtained by standard methods.

A potentiometric sensor for the determination of betaxolol hydrochloride in plasma and urine was developed by Alturiki [33]. The electrode employed PVC membrane with betaxolol-tetraphenylborate (BT-TPB) ion pair complex with di-butyl phosphate (DBP) as a plasticizer. The calibration curve was linear in the range from 1.0×10^{-6} to 1.0×10^{-2} mol/L with a correlation coefficient of 0.997 and slope of 59.0 mV/decade, which was close to the Nernstian slope. Selectivity coefficients of betaxolol related to a number of interfering cation, sugars and amino-acids were investigated. The influence of pH of the tested solutions on the formulation of the electrode was not as considerable since the electrode worked correctly in the pH range 3.20–8.30. The electrode was successfully used for potentiometric determination of betaxolol hydrochloride both in pure solutions and in biological fluids.

4.2.3.2 Voltammetric method

A differential pulse voltammetry coupled with multivariate curve resolution-alternating least squares (MCR-ALS) for simultaneous determination of betaxolol (Bet) and atenolol (Ate) was developed by Khoobi et al. [34]. The proposed method was performed in 0.20 M Britton–Robinson (B-R) buffer solution at the surface of a multi-walled carbon nanotube modified carbon paste electrode (MWCNT/CPE). Characterization of the modified electrode was carried out by electrochemical impedance spectroscopy (EIS) and cyclic voltammetry (CV). A strategy based on experimental design was followed. Operating conditions were improved with central composite rotatable design (CCRD) and response surface methodology (RSM), involving several chemical and instrumental parameters. Then second order data were built from variable pulse heights of DPV and after correction in potential shift analyzed by MCR-ALS. Analytical parameters such as linearity, repeatability, and stability were also investigated and a detection limit (DL) of 0.19 and 0.29 μM for Bet and Ate was achieved, respectively. The proposed method was successfully applied in simultaneous determination of the two analytes in human plasma.

A differential pulse voltammetric method was described by Khoobi et al. [30] for the detrmination of betaxolol in metabolite and plasma samples. In this method, the oxidation peak current of the betaxolol was optimized in a 0.20 M Britton–Robinson buffer solution by experimental design for its determination in the presence of acetaminophen. A central composite rotatable design was used to evaluate the effects of the variables by the differential pulse voltammetry method. These variables include scan rate, step potential, modulation amplitude, gold nanoparticles content and pH. Then, under the optimized conditions, the dynamic range for betaxolol in the presence of acetaminophen was from 0.50 to 125.00 μM and the detection limit was found to be 46.00 nM. The characterization of the gold nanoparticle modified carbon paste electrode was investigated by cyclic voltammetry and electrochemical impedance spectroscopy. Other voltammetric methods such as double-potential step chronoamperometry and linear sweep voltammetry were also applied. Using these methods, the diffusion coefficient ($D = 7.43 \times 10\text{–}5\ cm^2\ s^{-1}$) and the kinetic parameters such as the electron transfer coefficient ($\alpha = 0.60$) and exchanging c.d. ($j_0 = 4.6 \times 10^{-4}\ \mu A\ cm^{-2}$) for BET were determinate. The results indicated that the gold nanoparticle modified carbon paste electrode could be employed for the determination of betaxolol in real samples such as pharmaceutical formulations and plasma samples.

4.2.4 *Chromatography*

4.2.4.1 Thin-layer chromatography

Various analytical techniques, namely, TLC, HPLC and ^{1}H NMR, were used for recognizing chirality and determining the absolute configuration of the diastereomers of (RS)-Betaxolol, [(RS)-Bet] by Vashistha et al. [35]. The diastereomers of (RS)-Bet were separated and isolated using preparative TLC and a reverse phase HPLC method was developed and validated for the enantioseparation of (RS)-Betaxolol. Diastereomers were synthesized, using a naproxen based chiral derivatizing reagent, under microwave irradiation. The absorbance configuration of both diastereomers was established with the help of ^{1}H NMR spectroscopy and by developing the lowest energy optimized structures of the diastereomers based on density functional theory. The configuration of the diastereomers was established as [(S)-Nap-(R)-Bet] and [(S)-Nap-(S)-Bet]. The determination of the absorbance configuration of the two diastereomers lent support to elution order and separation mechanism. The limits of quantitation were found to be 0.80 ng/mL for the diastereomers.

A TLC method for determination of timolol maleate (Timohexal 0.1%, Timoptic 0.25%) and betaxolol hydrochloride (Betoptic 0.5%) in ophthalmic solution was developed by Hopkala et al. [36]. The selectivity of the separation of these drugs were investigated by TLC on silica gel 60 F_{254}, silanized C_8 silica gel, and aluminum oxide 60 F_{254} plates with ascending and horizontal development. The substances were visualized by UV irradiation at $\gamma = 254$ nm or by spraying with suitable reagents. Detection and quantification of timolol maleate and betaxolol hydrochloride were performed densitometrically at $\gamma = 300$ nm and $\gamma = 218$ nm, respectively, and videodensitometrically at $\gamma = 254$ nm. Results from quantitative analysis were statistically assessed by use of the calibration procedure, a calibration equation, the correlation coefficient, and relative standard deviation. The average value of the coefficient of variation was 2.93%. The best separation of the analyzed drugs was achieved by ascending or horizontal development on silica gel with ethyl acetate–methanol–25% aqueous ammonia, 80 + 18 + 2 (v/v), as mobile phase. The spots obtained by use of this mobile phase were sharp and oval. The time for development to a distance of 10 cm was 24 ± 1 min (mean ± SD). The detection limits were determined visually by establishing the minimum levels of analyte that could be reliably detected.

A stability indicating TLC-densitometric method for the determination of betaxolol (Bx) in pharmaceutical preparations was developed by Kwiecie et al. [37]. The stability of betaxolol in acidic solutions was studied. The method was developed on TLC aluminum plates precoated with silica gel F_{254} using the mobile phase chloroform-methanol-ammonia 25% (18:4:0.2 v/v/v) which gives compact spots for betaxolol (R(f) approximately equal to 0.64) and its degradation product (R(f) approximately equal to 0.39). Densitometric analysis was carried out in UV at 280 nm. The LOD and LOQ were 66.6 ng/spot and 200 ng/spot, respectively. The (RSD = 2.73%) and the mean recovery was 100.28% at 100% level. Betaxolol was subjected to acidic and alkaline hydrolysis but degradation was observed only in acidic solutions. The degradation process was described with kinetic and thermodynamic parameters. Based on LC-ESI/MS analysis, it was found that betaxolol decomposed in acidic solution to produce ethoxyphenoxy-3-[(1-methylethyl)amino]propan-2-ol.

4.2.4.2 Capillary electrophoresis

A capillary electrophoresis (CE) method for determination of betaxolol hydrochloride tablets was developed by Zhang et al. [38]. The proposed method was performed on the following conditions: fused-silica capillary of 37 cm × 75 μm (30 cm of effective length); running buffer: 30 mmol/L acetate buffer (pH 4.0); sample injection method (10 kV × 5 s); running voltage: 20 kV; column temp.: 25°, and detection wavelength: 273 nm. The linear concentration range was 0.05–0.30 mg mL^{-1} of betaxolol hydrochloride ($r = 0.9995$, $n = 6$). The average recovery was 99.2%, and RSD = 0.63%. The method could be used to detect the content of betaxolol hydrochloride tablets.

Servais et al. [39] described a non-aqueous capillary electrophoresis (NACE) method for the separation of four β-blocking drug substances (atenolol, sotalol, betaxolol, and metoprolol) selected as model basic analytes using the principle of ion-pair formation. Camphor sulfonate was selected as the counter-ion in a background electrolyte (BGE) made up of formate buffer in a mixture of acetonitrile/methanol or acetonitrile/ethanol. The influence on resolution of the concentration of the counter-ion, the nature and proportion of the organic solvents, as well as the concentration of the ionic components of the BGE were studied by a univariate approach. An experimental design was then applied to establish possible quadratic effects and first-order interactions. A two-level fractional factorial design with 16 experimental points was applied as a screening test to identify the most important factors affecting the separation of the four β-blockers. A three factor Box-Behnken design with 12 experimental points was used to predict and optimize the selectivity. Finally the optimal conditions obtained by the univariate and the multivariate approaches were compared.

4.2.4.3 HPLC methods

A stability-indicating reversed-phase high performance liquid chromatography (RP-HPLC) method was developed for the determination of betaxolol hydrochloride by Auvity et al. [40]. The chromatographic separation was achieved on a Nucleosil C18, 4 μm (150 × 4.6 mm) column, using isocratic elution at a 220 nm detector wavelength. The optimized mobile phase consisted of a 0.02 M potassium dihydrogen phosphate: methanol (40:60 v/v, pH 3.0 adjusted with o-phosphoric acid) as solvent. The flow rate was 1.6 mL/min and the retention time of betaxolol hydrochloride was 1.72 min. The linearity for betaxolol hydrochloride was in the range of 25–200 μg/mL. Recovery for betaxolol hydrochloride was calculated as 100.01%–101.35%. The stability-indicating capability was established by forced degradation experiments and the separation of unknown degradation products. The developed RP-HPLC method was validated according to the International Conference on Harmonization (ICH) guidelines. This validated method was applied for the determination of betaxolol hydrochloride in tablets.

A high-performance liquid chromatographic (HPLC) method for the separation and determination of S- and R- enantiomers of betaxolol in tablets and ophthalmic preparations was developed by Hefnawy et al. [41]. Baseline resolution was achieved by using teicoplanin macrocyclic antibiotic chiral stationary phase (CSP) known as Chirobiotic T with fluorescence detection at excitation/emission wavelengths 275/305 nm. The polar ionic mobile phase consisted of methanol-glacial acetic acid-triethylamine, (100:0.020:0.025, v/v/v) at a flow rate of 1.5 mL/min. In this study, S-(-)-atenolol was used as internal standard and conducted at ambient temperature. The stability of betaxolol enantiomers under different degree of temperature were also studied. The results showed that it is stable for at least 7 days at 70 °C. The method was validated for its linearity, accuracy, precision and robustness. Using the chromatographic conditions described, S- and R-betaxolol were well resolved with mean retention times of 11.3 and 12.6 min, respectively. Linear response ($r > 0.997$) was observed over the range of 10–500 ng/mL of betaxolol enantiomers, with detection limit of 5 ng/mL. The recoveries of S- and R-betaxolol from tablets and ophthalmic preparation ranged from 97.4% to 101.4% and 98.0% to 102.0%, respectively. The mean relative standard deviation (RSD%) for both enantiomers were 1.1%–1.4% and 1.3%–1.7% in tablets and ophthalmic solution, respectively.

Darmon et al. [42] described an HPLC method for the determination of (R)-[91878-53-4] and (S)-betaxolol [(S)-I] [93221-48-8] in blood and other biological fluids. Separation of the enantiomers was performed after preparation of diastereomeric derivatives with the chiral reagent R(-)-naphthylethylisocyanate by reversed-phase HPLC. Fluorimetric detection was used for the quantification of betaxolol enantiomers up to 0.5 ng/mL. This method was used to evaluate the pharmacokinetic profile of the betaxolol enantiomers in subjects following a single oral dose (20 mg) of racemic betaxolol [63659-18-7]. No significant difference was observed in blood levels of the isomers.

Zhou et al. [43] developed a new liquid chromatography method for the chiral separation of the enantiomers of betaxolol hydrochloride. The enantiomers were separated by a Chiralpak IB (250 × 4.6 mm, 5 mm) at 30 °C and detected at 274 nm, the mobile phase used was *n*-hexane-ethanol (95/5, V/V), 0.2% diethylamine (DEA), with a flow rate of 0.8 mL/min and the resolution was 5.26. With the fluorescence detection ($\lambda ex = 227$ nm, $\lambda em = 305$ nm), the method was applicable for the determination of the enantiomers of betaxolol hydrochloride in rabbit plasma.

A method for the separation of betaxolol hydrochloride by chiral high performance liquid chromatography was also described by Ke et al. [44]. The stability of optical purity for levobetaxolol hydrochloride was investigated. A normal phase HPLC was used to separate the isomer by column of Daicel OJ-H and OD-H (250 × 4.6 mm, 5 mm). The mobile phases were hexane-isopropanol-diethylamine (96:4:0.5 v/v) for Daicel OJ-H column and hexane-isopropanol-diethylamine (70:30:0.5 v/v) for OD-H column respectively. The column temperature was fixed at 20–40°. The flow rate was 0.5–1.5 mL/min and detection wavelength was 190–400 nm. The stability of optical purity for levobetaxolol was determined by column OD-H. The method for rapid separation of betaxolol hydrochloride was established by chiral column. The optical conditions for OJ-H column were as follows: wavelength at 274 nm, flow rate at 0.8 mL/min, column temperature at 25°, and hexane-isopropanol-diethylamine (96:4:0.5 v/v) as mobile phase. The test of stability showed that the optical purity for levobetaxolol hydrochloride was close to 100% in 10 days with high temperature, high humidity and strong light. This method had strong specificity. This method can be used to separate the racemate of betaxolol hydrochloride. The optical purity for levobetaxolol hydrochloride was stable.

A high performance liquid chromatography method with fluorescence detection for determination of betaxolol hydrochloride was established and its pharmacokinetics in human plasma was described by Du et al. [45]. After a single oral dose of 40 mg betaxolol hydrochloride tablets was given to 10 volunteers in randomized cross over study, the betaxolol concentration in serum was determined by HPLC-Fluorescence method. The Hypersil C_{18} column (4.6 × 250 mm, 5 μm) was used with mobile phase consisting of 0.2% phosphate buffer solution-acetonitrile-methanol (60:37:3). The flow rate was 1.0 mL/min with fluorescence detection set at λex = 227 nm, λem = 301 nm. The study curve was linear in the range of 2.5–80 μg/mL (r = 0.9998). The minimum detection concentration was 0.5 ng/mL (S/N > 3) and the absorbance recoveries of betaxolol from plasma were larger than 66%. The assay method was convenient for determination of betaxolol hydrochloride. The pharmacokinetics parameters were T_{max} = (4.4 ± 1.1) h, C_{max} = (52.39 ± 11.14) μg/L, T1/2ke = (15.78 ± 2.43) h, T1/2ka = (1.07 ± 0.66) h, V/F = (821.5 ± 146.7) L, AUC0-t = (1088 ± 178) μg/h/L, AUC0-∞ = (1219 ± 241) μg/h/L. The presented method was suitable for pharmacokinetic studies of betaxolol in human plasma.

A reversed-phase HPLC method for the determination of betaxolol in human aqueous humor was described by Dulger et al. [46]. Betaxolol and the internal standard was extracted with cyclohexane and separation was performed on a reversed-phase column (Luna C18, 250 × 4.6 mm, 5 mm) with a mobile phase containing acetonitrile-phosphate buffer (40:60, v/v) at a flow-rate of 0.8 mL/min. The column effluent was monitored with a fluorescence detector at 227 nm (excitation) and 301 nm (emission). The retention times for metoprolol and betaxolol were 3.55 and 5.63 min, respectively. The recovery from aqueous humor was 71.6% for betaxolol at 1.25 μg/mL. The within-day and day-to-day accuracy values were at 96.17%–105.2% for betaxolol at 0.1, 4 and 12 mg/mL ($n = 6$), within-day and day-to-day precision values were <10% for betaxolol at the concentrations given above. The detection limit corresponding to the signal-to-noise ratio of 3:1 was 15 ng/mL. The presented method was suitable for measuring betaxolol levels in human aqueous humor samples obtained from patients after topical administration.

An HPLC method for measurement of betaxolol in blood was described by Bhamra et al. [47]. A direct extraction of 200 mL of sample at alkaline pH into an organic solvent (Methyl *tert*-butyl ether) followed by HPLC analysis with fluorescence detection were performed. The HPLC column used was Spherisorb S5W silica, the mobile phase 1 mM camphor sulfonic acid in methanol, and the internal standard was benzimidazole. The limit of accurate measurement of betaxolol in plasma was 5 mg/L and the recovery was 97.41%. No endogenous sources of interference were absorbed and no potential metabolites were detected. Of the drugs studied, only propranolol co-eluted with betaxolol.

An HPLC method with fluorometric detection for determination of betaxolol in blood and urine was described by Caqueret et al. [48]. The method described was suitable for routine monitoring of betaxolol. The time required for the preparation of 40 samples for the injection was less than 2 h. Moreover, using automatic injection and an appropriate data system, manual intervention by the operator was minimal. A calibration curve prepared with blood (or urine) spiked with 10, 20, 40 and 80 ng (from a methanolic solution of 1 μg/mL) and 50 ng of metoprolol as internal standard, was run with each series of samples. The chromatographic column had Spherisorb CN. The mobile phase was acetonitrile-acetate buffer (40:60, v/v, pH 5.6). The blank blood extract showed no endogenous interfering peaks. The retention times of metoprolol and betaxolol were 6.4 and 7.9 min, respectively. The absolute sensitivity (signal to noise ratio = 3) of this method, checked using standard solutions, was 0.5 ng/mL. The linearity range for betaxolol was 5–500 ng/mL. The sensitivity of the proposed method was adequate for the quantification of the drug in plasma samples.

An HPLC method with fluorescence detection to determine betaxolol in human plasma was proposed by Wu and Yankai [49]. The excitation and emission wavelength were 227 and 301 nm, respectively. The column was

Hypersil C18 column (4.6 × 250 mm, 5 μm) using a mobile phase consisting of acetonitrile:methanol:0.2% potassium dihydrogen phosphate (37:3:60, v/v) after optimization of chromatographic conditions, adjusted with 0.025 M phosphoric acid to pH 3.0 and a flow rate of 1.0 mL/min. The standard curve was linear over a range of 2.5–80 ng/mL. The lower LOD was 0.5 ng/mL at S/N = 3 and the LOQ was 2.5 ng/mL (RSD = 3.4%, *n* = 5). The recovery from human plasma was found to be 88% of betaxolol. The method was accurate, precise and reliable to determine the betaxolol hydrochloride in human plasma.

A method for the identification and quantification of beta-adrenergic receptor antagonists betaxolol, carteolol, metipranolol, and timolol in plasma and urine was described by Tracqui et al. [50]. An extraction method was described using pindolol as an internal standard. An RSIL 10 μm column was used. The lower detection limits of the beta-blockers were found to be 4–27 ng/mL. An RSiL CN (Alltech) 10 μm column (300 × 4.1 mm, i.d.) was used at ambient temperature. The mobile phase consisted of water, acetonitrile, and 0.1 M buffered.

NaH_2PO_4, solution (60:30:10 v/v) adjusted to pH 3.0 with orthophosphoric acid. Flow rate was 1.0 mL/min with an average operating pressure of 1000 lb/in^2. The results were reproducible and linear over range of 50–1000 ng/mL (correlation coefficient value >0.98). Within-run precision was evaluated by multiple analyses ,of blank plasma spiked to contain two different concentrations (100 and 250 ng/mL) of drug tested. The coefficients of variation were 5.0%–7.8% at 100 ng/mL and 4.2%–8.3% at 250 ng/mL. The coefficient of variation for Day-to-day precision was found to be about 7.5% (Table 2). The lower limits of detectability in plasma were 4, 12, 15, and 27 ng/mL for carteolol, metipranolol, betaxolol and timolol, respectively. This method is simple, rapid and sensitive; moreover, it allows the determination of 8 other beta-blockers.

Gao et al. [51] a HPLC-fluorescence method for determination of betaxolol hydrochloride in plasma samples. Pharmacokinetics and relative bioavailability of betaxolol hydrochloride in young healthy volunteers were studied. A single oral dose (40 mg) of the study formulation and the test formulation were given to 18 volunteers in randomized crossover test. The concentration-time curves of two formulations fitted to a one-compartment open model. The T_{max} were (4.6 ± 1.4) h and (4.1 ± 1.1) h, the C_{max} were (51.45 ± 9.21) μg/L and (53.69 ± 10.58) μg/L, $t_{1/2}$ were (17.16 ± 2.27) h and (16.47 ± 2.80) h, $AUC_{0\text{-}60h}$ were (1007 ± 165) μg/h/L and (1145 ±218) μg/h/L, $AUC_{0-\infty}$ were (1165 ± 214) μg/h/L and (1281 ± 250) μg/h/L. The pharmacokinetic parameters obtained from our studies showed no significant difference between two formulations ($P > 0.05$). The relative bioavailability F_{0-t} was (89.4 ± 15.0)%, $F_{0-\infty}$ was (92.6 ± 18.0)% of tested formulation. The results of statistical analysis suggested that the two formulations were bioequivalent.

A high-performance liquid chromatography (HPLC) method for the simultaneous identification and quantification in human plasma of the 13 most commonly prescribed β-blockers atenolol, sotalol, diacetolol, carteolol, nadolol, pindolol, acebutolol, metoprolol, celiprolol, oxprenolol, labetalol, propranolol, tertatolol, betaxolol and one active metabolite was developed by Delamoye et al. [52]. It involved liquid-liquid extraction procedures followed by liquid chromatography coupled to photodiode-array UV detection with a fixed wavelength at 220 nm for quantification. Compounds were separated on a 5 μm Hypurity C18 (ThermoHypersil) analysis column (250 × 4.6 mm) using a gradient of acetonitrile-phosphate buffer pH 3.8 at a flow rate of 1.0 mL/min. The total analysis time was 26 min per sample. Extraction recoveries were between 74% and 113% for the polar compounds and between 20% and 56% for the most polar compounds. Calibration lines were linear in the range from 25 to 1000 ng/mL for all compounds except carteolol and nadolol (50–1000 ng/mL), all of them with coefficients of determination (r^2 values) ≥0.994. Limits of detection (LODs) ranged from 5 to 10 ng/mL. Intra-assay and inter-assay precision and accuracy were studied at two concentration levels (100 and 500 ng/mL). The intra-assay coefficients of variation (CVs) for all compounds were ≤8.3% and all inter-assays CVs were below 12.6%. The intra-assay and inter-assay accuracies for all compounds were found to be within 91.4% and 105.6% at 100 ng/mL and within 94.1 and 107.4% at 500 ng/mL. Thus, the performance of the method described allows its use in toxicological screening and in quantification of the most prescribed β-blockers drugs.

A method for separation of betaxolol, bevantolol, metoprolol, bisoprolol by HPLC with bonded chiral stationary column(3,5-dimethylcabamate cellulose) was developed by Zhou et al. [53].The optimal mobile phase compositions of betaxolol, bevantolol, metoprolol, bisoprolol were n-hexane/ethanol (95/5, v/v; 0.2% diethylamine); n-hexane/ethanol (90/10, v/v; 0.1% diethylamine); hexane/ethanol (60/40, v/v; 0.1% diethylamine); n-hexane/ethanol (95/5, v/v; 0.2% diethylamine), respectively. The optimal temperature was 30 °C with a flow rate of 0.8 mL min^{-1}. The detection wavelength, by 1200 V D UV detector, of betaxolol, bevantolol, metoprolol, bisoprolol was 259, 274, 224 and 223 nm, respectively. Experimental results demonstrated that baseline separation (Rs > 1.5) of the betaxolol, bevantolol, metoprolol, bisoprolol enantiomers.

A method for enantioseparation of orciprenaline, betaxolol, and propranolol using HPLC and new chiral reagents based on 1,5-difluoro-2,4-dinitrobenzene was proposed by Bhushan and Nagar [54]. Diastereomers of three β-blockers (orciprenaline, betaxolol, and propranolol) were synthesized using three new chiral derivatizing reagents that were prepared by substituting one fluorine atom in 1,5-difluoro-2,4-dinitrobenzene with two chirally pure amines and one L-amino acid. The reagents were characterized using ultraviolet-visible spectroscopy, infrared spectroscopy, elemental analysis, and proton nuclear magnetic resonance. Diastereomers were synthesized under microwave irradiation at 50 s at 80% of 800 W and also by stirring for 50 min at 45 °C. Chromatographic conditions: LiChrospher C18 (250 × 4.6 mm i.d., 5 μm particle size); mobile phase 2, linear gradient (35%–65%) of acetonitrile with 0.10% trifluoroacetic acid for diastereomers of propranolol and betaxolol, and linear gradient (30%–75%) of acetonitrile with 0.10% trifluoroacetic acid for diastereomers of orciprenaline in 45 min; flow rate, 1.0 mL/min; detection, 340 nm. t R1 is retention time, k1 is retention factor of [(R)-L]-diastereomers; α, separation factor; R s, resolution. Optimization of chromatographic conditions was carried out by varying the concentration of buffer (triethylamine, 5–15 mM and trifluoroacetic acid 0.05%–0.2%), pH (in the range 2.5–5.5), and flow rate (0.5–2.0 mL/min) of the mobile phase. The regression equations were $y = 0.942x - 11.42$ (R2 = 0.999) and $y = 0.943x - 10.37$ (R2 = 0.999) for the diastereomers of (R)- and (S)-propranolol, respectively. The values of relative standard deviation for [(R)-L]- and [(S)-L]-diastereomers, respectively, were 0.49%–1.28% and 0.59%–1.30% for intra-day precision, and 0.58%–1.39% and 0.65%–1.26% for inter-day precision. The recovery values for the [(R)-L]- and [(S)-L]-diastereomers were 98.87%–101.13% and 99.25%–101.06% for intra-day assay and 99.37%–100.77% and 99.24%–101.25% for inter-day assay, respectively. The limit of detection (LOD), corresponding to signal-to-noise ratio of 3, was found to be 12 pg/mL and 13 pg/mL for [(R)-L]- and [(S)-L]-diastereomers, respectively and limit of quantitation (LOQ) corresponding to signal-to-noise ratio of 10, was found to be 36 and 39 pg/mL for [(R)-L]- and [(S)-L]-diastereomers.

An HPLC method for direct enantiomeric separation of betaxolol was proposed by Krstulović et al. [55]. The column Chiralcel OD

(250 × 4.6 mm I.D., 10 pm average particle size) from Daicel Chemical Industries (Sochibo, Velizy Villacoublay, France) was used in the analysis. The column temperature was ambient (22 ± 1 °C). The mobile phase were mixtures of hexane, 2-propanol and/or ethanol or hexane-2-propanol-diethylamine (87: 13:0.05, v/v/v), and the flow rate was 1.5 mL/min. The method is simple and precise. The column equilibration times are short and the analysis time is approximately 7 min. By changing the nature or the concentration of the polar modifier, enantiomeric separations of other β-adrenergic blocking agents can be obtained. This method was applied for the routine determination of the enantiomeric excess of (S)-betaxolol and also demonstrated its applicability to the measure betaxolol in rat hepatocytes.

A HPLC method with spectrofluorometric detection for determination of betaxolol in aqueous humor was developed by Dulger et al. [46]. Betaxolol and the internal standard metoprolol were extracted with cyclohexane and separated on a reversed-phase column (Luna C18, 250 × 4.6 mm, 5 μm) with a mobile phase containing acetonitrile–phosphate buffer (40:60, v/v) at a flow-rate of 0.8 mL/min. The column effluent was monitored with a fluorescence detector at 227 nm (excitation) and 301 nm (emission). The retention times for metoprolol and betaxolol were 3.55 and 5.63 min, respectively. The recovery from aqueous humor was found to be 71.6% for betaxolol at 1.25 μg/mL. The within-day and day-to-day accuracy values were in the range of 96.17%–105.2% for betaxolol at 0.1, 4 and 12 μg/mL ($n = 6$), within-day and day-to-day precision values were less than 10% for betaxolol at the concentrations given above. The detection limit corresponding to the signal-to-noise ratio of 3:1 was 15 ng/mL. The presented method measured betaxolol levels in human aqueous humor samples obtained from patients after topical administration.

An ion sensitive and thermosensitive in situ gel of betaxolol hydrochloride for ophthalmic use was quantified by Yuan et al. [56]. The methods were developed for assay the drug content and evaluating the preparation in vitro. The deacetylated gellan gum and poloxamer 407 were used as gel base. HPLC method was established to determinate drug content of betaxolol in the gel. A membrane less model and corneal penetration test were used to study the drug release. Formula of gel was principal agent 0.28%, 0.5% deacetylated gellan gum and 9% poloxamer 407 as gel base, 5% mannitol as osmotic pressure regulation and 0.03% ethylparaben as antibacterial agent. Compared with betaxolol hydrochloride solution agent, the drug in in situ gel penetrated slower. The preparation was simple and it had sustained-release characteristic.

A reversed-phase HPLC (RP-HPLC) method for the simultaneous determination of betaxolol and hydrochlorothiazide in tablets was developed by Erram et al. [57]. The chromatographic separation was achieved on a Chromegabond MC 18 column with a mobile phase of 50 mM phosphate buffer-acetonitrile (70:30) adjusted to pH 3.5 with orthophosphoric acid. The flow rate was 1 mL/min and the effluent was monitored at 220 nm. Samples as low as 200 ng/mL betaxolol and 40 ng/mL hydrochlorothiazide could be determinated. Labetalol was used as internal study. The order of elution was 3.36, 5.58, 7.78 min for hydrochlorothiazide, labetalol, betaxolol, respectively. The method was applicable for in vitro analysis and the results obtained were statistically comparable.

4.2.4.4 HPLC tandem mass spectrometry methods

An HPLC and tandem mass spectrometry (MS/MS) method for determination of betaxolol in plasma, ocular tissues in human and normal monkey, was described by Hollo et al. [58]. In humans, mean betaxolol concentrations (excluding the aphakic patient) were 71.4 ± 41.8 ng/g in the retina, 31.2 ± 14.8 ng/g in the optic nerve head, and 1290 ± 1170 ng/g in the choroid. Mean concentrations in the iris and ciliary body were 73,200 ± 89,600 and 4250 ± 3020 ng/g, respectively. Betaxolol concentration was higher in all ocular tissues than in the plasma (0.59 ± 0.32 ng/mL). In the monkeys the concentrations in the posterior tissues of the treated eyes were higher than in the untreated eyes, with mean differences in the retina and optic nerve head of 121 and 130 ng/g, respectively. Topically applied betaxolol was bioavailable to posterior ocular tissues, including the retina and optic nerve head, of patients with glaucoma and of normal cynomolgus monkeys. The higher betaxolol levels in the treated versus untreated monkey eyes are consistent with betaxolol's reaching posterior tissues by local absorption and distribution.

An extraction procedure for the HPLC-MS determination of 5 sedatives and 14 β-adrenergic blockers in pig kidney samples was developed by Oliveira et al. [59]. The sedative analytes were xylazine, chlorpromazine, acepromazine, azaperol, and azaperone. The β-blocker analytes were pindolol, propranolol, atenolol, metoprolol, sotalol, penbutolol, carazolol, betaxolol, nadolol, bisoprolol, labetalol, acebutolol, nebivolol, and carvedilol. Three extraction procedures were evaluated in analyte recovery studies. The procedure using acetonitrile for extraction and cleanup with freezing at low temperature and dispersive solid-phase extraction using 500 mg Celite 545 before the concentration step gave the better results. The dried extracts were redissolved in methanol and analyzed by

HPLC-MS/MS system with electrospray ionization (ESI) in pos. MRM mode. The analyte recovery values were in the 75%–88% range. The robustness of the method was tested against small variations.

Boonjob et al. [60] presented a liquid chromatography-electrospray ionization tandem mass spectrometry (SIA-BI-mSPE-LC-ESI-MS/MS) method for determination of 10 beta-blockers, namely atenolol, sotalol, pindolol, acebutolol, timolol, metoprolol, labetalol, carazolol, propranolol and betaxolol. These compounds yielded the same product ions, therefore were affected in terms of quantification when flow injection anal.-mass spectrometry (FIA-MS) was used. Thus, analytes and matrix present in the sample travel together into the ionization source which can seriously affect the ionization efficiency and analyte signals due to monitoring over a short time period. Limits of detection ($S/N=3$) and quantification ($S/N=10$) for β-blockers in effluent wastewater were found in the range of 0.005–0.07 μg/L and 0.01–0.23 μg/L, respectively. Repeatability of the overall procedure provided RSDs ranging from 7 to 12%. The correlation coefficient for all of the target compounds exhibited excellent linearity ($r>0.9921$) over the range of 0.01–10 μg/L. The absolute recovery percentages within heart-cutting elution of 10 mL sample loading volume for the overall β-blockers in effluent waste waters ranged from 91% to 117%.

A LC-MS/MS method for the simultaneous extraction and determination of eight betablockers (atenolol, sotalol, nadolol, pindolol, timolol, metoprolol, betaxolol and propranolol) in various bovine eye tissues including sclera, choroid-RPE, retina, and vitreous was developed by Kadam et al. [61]. The analytes were extracted by liquid-liquid extraction after samples were alkalinized with 2% NaOH solution in water. The chromatography separation was performed on a Hypersil-ODS C18 column (100 × 2.1 mm, 3.9 mm) using a gradient mixture of (A) 5 mM ammonium formate in water (pH 3.5 adjusted with formic acid) and (B) acetonitrile: methanol (75:25) containing 0.02% triethyl amine (pH 4.0; adjusted with formic acid) as mobile phase at a flow rate of 0.4 mL/min. The compounds were ionized in the positive electrospray ionization (ESI) mode and detected in the multiple reaction monitoring (MRM) mode. The average recoveries in all four eye tissues for all beta-blockers were >82%, except for sotalol (>51%). The matrix effect for beta-blockers ranged from 81% to 110% in the four eye tissues. This method was validated and applied successfully for simultaneous quantification of the beta-blockers in sclera after tissue exposure using cassette dosing method. The calibration curve was linear in the range of 10–2000 ng/mL for all analytes, with the correlation coefficient >0.996. Intra-day and inter-day precision (% CV) was less than 15%, and accuracy ranged from 85% to 110% for all analytes. Scleral uptake was the lowest for sotalol and atenolol, two hydrophilic beta-blockers, and the highest for propranolol, a lipophilic beta-blocker.

Hernando et al. [62] described a multiresidue method for the extraction and determination of two therapeutic groups of pharmaceuticals, lipid regulating agents (clofibric acid, bezafibrate, gemfibrocil, fenofibrate) and b-blockers (atenolol, sotalol, metoprolol, betaxolol) in waters by solid-phase extraction followed by liquid chromatography-electrospray ionization tandem mass spectrometry (LC-ESI-MS-MS). Recoveries obtained from spiked HPLC water, as well as, from spiked real samples (sewage treatment plants influent and effluents, river and tap water) were all >60%, with the exception of betaxolol with a 52% recovery. The quantitative MS analysis was performed using a multiple reaction monitoring. The LC-MS-MS method gave detection limits ranging from 0.017 to 1.25 mg/L in spiked effluent. Precision of the method, calculated as relative study deviation, ranged from 3.7 to 18.5%. Individual and combined effects on *Daphnia magna* were evaluated for both therapeutic groups. Individual effects in culture medium showed these compounds as not harmful and not toxic, an exception is fenofibrate that is harmful, but at high, in the environment unrealistic concentrations (EC50 of 50 mg/L). Combined effect in wastewater showed synergistic toxic effects at low concentration level (2 mg/L).

4.2.4.5 Capillary column gas chromatography (CCGC) methods

Ganansia et al. [63] performed a gas chromatographic method with an OV-101 phase quartz capillary column at 225 °C with Ar-methane (95:5) as the carrier gas and a 63 Ni linear electron-capture detector. The retention times of betaxolol and the internal standard were 7.8 and 9.5 min, respectively. Of the compounds tested, only propranolol interfered with the analysis. The relative study deviation ranged from 10% at lower concentration to 4% at higher concentration. The limit of sensitivity is 0.5 ng I/mL of blood or urine. The method was used to determinate concentration in human blood, urine, or tissues.

5. Stability

A method for the separation of betaxolol hydrochloride by chiral high performance liquid chromatography by Ke et al. [44] established the stability of betaxolol. The stability of optical purity for levobetaxolol hydrochloride was investigated. Normal phase HPLC was used to separate the isomer by column of Daicel OJ-H and OD-H (250 × 4.6 mm, 5 mm). The mobile phases were hexane-isopropanol-diethylamine (96:4:0.5, v/v) for Daicel OJ-H column and hexane-isopropanol-diethylamine (70:30:0.5, v/v) for OD-H column respectively. The column temperature was fixed at 20–40 °C. The flow rate was 0.5–1.5 mL/min and detection wavelength was 190–400 nm. The stability of optical purity for levobetaxolol was determined by column OD-H. The method for rapid separation of betaxolol hydrochloride was established by chiral column. The optical condition for OJ-H column were detection wavelength at 274 nm, flow rate at 0.8 mL/min, column temperature at 25°, and hexane-isopropanol-diethylamine (96:4:0.5, v/v) as mobile phase. The test of stability showed that the optical purity for levobetaxolol hydrochloride was close to 100% in 10 days with high temperature, high humidity and strong light. This method has strong specificity. This method can be used to separate the racemate of betaxolol hydrochloride. The optical purity for levobetaxolol hydrochloride was stable.

A delivery system for ophthalmic drugs using an antiglaucoma agent betaxolol hydrochloride as a model developed by Jani et al. [64]. The new delivery system involved both the binding and release of drug from ion exchange resin particles. Betaxolol was studied in-vitro via a release model analysis. The ocular comfort of betaxolol was greatly enhanced by reducing the availability of free drug molecules in the pre-corneal tear film. The amount of resin concentration was selected to obtain optimum binding of the drug. The zeta potential of suspended particles was adjusted to produce flocculated suspension. Drug resin particles were then incorporated into the structured vehicle, containing Carbomer 934P as a polymer, to enhance the physical stability and ease of resuspendability of the product. This delivery system also optimized the bioavailability of betaxolol, reducing the total drug concentration in half to 0.25% Betaxolol in 0.25% BETOPTIC S Ophthalmic Suspension as compared with 0.5% Betaxolol in BETOPTIC 0.5% Sterile Ophthalmic Solution dosage form. Increased comfort of 0.25% BETOPTIC S Ophthalmic Suspension, as well as its bioequivalency data in animal models (rabbits), was confirmed in actual clinical trials of the product 0.25% BETOPTIC S Ophthalmic Suspension. The 0.25% BETOPTIC S Ophthalmic Suspension formulation has an increased bioavailability (equivalent to BETOPTIC 0.5% Sterile Ophthalmic Solution at half the concentration of drug); and pharmaceutically, is an elegant suspension product which settles slowly providing uniform dosage and increased ocular comfort.

A stability-indicating reversed-phase high performance liquid chromatography (RP-HPLC) method was developed for the determination of betaxolol hydrochloride, a drug used in the treatment of hypertension and glaucoma was described by Auvity et al. [65]. The desired chromatographic separation was achieved on a Nucleosil C18, 4 µm (150 × 4.6 mm) column, using isocratic elution at a 220 nm detector wavelength. The optimized mobile phase consisted of a 0.02 M potassium dihydrogen phosphate: methanol (40:60, v/v, pH 3.0 adjusted with o-phosphoric acid) as solvent. The flow rate was 1.6 mL/min and the retention time of betaxolol hydrochloride was 1.72 min. The linearity for betaxolol hydrochloride was in the range of 25–200 µg/mL. Recovery for betaxolol hydrochloride was calculated as 100.01%–101.35%. The stability-indicating capability was established by forced degradation experiments and the separation of unknown degradation products. A chromatogram of acid hydrolysis performed at room temperature for 24 h showed degradation of betaxolol hydrochloride with degradation product peak at retention time 0.950 and 2.092 min. A chromatogram of base hydrolysis performed at room temperature for 24 h showed degradation of betaxolol hydrochloride with a degradation product peak at a retention time of 0.892 min. The chromatogram of oxidized betaxolol hydrochloride with 3% hydrogen peroxide for 24 h showed degradation of betaxolol hydrochloride with a degradation product peak at a retention time of 1.383 min. No degradation of betaxolol hydrochloride exposed to sunlight at room temperature for 7 days was observed. The compound was particularly stable at sunlight condition.

A stability-indicating HPLC method for betaxolol HCl and its pharmaceutical dosage forms was developed by Mahalaxmi et al. [66]. Betaxolol HCl was chromatographed on a microbondapak C18 column utilizing a simple mixture of methano1:acetonitrile:0.1% diethylamine (pH 3.0 adjusted using orthophosphoric acid). It was detected at 222 nm. Under the proposed conditions betaxolol showed a linear response over a range of 0.1–1.0 µg/mL. The method is accurate and precise with a percent relative standard deviation of 0.11 based on 6 readings. A number of inactive ingredients present in the dosage forms (eye drop, tablet, gel) did not interfere in the assay procedure. The recovery from synthetic mixtures was quantitative. The extraction procedure from the dosage forms was very simple. The drug was found to be very sensitive to acids (such as sulfuric acid) since 100% of the drug decomposed on boiling for 5 min.

The Stability of Betaxolol Suspensions in Oral Syringes and Glass Bottles was described by Verdier [67]. There were no detectable changes in color, odor, or taste. The initial pH of the suspensions in deionized water, InOrpha, and Ora-Plus/Ora-Sweet was 5.68, 4.69, and 4.11, respectively. At 14 days, the pH had changed by +7.4%, +3.0%, and +3.0% as compared to the initial pH for the 3 suspensions. The concentration of betaxolol remained at least 95% of its initial level throughout the 14-day study period, regardless of the suspension vehicle, packaging, and storage temperature (4 or 20 °C). Since temperature does not appear to influence stability, it might be possible to prepare betaxolol suspensions in advance and store them at 4 °C to control microbial growth, either in plastic syringes or in glass bottles for the duration of the study period, until needed in pediatric departments. These suspensions (particularly those in InOrpha, which does not contain parabens [chemical preservatives with estrogenic properties]), can be considered suitable for pediatric patients. The liquid form would also permit the administration of accurate doses of betaxolol to infants and young children based on their body weight.

6. Pharmacology

6.1 Pharmcodynamics

The onset of action with betaxolol can generally be noted within 30 min and the maximum effect can usually be detected 2 h after topical administration. A single dose provides a 12-h reduction in intraocular pressure. In some patients, the intraocular pressure lowering responses to BETOPTIC S® may require a few weeks to stabilize. As with any new medication, careful monitoring of patients is advised.

No evidence of cardiovascular beta adrenergic-blockade during exercise was observed with betaxolol in a double-masked, crossover study in 24 normal subjects comparing ophthalmic betaxolol and placebo for effects on blood pressure and heart rate.

In subjects with normal renal function, betaxolol did not alter glomerular filtration rate, but produced a small increase in renal blood flow. Its effects on sodium and potassium excretion varied between studies, and need further clarification.

Respiratory function of normal subjects was not affected by betaxolol, while that of subjects with airways disease was less affected by betaxolol than by propranolol.

In non-diabetic healthy subjects or patients with cardiovascular disease, betaxolol did not affect either glucose metabolism, the reduction in blood pressure produced by insulin, or the time taken to recover from hypoglycemia (unlike propranolol which prolonged recovery time).

Similarly, insulin-induced tachycardia was minimally affected by betaxolol (but was reduced by propranolol). However, the effects produced by betaxolol on insulin-induced changes in free fatty acid and glycerol serum concentrations were similar to those produced by propranolol. Mean total cholesterol and triglyceride serum concentrations were unchanged or increased only slightly during long term treatment with betaxolol. There was no change in apoprotein B concentrations and no reduction in HDL cholesterol [13].

Ophthalmic betaxolol solution at 1% (one drop in each eye) was compared to placebo in a crossover study challenging nine patients with reactive airway disease. Betaxolol HCl had no significant effect on pulmonary function as measured by FEV1, Forced Vital Capacity (FVC), FEV1/FVC and was not significantly different from placebo. The action of isoproterenol, a beta stimulant, administered at the end of the study was not inhibited by ophthalmic betaxolol. Used topically in glaucoma and ocular hypertension, betaxolol 0.5% solution produces a reduction in intraocular pressure of between 13% and 30%, an effect comparable with that of ocular timolol. It may usefully be combined with other types of anti-glaucoma agents [68–70].

6.2 Mechanism of action

The mechanism of the antihypertensive effects of beta-adrenergic receptor blocking agents has not been established. Several possible mechanisms have been proposed, however, including: (1) competitive antagonism of catecholamines at peripheral (especially cardiac) adrenergic-neuronal sites, leading to decreased cardiac output, (2) a central effect leading to reduced sympathetic outflow to the periphery, and (3) suppression of rennin activity. Betaxolol selectively blocks catecholamine stimulation of beta (1)-adrenergic receptors in the heart and vascular smooth muscle. These results in a reduction of heart rate, cardiac output, systolic and diastolic blood pressure, and possibly reflex orthostatic hypotension. Betaxolol can also competitively block beta (2)-adrenergic responses in the bronchial and vascular smooth muscles, causing bronchospasm [71]. Betaxolol Hydrochloride, a cardioselective (beta-1-adrenergic) receptor blocking agent, does not have significant membrane-stabilizing (local anesthetic) activity and is devoid of intrinsic sympathomimetic action. Orally administered beta-adrenergic blocking agents reduce cardiac output in healthy subjects and patients with heart disease. In patients with severe impairment of myocardial function, beta-adrenergic receptor antagonists may inhibit the sympathetic stimulatory effect necessary to maintain adequate cardiac function. When instilled in the eye, Betaxolol Hydrochloride Ophthalmic Solution has the action of reducing elevated as well as normal intraocular pressure, whether or not accompanied by glaucoma. Ophthalmic Betaxolol has minimal effect on pulmonary and cardiovascular parameters [70].

Osborne et al. [72] determined the betaxolol is a neuro-protective agent and can therefore slow down the changes seen in the retina following ischemia/reperfusion.

6.3 Pharmacokinetics

Betaxolol is very lipid soluble but does not undergo extensive first-pass metabolism. Bioavailability after oral administration is about 80–89%, and is unaffected by the presence of food in the gut. Peak plasma concentrations after betaxolol show less interindividual variability than after propranolol. Betaxolol is distributed widely, including to the placenta and into milk (where levels are about 3 times higher than in blood). Little protein binding (about 50%) occurs. Betaxolol is metabolized into mainly inactive metabolites which are excreted in the urine; some unchanged drug (about 15% of a dose) is also excreted in the urine. Betaxolol's elimination half-life is long (14–22 h), and is increased by severe renal impairment but is not affected by hepatic dysfunction. Its half-life is also increased in elderly subjects and in day-old infants. Otherwise, its pharmacokinetic behavior is largely the same in children as in adults [13].

Teratogenic effects: Reproduction, teratology, and peri- and postnatal studies have been conducted with orally administered betaxolol HCl in rats and rabbits. There was evidence of drug related post implantation loss in rabbits and rats at dose levels above 12 and 128 mg/kg, respectively. Betaxolol HCl was not shown to be teratogenic, however, and there were no other adverse effects on reproduction at subtoxic dose levels. There are no adequate and well-controlled studies in pregnant women. BETOPTIC S® Ophthalmic Suspension should be used during pregnancy only if the potential benefit justifies the potential risk to the fetus.

Stagni et al. [73] described the pharmacokinetic behavior of betaxolol enantiomers in healthy male subjects. Betaxolol enantiomer concentration were determined in samples collected up to 48 h after i.v. administration of a 10-mg dose over a 30-min period by constant-rate infusion in 12 subjects and after oral administration of 40-mg capsules to 8 of the same subjects. Betaxolol excreted from whole blood was reacted with

(+) or (-)-1-naphthylethyl isocyanate. The resulting diastereoisomeric derivatives were analyzed by reversed-phase HPLC with fluorimetric detection. Following the i.v. dose, there were no differences in clearance or volume of distribution for the 2 enantiomers. Likewise, after the oral dose, there were no differences in the maximum concentration, time of maximum concentration, bioavailability, or apparent absorption rate constant. Thus, the pharmacokinetic behavior of racemic betaxolol accurately reflects the behavior of betaxolol enantiomers in this subject group.

The effects of an ocular b1-selective antagonist, betaxolol 0.5%, on plasma total cholesterol, high d. fraction, triglycerides and glucose during a treatment period of 6 months. Group A, consisting of 40 patients, received 0.5% betaxolol 1 drop in both eyes twice daily without nasolacrimal occlusion, group B, involving 35 patients, received only the vehicle of betaxolol (placebo), in the same doses without nasolacrimal occlusion. Cardiac parameters such as atrioventricular conduction, pulse rate and both systolic and diastolic blood pressure were controlled during the study. After monthly checks and at the end of the trial, betaxolol did not produce any significant variations in plasma lipid and glucose levels in the patients examined [74].

Nursing Mothers: It is not known whether betaxolol HCl is excreted in human milk. Because many drugs are excreted in human milk, caution should be exercised when BETOPTIC S® Ophthalmic Suspension 0.25% is administered to nursing women.

Pediatric Use: Safety and IOP-lowering effect of BETOPTICS® Ophthalmic Suspension 0.25% has been demonstrated in pediatric patients in a 3-month, multicenter, double-masked, active-controlled trial.

Geriatric Use: No overall differences in safety or effectiveness have been observed between elderly and younger patients [70].

6.3.1 Absorption

Betaxolol is extensively absorbed from the gastrointestinal tract following oral administration. Peak plasma concentrations were reached within 2–4 h of a single 20 mg dose in healthy volunteers. Administration of betaxolol with food did not affect the rate or extent of its absorption.

The mean peak plasma concentration after betaxolol 20 mg was comparable to that after propranolol 80 mg in 6 healthy subjects (approximately 50 and 36 μg/L, respectively). However, plasma concentrations after betaxolol showed less interindividual variability than after propranolol. Thus, the peak plasma concentrations after betaxlol varied from 48 to 75 μg/L while those after propranolol varied from 8 to 82 μg/L [13].

6.3.2 Distribution

Betaxolol was rapidly and widely distributed throughout the body of the rat in a two-compartment fashion, the highest concentrations occurring in the lung, kidneys, heart, brain and liver. Reported values for the volume of distribution of betaxolol in man include 4.9, 6, and 9.8 L/kg in a preliminary study. After repeated oral doses, the mean volume of distribution was 7.7–8.8 L/kg.

Betaxolol is not extensively bound to plasma proteins. In one study involving 14 healthy subjects it was found to be about 55% bound, 42% to albumin and the remainder to glycoprotein. No significant change in the extent of plasma binding was observed in patients with renal or hepatic disease. The extent of protein binding was not affected by other drugs such as aspirin, digoxin, frusemide (furosemide), chlorothiazide and chlorthalidone [13].

6.3.3 Metabolism

The metabolic pattern of betaxolol was similar in animals (rat and dog) and man and consisted of O-dealkylation followed by aliphatic hydroxylation. Betaxolol was more extensively metabolized in animals (97%–99%) than in man (84%), the remaining drug being excreted unchanged in the urine. The two major metabolic pathways of betaxolol in man yielded inactive metabolites. A minor pathway, involving hydroxylation of a carbon atom in the benzene ring, yielded a small and clinically insignificant amount of active metabolite (Beresford et al. [13]).

6.3.4 Excretion

Betaxolol and its metabolites are excreted in the urine. Its elimination half-life did not change with repeated administration, the presence of food or concomitant administration of hydrochlorothiazide. Its clearance rate was 0.28 L/h/kg [13].

6.3.5 Elimination half life

In pharmacokinetic studies, Betaxolol's elimination half-life is long (14–22 h), and is increased by severe renal impairment but is not affected by hepatic dysfunction. Its half-life is also increased in elderly subjects and in day-old infants. Otherwise, its pharmacokinetic behavior is largely the same in children as in adults [13].

References

[1] Betaxolol. [cited 2020 28 April]; Available from: http://www.chemspider.com/Chemical-Structure.2279.html?rid=51ec5bde-d3ec-4052-a90c-d8d663a221e3&page_num=0.
[2] Betaxolol. [cited 2020 28 April]; Available from: https://pubchem.ncbi.nlm.nih.gov/compound/Betaxolol.
[3] Betaxolol Hydrochloride. [cited 2020 28 April]; Available from: https://pubchem.ncbi.nlm.nih.gov/compound/Betaxolol-hydrochloride.
[4] Betaxolol Hydrochloride. [cited 2020 April 28]; Available from: http://www.chemspider.com/Chemical-Structure.97073.html?rid=87554247-d426-46f4-b56b-aba14e53fe13.
[5] Levobetaxolol. [cited 2020 28 April]; Available from: http://www.chemspider.com/Chemical-Structure.54669.html?rid=462492d8-ef1f-45c5-9947-2ee71d6a08f6.
[6] Levobetaxolol hydrochloride. [cited 2020 28 April]; Available from: http://www.chemspider.com/Chemical-Structure.54668.html?rid=f5a1d7b8-8cdf-4da1-8c76-889afa927dea.
[7] Levobetaxolol. [cited 2020 28 April]; Available from: https://www.drugbank.ca/drugs/DB09351.
[8] Clarke's Analysis of Drugs and Poisons in Pharmaceuticals, Body Fluids and Postmortem Material, fourth ed., Pharmaceutical Press, USA, 2011. 1 Lambeth High Street, London SE1 7JN, UK, 1559 St Paul Avenue, Gurnee, IL 60031.
[9] Betaxolol Brand Names. [cited 2020 28 April]; Available from: https://www.drugs.com/ingredient/betaxolol.html.
[10] Kerledex. [cited 2020 28 April]; Available from: https://www.rxlist.com/kerledex-drug.htm.
[11] S.C. Sweetman, Martindale: The Complete Drug Reference. Drug Monographs, Pharmaceutical Press, 2011.
[12] Percent Composition Calculator. [cited 2020 28 April]; Available from: http://www.chemhaven.org/che101/tools/percentcomp/percentcomp.cgi.
[13] R. Beresford, R. Heel, Betaxolol: a review of its pharmacodynamic and pharmacokinetic properties, and therapeutic efficacy in hypertension, Drugs 31 (1) (1986) 6–28.
[14] P.M. Manoury, et al., Synthesis of a series of compounds related to betaxolol, a new beta 1-adrenoceptor antagonist with a pharmacological and pharmacokinetic profile optimized for the treatment of chronic cardiovascular diseases, J. Med. Chem. 30 (6) (1987) 1003–1011.
[15] J. Allen, A. Tizot, The synthesis of betaxolol labelled with tritium at high specific activity, J. Label. Compd. Radiopharm. 25 (9) (1988) 931–937.
[16] G.K. Datta, et al., Selective terminal Heck arylation of vinyl ethers with aryl chlorides: a combined experimental-computational approach including synthesis of betaxolol, J. Org. Chem. 71 (10) (2006) 3896–3903.
[17] M. Muthukrishnan, et al., Concise synthesis of β-blockers (S)-metoprolol and (S)-betaxolol using hydrolytic kinetic resolution, Tetrahedron 63 (8) (2007) 1872–1876.
[18] R.A. Joshi, et al., Process for Preparation of S-(-)-betaxolol and Salts Thereof, 2006. Google Patents.
[19] N. Perišić-Janjić, et al., Study of the lipophilicity and retention behavior of some beta-adrenoceptor antagonists, JPC-J. Planar Chromat. 16 (5) (2003) 347–350.
[20] R. Kaliszan, A. Nasal, A. Buciński, Chromatographic hydrophobicity parameter determined on an immobilized artificial membrane column: relationships to standard measures of hydrophobicity and bioactivity, Eur. J. Med. Chem. 29 (2) (1994) 163–170.
[21] Betaxolol Solubility. [cited 2020 28 April]; Available from: https://pubchem.ncbi.nlm.nih.gov/compound/2369#section=Solubility.

[22] T.M. Maria, et al., Polymorphism and melt crystallisation of racemic betaxolol, a β-adrenergic antagonist drug, J. Therm. Anal. Calorim. 111 (3) (2013) 2171–2178.
[23] J. Canotilho, et al., The structure of betaxolol from single crystal X-ray diffraction and natural bond orbital analysis, J. Mol. Struct. 891 (1–3) (2008) 437–442.
[24] A. Habibi-Yangjeh, E. Pourbasheer, M. Danandeh-Jenagharad, Prediction of melting point for drug-like compounds using principal component-genetic algorithm-artificial neural network, Bull. Korean Chem. Soc. 29 (4) (2008) 833.
[25] United States Pharmacopeia, 36/National Formulary 31, US Pharmacopeial Convention, Inc., Rockville, MD, 2013.
[26] HM Patel, S Pancholi, N Jivani, Spectrophotometric method for simultaneous determination of betaxolol hydrochloride and chlorthalidone in tablet formulation by simultaneous equation method, Inventi Rapid: Pharm Analysis & Quality Assurance 2013 (2) (2013) Inventi:ppaqa/609/12.
[27] YN Zhuk, SO Vasyuk, Spectrophotometric determination of betaxolol in pure form and pharmaceuticals, Pharma Innov. J. 5 (6) (2016) 31–33.
[28] B. Suhagia, et al., Spectrophotometric estimation of Betaxolol Hydrochloride in bulk powder and its dosage forms, Indian J. Pharm. Sci. 68 (2) (2006).
[29] H. Salem, Spectrophotometric determination of β-adrenergic blocking agents in pharmaceutical formulations, J. Pharm. Biomed. Anal. 29 (3) (2002) 527–538.
[30] A.F. El Yazbi, et al., Kinetic spectrophotometric determination of Betaxolol, Clopidogrel and Imidapril in pharmaceutical preparations, Curr. Anal. Chem. 6 (3) (2010) 228–236.
[31] M. Sultan, Spectrofluoremetric determination of Betaxolol in dosage forms and spiked, Zagazig J. Pharma. Sci. (2005).
[32] A.A. Wassil, A.E.-F.B. Farag, F.A. Moukdad, Polymer membrane electrodes for sensitive potentiometric determination of β-blockers, Drug Dev. Ind. Pharm. 33 (1) (2007) 57–62.
[33] A.S. Alturiqi, Potentiometric sensor for the determination of Betaxolol hydrochloride in plasma and urine, Int. J. Pharm. Sci. Rev. Res. 41 (2) (2016) 332–336.
[34] A. Khoobi, et al., Multivariate curve resolution-alternating least squares assisted by voltammetry for simultaneous determination of Betaxolol and atenolol using carbon nanotube paste electrode, Bioelectrochemistry 94 (2013) 100–107.
[35] V.K. Vashistha, R. Bhushan, Chirality recognition for assessing the enantiomeric purity of Betaxolol, Tetrahedron: Asymmetry 26 (5) (2015) 304–311.
[36] H. Hopkała, et al., Densitometric and videodensitometric TLC determination of timolol and betaxolol in ophthalmic solutions, JPC-J. Planar Chromat. 16 (4) (2003) 280–285.
[37] A. Kwiecie, et al., Development and validation of stability-indicating TLC-densitometric method for determination of Betaxolol with LC-ESI/MS analysis of degradation product, Acta Pol. Pharm. Drug Res. 70 (4) (2013) 643–652.
[38] H.-l. Zhang, H. Shao, Z.-z. Zhang, Capillary electrophoresis determination of betaxolol hydrochloride tablets, Chin. J. Pharma. Anal. 27 (10) (2007) 1630–1632.
[39] A.C. Servais, et al., Optimization of the separation of β-blockers by ion-pair capillary electrophoresis in non-aqueous media using univariate and multivariate approaches, J. Sep. Sci. 25 (15–17) (2002) 1087–1095.
[40] S. Auvity, et al., A rapid stability-indicating RP-HPLC method for the determination of betaxolol hydrochloride in pharmaceutical tablets, Anal. Chem. insights 8 (2013) 1.
[41] M.M. Hefnawy, M.A. Sultan, M.M. Al-Shehri, Direct enantiomeric resolution of betaxolol with application to analysis of pharmaceutical products, Anal. Chem. insights 1 (2006) 13.
[42] A. Darmon, J. Thenot, Determination of betaxolol enantiomers by high-performance liquid chromatography: application to pharmacokinetic studies, J. Chromatogr. B Biomed. Sci. Appl. 374 (1986) 321–328.

[43] J. Zhou, et al., Separation of betaxolol hydrochloride with new bonded cellulose chiral stationary phase and determination of the enantiomers in plasma by HPLC, J. Liq. Chromatogr. Relat. Technol. 35 (13) (2012) 1767–1778.
[44] Y. Ke, D. Yuan, H. Liu, Investigation on enantiomeric separation of betaxolol hydrochloride by normal phase chiral HPLC and stability, Zhengzhou Daxue Xuebao, Yixueban (2010) 308–311.
[45] B. Du, et al., HPLC determination of betaxolol hydrochloride in human plasma, Chin. J. Pharma. Anal. 26 (9) (2006) 1207–1210.
[46] B. Dulger, et al., Determination of betaxolol in human aqueous humour by high-performance liquid chromatography with fluorescence detection, J. Chromatogr. B 772 (1) (2002) 179–183.
[47] R. Bhamra, A. Ward, D. Holt, Column liquid chromatographic measurement of betaxolol in plasma or serum, J. Chromatogr. B Biomed. Sci. Appl. 417 (1987) 229–232.
[48] H. Caqueret, G. Bianchetti, Simple method for routine determination of betaxolol in blood and urine by automated high-performance liquid chromatography with fluorimetric detection, J. Chromatogr. B Biomed. Sci. Appl. 311 (1984) 199–205.
[49] C. Wu, Y. Liu, Determination of betaxolol in human plasma by high-performance liquid chromatography with fluorescence detection, Life Sci. J. 4 (3) (2007).
[50] A. Tracqui, et al., A specific HPLC method for determination of β-blockers topically used in ophthalmological diseases, Forensic Sci. Int. 38 (1) (1988) 37–41.
[51] N. Gao, et al., Pharmacokinetics and relative bioavailability of betaxolol hydrochloride tablets in Chinese healthy volunteers, Chin. J. Clin. Pharm. 23 (3) (2007) 191.
[52] M. Delamoye, et al., Simultaneous determination of thirteen β-blockers and one metabolite by gradient high-performance liquid chromatography with photodiode-array UV detection, Forensic Sci. Int. 141 (1) (2004) 23–31.
[53] J. Zhou, et al., Separation of four-lol drugs by HPLC with new bonded chiral stationary column, Chin. Sci. Bull. 58 (11) (2013) 1267–1273.
[54] R. Bhushan, H. Nagar, Enantioseparation of orciprenaline, betaxolol, and propranolol using HPLC and new chiral reagents based on 1,5-Difluoro-2, 4-dinitrobenzene, Anal. Lett. 47 (2) (2014) 202–219.
[55] A. Krstulović, et al., Direct enantiomeric separation of betaxolol with applications to analysis of bulk drug and biological samples, J. Chromatogr. A 452 (1988) 477–483.
[56] J.W. Yuan, Hui-juan, D.-g. Guo, Y. Zhou, Preparation and evaluation of ion sensitive & thermosensitive in situ gel of betaxolol hydrochloride for ophthalmic use, Zhongguo Yiyuan Yaoxue Zazhi (2013) 1137–1141.
[57] S. Erram, H. Tipnis, Simultaneous estimation of betaxolol hydrochloride and hydrochlorothiazide from tablets by RP-HPLC, Indian Drugs-Bombay 29 (1992) 553.
[58] G. Holló, et al., Concentrations of betaxolol in ocular tissues of patients with glaucoma and normal monkeys after 1 month of topical ocular administration, Invest. Ophthalmol. Vis. Sci. 47 (1) (2006) 235–240.
[59] L.G.d. Oliveira, et al., Development of extraction procedure for analysis of sedatives and β-blockers in swine kidney, Quím. Nova 37 (8) (2014) 1342–1350.
[60] W. Boonjob, et al., Renewable sorbent material for solid phase extraction with direct coupling of sequential injection analysis-bead injection to liquid chromatography-electrospray ionization tandem mass spectrometry, Anal. Bioanal. Chem. 407 (19) (2015) 5719–5728.
[61] R.S. Kadam, U.B. Kompella, Cassette analysis of eight beta-blockers in bovine eye sclera, choroid–RPE, retina, and vitreous by liquid chromatography–tandem mass spectrometry, J. Chromatogr. B 877 (3) (2009) 253–260.
[62] M.D. Hernando, et al., Analysis by liquid chromatography–electrospray ionization tandem mass spectrometry and acute toxicity evaluation for β-blockers and lipid-regulating agents in wastewater samples, J. Chromatogr. A 1046 (1) (2004) 133–140.

[63] J. Ganansia, et al., Improved determination of betaxolol in biological samples by capillary column gas chromatography, J. Chromatogr. B Biomed. Sci. Appl. 275 (1983) 183–188.
[64] R. Jani, et al., Ion exchange resins for ophthalmic delivery, J. Ocul. Pharmacol. Ther. 10 (1) (1994) 57–67.
[65] S. Auvity, et al., A Rapid Stability-Indicating RP-HPLC Method for the Determination of Betaxolol Hydrochloride in Pharmaceutical Tablets, SAGE Publications, London, England, 2013.
[66] D. Mahalaxmi, et al., Stability-indicating HPLC method for betaxolol HCl and its pharmaceutical dosage forms, Drug Dev. Ind. Pharm. 22 (9–10) (1996) 1037–1039.
[67] J. Verdier, Stability of betaxolol suspensions in oral syringes and glass bottles, Ann. Pharmacother. 47 (9) (2013) 1237–1238.
[68] M.M.-T. Buckley, K.L. Goa, S.P. Clissold, Ocular betaxolol, Drugs 40 (1) (1990) 75–90.
[69] G.R. Reiss, R.F. Brubaker, The mechanism of betaxolol, a new ocular hypotensive agent, Ophthalmology 90 (11) (1983) 1369–1372.
[70] Betaxolol Prescribing Information. 2007 [cited 2020 28 April]; Available from: https://www.accessdata.fda.gov/drugsatfda_docs/label/2007/019845s020lbl.pdf.
[71] A. Miki, et al., Betaxolol-induced deterioration of asthma and a pharmacodynamic analysis based on beta-receptor occupancy, Int. J. Clin. Pharmacol. Ther. 41 (8) (2003) 358–364.
[72] N. Osborne, et al., In vivo and in vitro experiments show that betaxolol is a retinal neuroprotective agent, Brain Res. 751 (1) (1997) 113–123.
[73] G. Stagni, P.J. Davis, T.M. Ludden, Human pharmacokinetics of betaxolol enantiomers, J. Pharm. Sci. 80 (4) (1991) 321–324.
[74] A. Caccavale, et al., Effects of ocular betaxolol on glucose and lipid metabolism after long-term treatment, Clin. Drug Investig. 14 (5) (1997) 363–368.

CHAPTER FOUR

Rabeprazole: A comprehensive profile

Ahmed H. Bakheit[a,b,*], Hamad M. Al-Kahtani[a], and Salem Albraiki[a]
[a]Department of Pharmaceutical Chemistry, College of Pharmacy, King Saud University, Riyadh, Kingdom of Saudi Arabia
[b]Department of Chemistry, Faculty of Science and Technology, Al-Neelain University, Khartoum, Sudan
*Corresponding author: e-mail addresses: abujazz76@gmail.com; abakheit@ksu.edu.sa

Contents

Profiles of Drug Substances, Excipients, and Related Methodology, Volume 46
ISSN 1871-5125
https://doi.org/10.1016/bs.podrm.2020.07.003

1. Description

Rabeprazole is a benzimidazole substitution, proton pump inhibitor. Sodium salt of rabeprazole is a proton pump inhibitor in the stomach, which can suppress gastric acid secretion by inhibiting H^+, K^+-ATPase on the secretory surface of the partial cellular cells without affecting cholinergic or histamine H_2-receptors. Sodium rabeprazole is usually used in combination prescriptions to eliminate *Helicobacter pylori*. Moreover, rabeprazole is one of the treatments for duodenal ulcers. It is also used in the treatment of gastroesophageal reflux and Zollinger-Ellison syndrome, which is the conditions in which the stomach produces much acid. It is also used in cases of gastric ulcers caused by bacteria where it is used with antibiotics [1].

1.1 Nomenclature

1.1.1 Systematic chemical names

1.1.1.1 Rabeprazole base

- 2-[[4-(3-Methoxypropoxy)-3-methylpyridin-2-yl]methylsulfinyl]-1*H*-benzimidazole.
- 2-[[[4-(3-Methoxypropoxy)-3-methyl-2-pyridinyl]methyl]sulfinyl]-1*H*-benzimidazole [2].

1.1.1.2 Rabeprazole sodium salt

- 1*H*-Benzimidazole, 2-[[[4-(3-methoxypropoxy)-3-methyl-2-pyridinyl] methyl]sulfinyl]-, sodium salt [3].
- 2-[[[4-(3-Methoxypropoxy)-3-methyl-2-pyridyl]methyl]sulfinyl]benzimidazole sodium salt [117976-90-6] [3].

1.1.2 Non-proprietary names

Generic:

Rabeprazole sodium, rabeprazole sodium hydrate.

1.1.3 Proprietary names (brand names)

Pareit; acephex, abbot-rabeprzole, apo-rabeprazole [4].

Aciphex; Clofezone; Pariet; Q9Z21M00M9; LY307640; C063129; Chemistry 2153; DB01129.

E3810 [5].

1.2 Formula

1.2.1 *Empirical formula, molecular weight, CAS number*

Empirical formula, molecular weight, and CAS number are tabulated in Table 1.

1.2.2 *Structural formula*

The structural formula of rabeprazole base and rabeprazole sodium salt is performed in Fig. 1.

1.3 Elemental analysis

The elemental composition of rabeprazole base and sodium salt is written in Table 2.

Table 1 Empirical formula, molecular weight, CAS number [6].

Compounds	Empirical formula	Molecular weight	CAS number
Rabeprazole base	$C_{18}H_{21}N_3O_3S$	359.44	117976-89-3
Rabeprazole sodium salt	$C_{18}H_{20}N_3NaO_3S$	381.42	117976-90-6

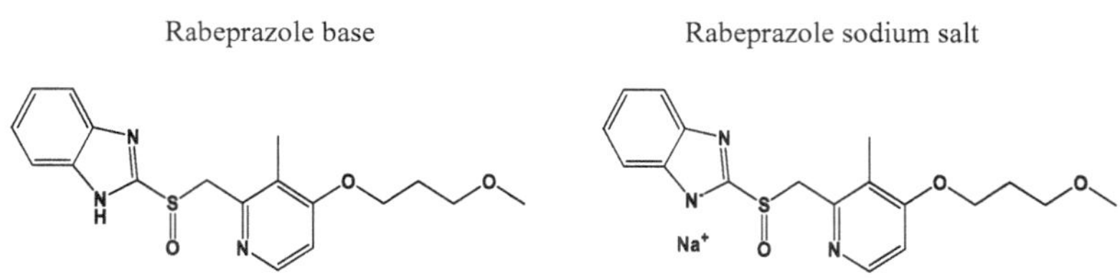

Fig. 1 Rabeprazole base and salt structure.

Table 2 The elemental composition of rabeprazole base and sodium salt [6].

Elements	Rabeprazole base (%)	Rabeprazole sodium salt (%)
C	60.15	56.68
H	5.89	5.29
N	11.69	11.02
O	13.35	12.58
S	8.92	8.41
Na	0	6.03

2. Methods of preparation

The Rabeprazole and many other substituted benzimidazole-style compounds with anti-ulcer activity are reported by Souda et al. [7]. This patent further reveals the method of Rabeprazole preparation using 85% m-chloroperabenzoic acid for oxidation of Rabeprazole sulfide **2**, in a mixture of dichloromethane and diethyl ether, after that, work is performed to get product as oil. A dichloromethane/ether mixture is used to crystalize the oil produced. In aqueous solution of sodium hydroxide the oily crude is optionally dissolved. The solution is exposed to distillation of azeotropic via ethanol to eliminate water and adding ether to give crystalline Rabeprazole base (Scheme 1).

Ray et al. [4] prepared rabeprazole from reaction of the free base 2-(dichloraneyl)methyl)-4-(3-methoxypropoxy)-3-methylpyridine **2** with m-chloroperabenzoic acid (mCPBA) in chloroform medium at 58 °C to give 2-(chloromethyl)-4-(3-methoxypropoxy)-3-methylpyridine **3**. The compound **3** was condensed with 2-(chloromethyl)-4-(3-methoxypropoxy)-3-methylpyridine 1-oxide **4** in *N,N*-dimethylformamide containing potassium carbonate to yield a chlorine-free compound as 2-(((1H-benzo[*d*]imidazol-2-yl)thio)methyl)-4-(3-methoxypropoxy)-3-methyl pyridine 1-oxide **6.** The compound **6** was oxidized with mCPBA in dichloromethane at 0 °C to give mixture of two compounds **7a** and 7**b**, compound **7b** was

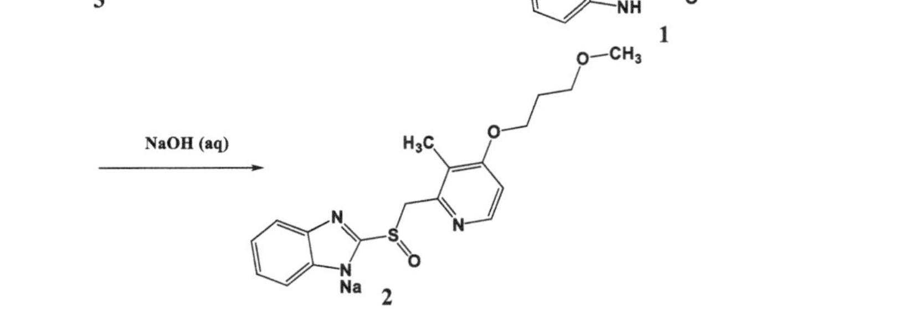

Scheme 1 Synthesis of rabeprazole by Souda method.

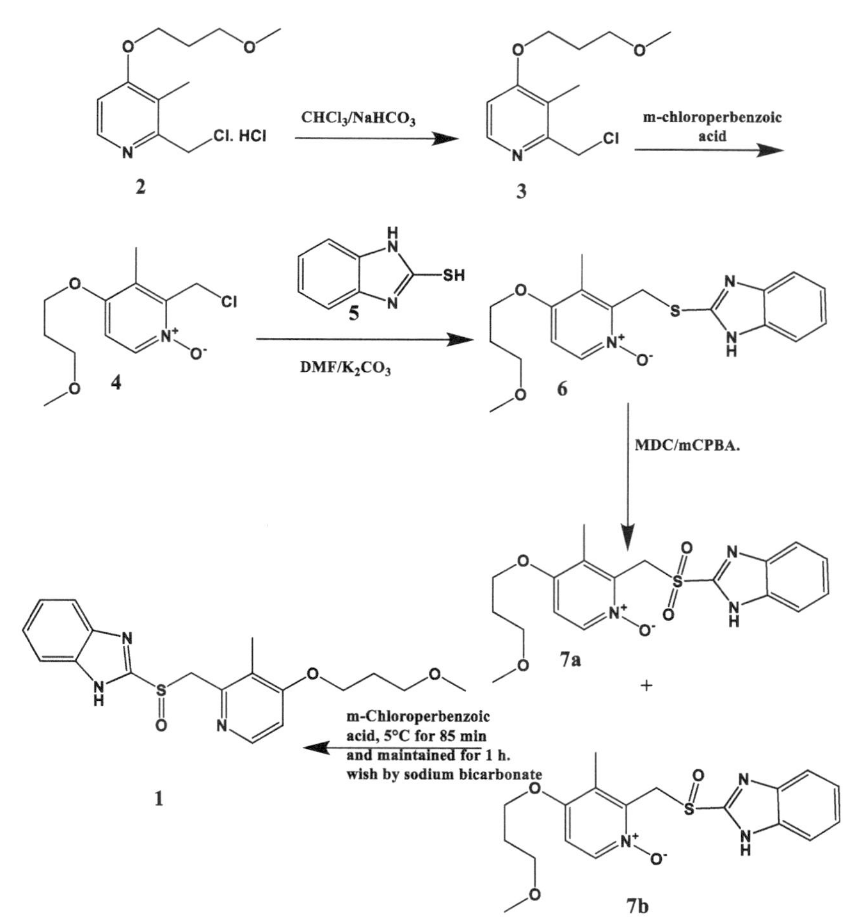

Scheme 2 Synthesis of rabeprazole by Ray method.

treated by *m*-Chloroperabenzoic acid at 58 °C for 85 min and maintained for 1 h to give compound **1** (Scheme 2).

Vardanyan and Hruby [8] synthesized rabeprazole **1**. This method was started from 4-nitro-2,3-dimethylpyridine-*N*-oxide **2**, which on heating with concentrated hydrochloric acid gave 4-chloro-2,3-dimethylpyridine-*N*-oxide **3**. A reaction with sodium 3-methoxypropan-1-olate in DMSO yields the corresponding 4-alkoxy intermediate **4**. The 4-alkoxy intermediate **4**, on further reaction with acetic anhydride, underwent Boekelheide rearrangement to produce *O*-acetyl intermediate **5**, which, after alkaline hydrolysis, produced pyridin-2-ylmethanol **6**, which was chlorinated using thionyl chloride to furnish a chloro intermediate **6**. A condensation of the

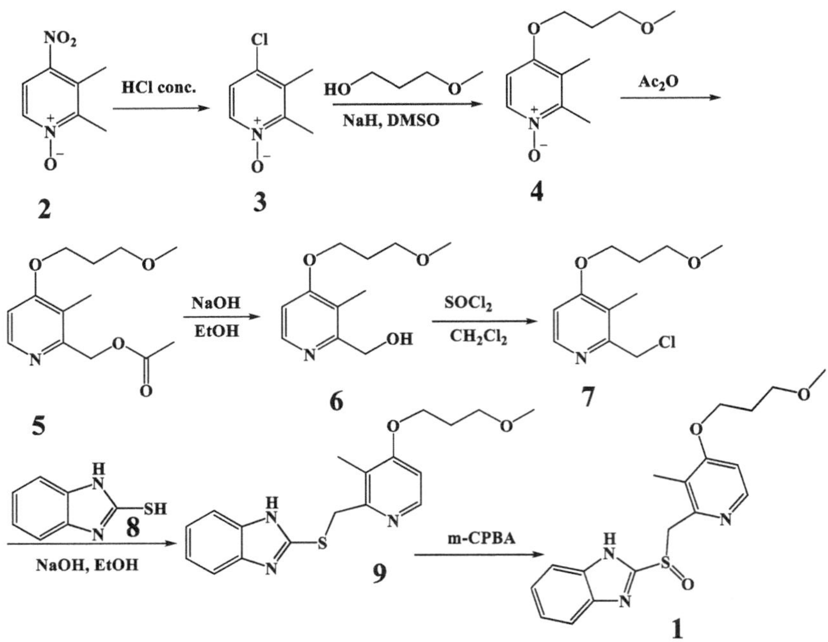

Scheme 3 Synthesis of rabeprazole by Vardanyan method.

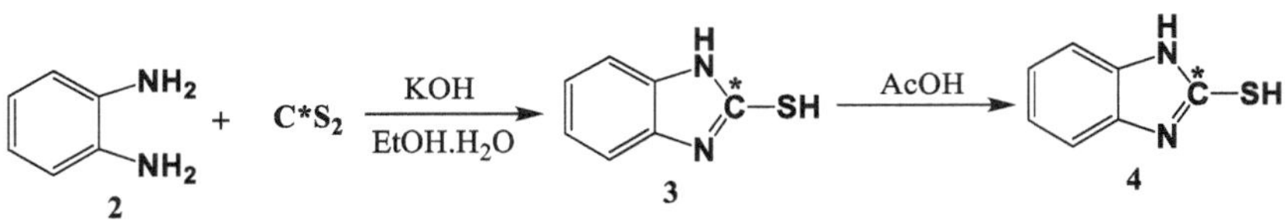

Scheme 4 synthesis of 2-Mercapto[2-^{14}C]-benzimidazole 4 by Amersham International plc.

obtained compound with 2-mercapto benzimidazole **8** in the presence of base produced a 2-((pyridin-2-ylmethyl)thio)-1*H*-benzo[*d*]imidazole derivative **9**, which oxidized with meta-chloroperabenzoic acid, gave the final rabeprazole **1** (Scheme 3).

Tagami et al. [9] synthesized rabeprazole as illustrated in Scheme 5. 2-Mercapto[2-^{14}C]-benzimidazole **4**, the starting material, was synthesized by Amersham International plc, as shown in Scheme 4. [^{14}C] carbon disulfide was treated with phenylenediamine **2** in a solution of potassium hydroxide followed by desalination via acetic acid to give compound **4**. An equivalent amount of the product from nonlabeled 2-mercapto-benzimidazole was used for dilution.

A mixture of an ethanolic solution containing of 2-mercapto[^{14}C] benzimidazole **4**, *Z*-mercapto-benzimidazole, 2-chloromethyl-4-(3-methoxypropoxy)-3-methylpyridine hydrochloride **5** and sodium hydroxide were refluxed with stirring for 1 h. Then the mixture was cooled to room temperature and filtered under vacuum to eliminate inorganic salt. The compound **6** was extracted from the residue in the aqueous phase with dichloromethane organic phase. Furthermore, compound **6** was isolated form the residue with column chromatography on silica (first eluting with 66.7% *n*-hexane/ethyl acetate, and then with 50% *n*-hexandethyl acetate), and finally with 33.3% *n*-hexandethyl acetate as white crystals.

After cooling to −68 °C, the compound **6** dissolved in dichloromethane was treated with ethyl ether and m-chloroperabenzoic acid followed by stirred the mixture for 1 h. After that the mixture was treated by triethylamine and stirred for 10 min, and then with 1N sodium hydroxide aqueous solution. The solution was stirred for 30 min at room temperature. In addition, dichloromethane was used to remove unreacted thioether **6** from the aqueous layer which was treated with 2 M ammonium acetate solution and dichloromethane. The collective organic layers were dried over magnesium sulfate and the solution filtered. The resulting solution was treated with ethyl ether and the solid produced was filtered off and air-dried at room temperature to give compound **7**, which was dissolved in alcoholic sodium hydroxide and the solution was filtered and washed with ethanol and dichloromethane. The solvent was removed under vacuum. The residue was treated with Ethanol and the solvent was again removed. The residue was dried under vacuum and then treated with ethyl ether to afford compound **1** as a pale yellowish solid (Scheme 5).

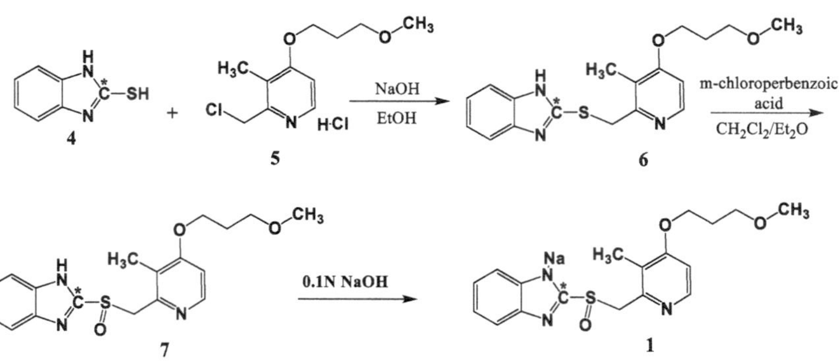

Scheme 5 Synthetic process for sodium [^{14}C] rabeprazole.

3. Physical characteristics

3.1 Physical description

A white to yellowish-white solid in its pure form [10].

3.2 Dissociation constants

$pK_{a1} = 1.9$.
$pK_{a2} = 3.2$.
$pK_{a3} = 4.2$ [11].

3.3 Flash point, and vapor pressure

Flash point: $319.1 \pm 34.3\,°C$.
Vapor pressure (mm Hg): 0.0 ± 1.7 mmHg at 25 °C [12].

3.4 Solubility characteristics

Rabeprazole is highly soluble in water and methanol, freely soluble in ethanol and chloroform; ether and insoluble in *N*-hexane. [13].

3.5 Partition coefficient

The *n*-octyl alcohol/water partition coefficient is 214 [14].

3.6 Particle morphology

The powder of rabeprazole was characterized structurally by high-resolution X-ray diffraction (XRD) using Rigaku Ultima IV diffractometer with scintillation detector with reflection mode, Cu K{acute over (α)} 19.400 radiation (1.5406 Å), Scanning range: 3–60°2θ, Step size: 0.02° 2θ and Time per step: 1 s, monochromatized with a graphite crystal. The pattern was collected at 40 kV of tube voltage and 40 mA of tube current in step scan mode. The peaks (reflections) of Rabeprazole powder are obtained at Table 3 and shown in Fig. 2.

Table 3 The X-ray powder diffraction pattern of rabeprazole.

Peak no.	2theta	Flex width	d-Value	Intensity	I/Io
1	5.500	0.353	16.0548	147	9
2	19.400	0.353	4.5717	74	5
3	38.100	0.235	2.3600	1641	100
4	44.300	0.235	2.0430	571	35

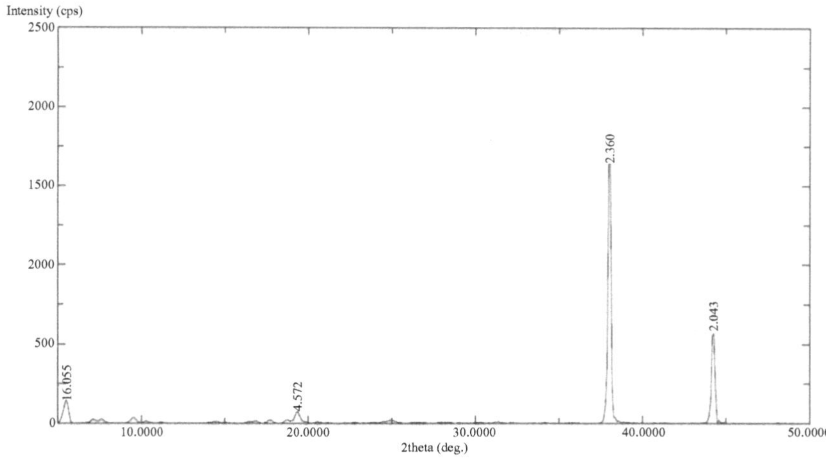

Fig. 2 The X-ray powder diffraction pattern of rabeprazole.

4. Characterization and identification

4.1 Thermal methods of analysis

4.1.1 Melting behavior

Melting points of rabeprazole salt were measured using a Büchi melting point B-545 apparatus, and the melting point range was found at 140–141 °C and decompose range at 153.2–154 °C which was agree with Ref. [13].

4.1.2 Differential scanning calorimetry

Rabeprazole differential scanning (DSC) data were obtained by a differential calorimeter Perkin-Elmer DSC-7 with a TAC 71DX thermal analysis controller (Perkin-Elmer, Norwalk, CT). The samples were precisely weighed (3.6 mg) in aluminum plates, and thermograms acquired at heating range of 10 °C per minute for a temperature range of 40–300 °C. The research was carried out with Pyris Program 2.04 (Perkin-Elmer). Rabeparazole differential scanning calorimetry (DSC) (Fig. 3) shows a 222 °C exothermic process. An exothermic process is followed immediately by weight loss due to decomposition (mp 198–201 °C).

4.1.3 Thermogravimetric analysis

Thermal gravimetric analysis (TGA) of rabeprazole was obtained using a Perkin Elmer pyris 1 apparatus. The samples (2.042 mg) were placed in aluminum pan, pierced prior to scan, and temperature profile 50–550 °C at a

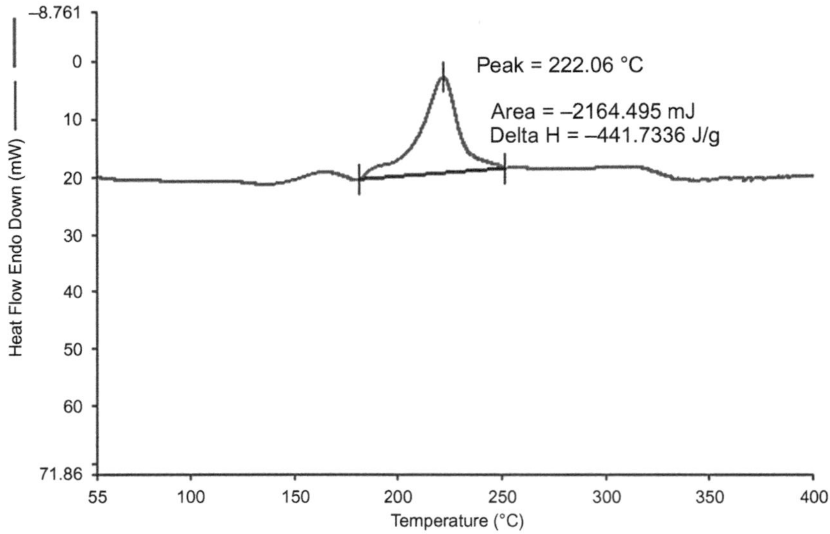

Fig. 3 DSC curves of rabeprazole RS.

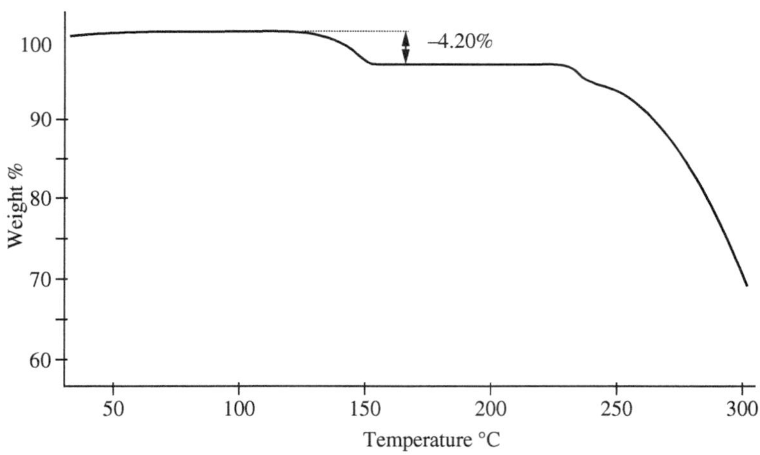

Fig. 4 Thermogravimetric analysis (TGA) of rabeprazole.

rate of 10 °C/min under nitrogen purge (50 mL/min). The results show that the powder of rabeprazole is chemically unstable above the melting point temperatures and the compound was decompose at 153.2–154 °C as shown in Fig. 4.

4.2 UV/vis spectroscopy

A Shimadzu UV-spectrophotometer (model no. UV-1800) was used to record an ultraviolet absorption spectrum of methanolic solution of Rabeprazole

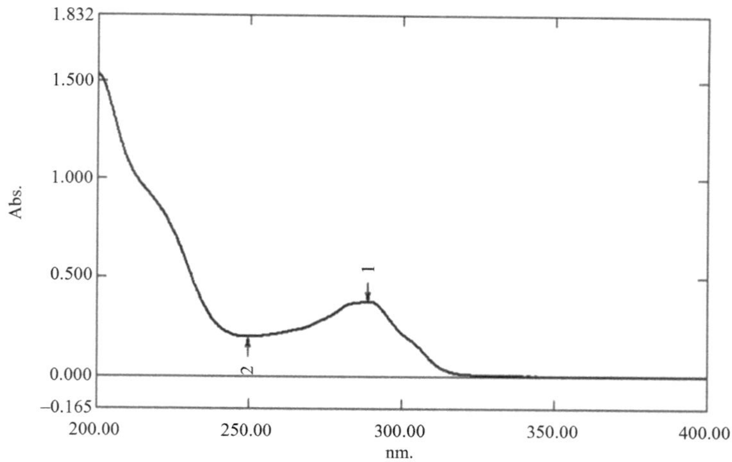

Fig. 5 The UV-absorption spectrum of 20 μg/mL of rabeprazole in acetonitrile.

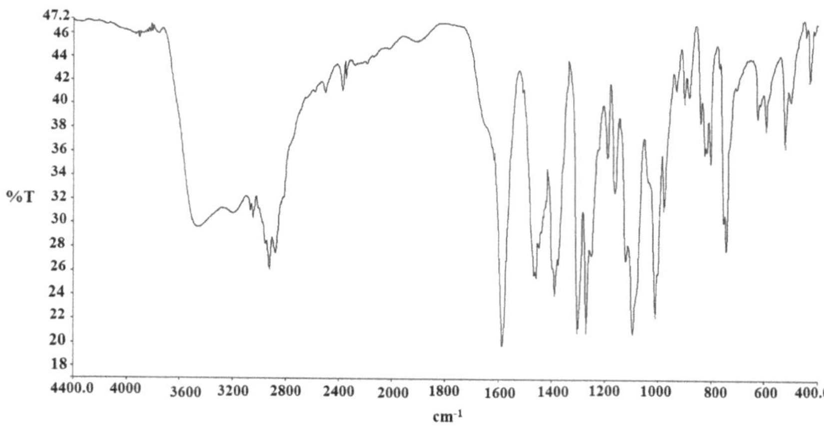

Fig. 6 IR spectrum of rabeprazole.

(20 μg/mL) in methanol. The absorption spectra are measured in a 1 cm quartz cell within 200–400 nm range. The ultraviolet spectrum is displayed in Fig. 5 with a two absorption maxima of Rabeprazole at 275 and 205 nm.

4.3 Infrared spectroscopy

A Fourier-transform infrared (FT-IR) absorption spectrum of rabeprazole was recorded (as KBr disc) using the Perkin Elmer FT-IR Spectrum BX apparatus. Fig. 6 shows the FT-IR of rabeprazole. The characteristic absorption band of rabeprazole is shown in Tables 4 and 5.

Table 4 Interpretation of rabeprazole FT-IR spectra.

No.	Functional group	Standard value (cm^{-1})	Obtained value
1	C=C group	1600–1475	1584.99
2	C—N group	1350–1000	1094.79
3	C—O group	1000–1300	1010.69
4	S=O group	1050	1161.73
5	C—H group	2850–3000	2923.03
6	N—H group	3300–3500	3382

4.4 Mass spectrometry

The mass spectrum of Rabeprazole was obtained using a mass spectrometer Agilent 6320 ion-trap (Agilent Technologies, USA) fitted with the electrospray ionization interface (ESI). Instead of a column, a connector was used. The mobile phase consisted of a combination of HPLC grade water and acetonitrile (50:50). The rabeprazole powder (10 mg) was dissolved in acetonitrile and diluted to 10 mL with a mobile phase. Based on the intensities of the ions the test solution was diluted to 10–30 $\mu g\,mL^{-1}$ with the mobile-phase. Rate of flow was 0.4 $mL\,min^{-1}$ and 5 min is runtime. The parameters of mass spectrometry were optimized. The scan was in ultra-scan mode. In the mass range 50–1000 m/z, MS2 scans were performed. The ESI was in a positive mode. The source temperatures were set to 350 °C. The molecular ion peak at $m/z = 360$ $[M+1]^+$ was observed with a nebulizer gas pressure of 55,00 psi, dry gas flow rate of 12.00 L min. The full scan mass spectra of rabeprazole demonstrated that the main rabeprazole peak was present at m/z 351.2 (Fig. 7).

The spectra of product ion mass by these protonated molecular ions (Fig. 8) demonstrated that the main rabeprazole product ion was present at m/z 194.8, 73.0, 119, 169.9, 150 and 242.

The resulting MS data are similar to that of described in the literature [15–18]. The ESI negative ion mode is also reported; the important fragments (m/z) are 434.25 ($M+H^+$), 390.1, 350.1, and 179.2 [19,20]. Both positive and negative ion modes are used for determination of Rabeprazole by LC-MS/MS.

The EI-MS of rabeprazole was reported by Hishinuma et al. [21]. The important fragments (m/z) were 242.1 [M–118], 342 [M–18], 210 [M–150], 195.3 [M–164.7], and 73.2 [M–286.8].

Table 5 Summary of chromatographic methods of rabeprazole.

<table>
<tr><th>Sample</th><th>Method</th><th>Method mobile/flow rate</th><th>Detection (λ, nm)</th><th>Stationary phase</th><th>Linearity range</th><th>References</th></tr>
<tr><td>Pure Rabeprazole Sodium RS</td><td>HPLC</td><td>Mobile phase composition:
—mobile phase A: mix 5 volumes of acetonitrile R and 95 volumes of a 4.35 g/L solution of dipotassium hydrogen phosphate R previously adjusted to pH 7.0 with phosphoric acid R
—mobile phase B: methanol R
—mobile phase C: acetonitrile R
<table>
<tr><th>Time (min)</th><th>0</th><th>2</th><th>7</th><th>27</th><th>32</th></tr>
<tr><td>A %</td><td>100</td><td>100</td><td>85</td><td>30</td><td>15</td></tr>
<tr><td>B %</td><td>0</td><td>0</td><td>0</td><td>40</td><td>55</td></tr>
<tr><td>C%</td><td>0</td><td>0</td><td>15</td><td>30</td><td>30</td></tr>
</table></td><td>280 nm</td><td>size: $l = 0.25$ m, $Ø = 4.6$ mm</td><td>–</td><td>[23]</td></tr>
<tr><td>-USP Rabeprazole Sodium RS
-Ssample of Rabeprazole Sodium</td><td>HPLC</td><td>Flow rate: 1 mL/min
Mobile phase composition:
Buffer 1: 4.35 g/L of dibasic potassium phosphate. Adjust with phosphoric acid to a pH of 7.0
Solution A: Acetonitrile and buffer 1 (5:95)
Solution B: Methanol
Solution C: Acetonitrile
<table>
<tr><th>Time (min)</th><th>0</th><th>2</th><th>7</th><th>27</th><th>32</th></tr>
<tr><td>Solution A %</td><td>100</td><td>100</td><td>85</td><td>30</td><td>15</td></tr>
<tr><td>Solution B %</td><td>0</td><td>0</td><td>0</td><td>40</td><td>55</td></tr>
<tr><td>Solution C %</td><td>0</td><td>0</td><td>15</td><td>30</td><td>30</td></tr>
</table></td><td>280 nm</td><td>4.6-mm × 25-cm; 5-μm packing L1</td><td>–</td><td>[22]</td></tr>
</table>

Continued

Table 5 Summary of chromatographic methods of rabeprazole.—cont'd

Sample	Method	Method mobile/flow rate	Detection (λ, nm)	Stationary phase	Linearity range	References
The rabeprazole and mosapride in their combined Pharmaceutical dosage forms	HPLC	0.01 M 6.5 pH ammonium acetate buffer–methanol–acetonitrile (30:40:30, v/v) adjusted to pH 5.70 ± 0.02 with acetic acid–ammonia. (Flow rate of 1.0 mL/min)	276 nm	Phenomenex C18 column (250 mm × 4.6-mm i.d., 5-μm particle size)	The range of 0.30–3.67% and 1.03–3.58%, respectively	[18]
Rabeprazole enantiomers in commercial tablets	HPLC	An isocratic elution: hexane:ethanol:ethylenediamine (30:70:0.05, v/v) at a flow rate of 1.0 mL/min	280 nm	Chiralpak IC (150 × 4.6 mm, 5 μm) column	10–100 μg/mL for each enantiomer	[44]
Rabeprazole sodium	RP-HPLC	The optimized mobile phase consisted of a 10 mM potassium dihydrogen phosphate buffer, acetonitrile and methanol (50:18:32, v/v)	280 nm	Pursuit C18 column, i.e., 4.6 mm × 25 cm, 5 μm	50–150 μg/mL	[45]
Tablet dosage form	RP-HPLC	phosphate buffer (pH 5.5), methanol (30:70) with flow rate at 0.9 mL/min at wavelength 284 nm	284 nm	Symmetry C18 (4.6 × 150 mm, 5 mm, Make: XTerra)	20–60 μg/mL	[46]
Human plasma	RP-HPLC	5 mM ammonium acetate buffer (pH 7.4)/acetonitrile/methanol (45/20/35, v/v)	284 nm	Waters symmetry® C18 column	20–1000 ng/mL	[47]
Human plasma	NP-HPLC–MS/MS	Diamonsil C_{18} analytical column	Mass detector	*n*-hexane–dichloromethane–isopropanol (20:10:1, v/v)	2.0–800 ng/mL	[48]
Marketed tablet samples	RP-HPLC	0.1 M sodium phosphate buffer (pH 6.5) and acetonitrile 65:35 with UV detection (285 nm) with a flow rate of 1.2 mL/min^{-1}	285 nm	Lichrosphere RP-100 C_8 column	0.025–150%	[49]
Pure powder and tablet dosage form	RP-HPLC	Phosphate buffer (pH 6): acetonitrile (65:35, v/v) flow rate of 1.5 mL/min	280 nm	Inertsil ODS 3V, 150 × 4.6 mm, 5 μ	80–120 μg/mL	[50]

Pharmaceutical dosage forms	RP-HPLC	Mixture of phosphate buffer (pH 7.4) and acetonitrile (65:35, v/v) A flow rate of 1.5 mL/min	290 nm.	C-18 column (4.6 mm × 250 mm, 5 μm)	0.1–1.0 μg/mL	[51]
Rabeprazole and domperidone in their combined dosage forms	RP-HPLC	0.01 M, pH 6.5 ammonium acetate buffer:methanol: acetonitrile (40:30:30, v/v, pH 7.44) flow rate of 1.0 mL/min	287 nm	Phenomenex C18 (2) column (250 mm × 4.6 mm i.d., 5 μm)	200–2000 ng/mL	[52]
Rabeprazole sodium in pharmaceutical formulations	RP-HPLC	Acetonitrile:phosphate buffer (70:30, v/v, pH 7.0) at a flow rate of 0.8 mL/min	228 nm	C18 column [Use Inertsil C18, 5 μ, 150 mm × 4.6 mm]	0.1–30 μg/mL	[53]
Rabeprazole sodium and itopride hydrochloride in solid dosage form	HPLC	Acetonitrile:buffer (35:65, v/v) and the pH 7	266 nm	Luna C18 (5 μM, 25 cm × 4.6 mm i.d) phenomenex	2–16 μg/mL	[54]
Rabeprazole in bulk and tablet dosage form.	HPLC	Methanol and water (65:35, v/v), the flow rate is 0.8 mL/min	284 nm	RP-C18 column (150 mm × 4.6 mm I.D, 5 μm particle size)	0.25–20 μg/mL	[55]
Bulk drugs and pharmaceutical dosage forms	RP-HPLC	Triethyl amine buffer (pH 5):acetonitrile (50:50, v/v) with flow rate 2 mL min^{-1}	284 nm	BDS Hypersil C8 (250 mm × 4.5 mm, 5 μm)	5–25 μg/mL	[56]
Rabeprazole sodium and Aceclofenac in a combined dosage form	RP-HPLC	Methanol:acetonitrile:water (60:10:30, v/v/v) at a flow rate of 1.0 mL/min	280 nm	C-18 column methanol: acetonitrile:water (60:10:30, v/v/v)	1–10 μg/mL	[57]
Rabeprazole Sodium (RS) in solid dosage	RP-HPLC	50 mM ammonium acetate in water (pH 8) with ammonia and methanol	PDA detector	YMC C18, ODS-AM stainless steel column (250 mm × 4.6 mm ID; particle size 5 m)	100–500 μg/mL	[58]

Continued

Table 5 Summary of chromatographic methods of rabeprazole.—cont'd

Sample	Method	Method mobile/flow rate	Detection (λ, nm)	Stationary phase	Linearity range	References
Rabeprazole, pantoprazole, and itopride in combined dosage form	RP-HPLC	10 mM potassium dihydrogen orthophosphate (adjusted to pH 6.8):acetonitrile (70:30, v/v) at a flow rate of 1.0 mL/min	288 nm.	Phenomenex C18 (Luna) column (250 mm × 4.6 mm, dp = 5 μm)	2.5–25 μg/mL	[59]
Rabeprazole sodium in tablet formulations	HPLC	Acetonitrile and phosphate buffer (pH 7.4) in the ratio of 35:65 (v/v). The flow rate was 0.4 mL/min	280 nm	a Waters Acquity BEH C18 (50 × 2.1 mm, particle size 1.7 μm)	0.03–30 μg/mL	[60]
Rabeprazole sodium in bulk and its tablet dosage forms	RP-HPLC	Methanol:water (50:50; v/v) at the flow rate of 1.0 mL/min	290 nm	Waters C18 (4.6 × 150 mm, 5 μm)C	10–200 μg/mL	[61]
Clidinium Bromide, Chlordiazepoxide, Dicyclomine HCl and Rabeprazole Sodium in combined solid dosage	RP-HPLC	Phosphate buffer (pH 4.5):acetonitrile (80:20, v/v) and pH adjusted with 0.5% orthophosphoric acid -At a flow rate of 1.0 mL/min	215 nm	ODS C18 RP column (250 mm × 4.6 mm i.d., 5 μm)	20–60 μg/mL	[62]
Naproxen and rabeprazole sodium in pure and pharmaceutical dosage	RP-HPLC	Sodium dihydrogen phosphate buffer (pH 7.5 ± 0.1) and acetonitrile in the ratio of 70:30 (v/v) -The flow rate 1 mL/min	275	A phenomenex ODS C18 column of dimension 4.6 mm × 250 mm, 10 μm	1.0–30.2 μg/mL	[63]
Rabeprazole and mosapride in bulk and tablet dosage forms	RP-HPLC	Methanol and buffer (55:45%) [pH 4.0 adjusted with triethylamine] The flow rate 1.0 mL/min	276	An Inertsil-C18 ODS column (250 × 4.6 mm; 5 μ)	20–80 μg/mL	[64]

<table>
<tr><td>Rabeprazole and related impurities in bulk</td><td>HPLC-PDA</td><td>0.02 M K_2HPO_4:acetonitrile:methanol (85:5:10, v/v), with gradient programmed at flow rate 1.0 mL/min
—mobile phase A: 0.02 M K_2HPO_4
—mobile phase A: Acetonitrile R
—mobile phase C: Methanol R
<table>
<tr><th>Time (min)</th><th>0</th><th>10</th><th>15</th><th>25</th><th>35</th><th>40</th></tr>
<tr><td>A %</td><td>85</td><td>60</td><td>30</td><td>30</td><td>85</td><td>85</td></tr>
<tr><td>B %</td><td>5</td><td>30</td><td>60</td><td>60</td><td>5</td><td>5</td></tr>
<tr><td>C%</td><td>10</td><td>10</td><td>10</td><td>10</td><td>10</td><td>10</td></tr>
</table></td><td>285 nm</td><td>(250 × 4.6 mm), 5.0 μm, Phenomenex C18 column</td><td></td><td>[65]</td></tr>
<tr><td>Rabeprazole sodium in raw material and tablet dosage form</td><td>RP-HPLC</td><td>Buffer, methanol (30:70%, v/v)
Flow rate 0.9 mL/min</td><td>284 nm</td><td>Column C18 (4.6 × 150 mm, 3.5 μm, XTerra)</td><td>20–60 μg/mL</td><td>[66]</td></tr>
<tr><td>Rabeprazole in plasma</td><td>HPLC</td><td>5.0 mM ammonium acetate buffer (pH adjusted to 7.4 with dilute ammonia solution) methanol and acetonitrile at the ratio 45:35:20 (v/v/v)
A flow rate of 1.0 mL/min</td><td>290 nm</td><td>Luna C18 column (5 mm, 250 mm × 4.6 mm i.d.; 5 μm Phenomenex, CA, USA) with a guard column (30 mm × 4.6 mm i.d.; Phenomenex, CA, USA)</td><td>20.0–1000.0 ng/mL</td><td>[67]</td></tr>
<tr><td>Rabeprazole sodium in raw material and tablet dosage form</td><td>RP-HPLC</td><td>pH 5 phosphate buffer, methanol (30:70%, v/v)
flow rate of 0.9 mL/min flow rate</td><td>284 nm</td><td>Column: C18 (4.6 × 150 mm, 3.5 μm, XTerra)</td><td>20–60 μg/mL</td><td>[68]</td></tr>
</table>

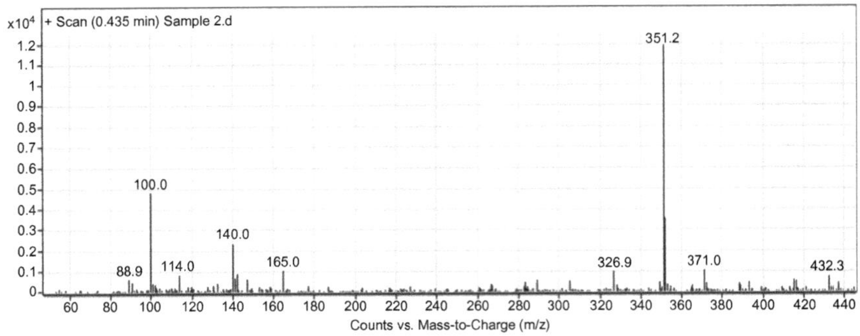

Fig. 7 Traces show full scan mass spectra of rabeprazole.

5. Methods of analysis

5.1 Compendial methods of analysis

5.1.1 Identification

USP 41 [22] described two methods for identification of Rabeprazole sodium bulk drug, i.e., IR absorption (197K), (197M), or (197A) may be used. LC system (compare to the retention time) and a TLC method were applied for identification of Rabeprazole. The sample and standard (0.2 mg/mL) were prepared in in methanol and 10 μL of sample and standard solution were added to thin layer chromatography plate. The mobile phase was a mixture of 10 mM ammonium acetate in water. Adjust with glacial acetic acid to a pH of 7.0:Acetonitrile (1:1). The R_f of the samples should be identical to that of the standard solutions.

British Pharmacopeia 2019 [23] described three methods (A, B, and C) for identification of Rabeprazole sodium: (A) Infrared absorption spectrophotometry (2.2.24), the sample was dissolved in methanol and evaporated to dryness then it was recorded the spectrum by the residue. The result spectrum was comparison with standard of rabeprazole sodium hydrate. (B) Loss on drying and (C) It gives reaction (A) of sodium (2.3.1). The identification must be conducted by using either methods A and B or A and C. Method B consist many tests, first test, 0.1 g of Rabeprazole sodium sample was dissolved in carbon dioxide-free water R and diluted to 10 mL with same solvent. The pH of this solution should be in range 9.5–11.5. Second test should be performed by using LC system (see Table 3). The sample used in this method contains impurity A and H. Finally method B tests, 0.200 g of sample was used to determine the water percentage in the sample and it should be in range 1.5–7.0%. For method C, addition of a 150 g/L

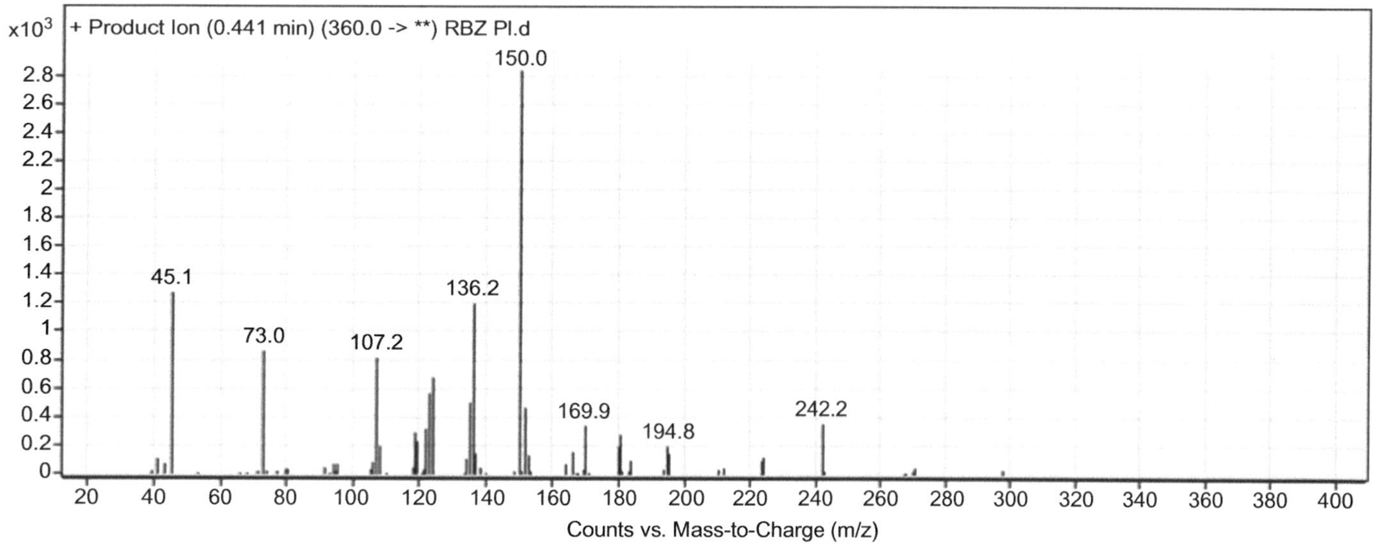

Fig. 8 Traces show product ion spectra of rabeprazole.

potassium carbonate R solution to a 0.05 g/mL of Rabeprazole sodium solution, then mixture solution was boiled after that, 4 mL of potassium pyroantimonate solution R was added to the mixture and heated to boiling. Next, the mixture was cold in ice water and A dense white precipitate was formed detection for sodium ion.

5.1.2 Related compounds

USP 36 [22] described two limit tests for Rabeprazole-related compounds using LC methods. Test 1 is the limit test for related compound A (RS A), while test 2 is for related compound B (RS B), related compound C (RS C), and other related compounds. The percentage of the related compounds can be calculated by using the formula:

$$\text{Result} = (r_u/r_s) \times (C_s/C_u) \times 100$$

in which C_s is the concentration (mg/mL) of the Rabeprazole-related compound A in the standard solutions, while C_u is the concentration of Rabeprazole (mg/mL) in the test solutions; r_u and r_s are the peak responses of RS A from the test and standard solutions, respectively. The acceptability limit is 1%. Details of LC methods are described in Table 3.

British Pharmacopeia 2019 [23] described an LC method for determination of the impurity in bulk drugs (see Table 3). The peak-to-valley ratio of Rabeprazole should be minimum 1.5, where the peak height above (Hp) the baseline due to impurity H and height above the baseline of the lowest point of the curve separating this peak from the peak due to rabeprazole.

5.1.3 Assay

USP 41 [22] described LC methods for the determination of Rabeprazole bulk drug. All LC methods used the same material for column (L1) but with different dimensions and particle sizes, while mobile phases are identical. A gradient elution was applied for analysis of Rabeprazole. Details of LC conditions are presented in Table 3.

British Pharmacopeia 2019 [23] used Liquid chromatography for assay of Rabeprazole in bulk drugs (see Table 3).

5.1.4 Impurities

USP [22] Rabeprazole Sodium RS

- USP Rabeprazole Related Compound **A RS** Sodium 1-(1H-benzimidazol-2-yl)-3-methyl-4-oxo-1,4- dihydropyridine-2-carboxylate ($C_{14}H_{10}N_3NaO_3$ 291.24).

- USP Rabeprazole Related Compound **B** RS 2-{[(1H-benzimidazol-2-yl)sulfinyl]methyl}-4-(3-methoxypropoxy)-3-methylpyridine 1-oxide ($C_{18}H_{21}N_3O_4S$ 375.44).
- USP Rabeprazole Related Compound **C** RS 1H-benzimidazole-2-thiol ($C_7H_6N_2S$ 150.20).
- USP Rabeprazole Related Compound **D** RS 2-({[4-G-meethotortor 3-methyl-2-pyridyl] reilly ay benzimidazole ($C_{18}H_{21}N_3O_4S$ 375.44).
- USP Rabeprazole Related Compound E RS 2-{[4-(3-methoxypropoxy)-3-methyl-2-pyridyl]methylthio}benzimidazole ($C_{18}H_{21}N_3O_2S$ 343.44).
- USP Rabeprazole Related Compound F RS 2-{[4-chloro-3-methyl-2-pyridyl]methyl]sulfinyl}benzimidazole ($C_{14}H_{12}ClN_3OS$ 305.78).

British Pharmacopeia 2019 [23] detected 10 impurities compound, the main impurities *specified impurities A*. Other detectable impurities include: *B, C, D, E, F, G, H, I, K*. It is not necessary to identify these impurities for demonstration of compliance.

A. 2-[[[4-(3-methoxypropoxy)-3-methylpyridin-2-yl]methyl]sulfonyl]-1*H*-benzimidazole,

B. 2-[[[4-(3-methoxypropoxy)-3-methylpyridin-2-yl]methyl]sulfanyl]-1*H*-benzimidazole,

C. 1-(1*H*-benzimidazol-2-yl)-3-methyl-4-oxo-1,4-dihydropyridine-2-carboxylic acid,

D. 2-[[(*RS*)-(1*H*-benzimidazol-2-yl)sulfinyl]methyl]-4-(3-methoxypropoxy)-3-methylpyridine 1-oxide,

E. 2-[(*RS*)-[(4-methoxy-3-methylpyridin-2-yl)methyl]sulfinyl]-1*H*-benzimidazole,

F. 1*H*-benzimidazole-2-thiol,

G. 2-[[(4-methoxy-3-methylpyridin-2-yl)methyl]sulfanyl]-1*H*-benzimidazole,

H. 2-[(*RS*)-[(4-chloro-3-methylpyridin-2-yl)methyl]sulfinyl]-1*H*-benzimidazole,

I. 2-[[(1*H*-benzimidazol-2-yl)sulfonyl]methyl]-4-(3-methoxypropoxy)-3-methylpyridine 1-oxide,

K. 1*H*-benzimidazol-2-ol.

5.2 Reported methods of analysis

5.2.1 Electrochemical method

5.2.1.1 Voltammetric

Radi et al. [24] studied the oxidative behavior of rabeprazole at a glassy carbon electrode in Britton–Robinson (BR) buffer solutions using cyclic, linear sweep and differential-pulse voltammetry. The process of oxidation was seen to be irreversible over the pH range (6.0–11.0) and managed diffusion–adsorption. For the determination of rabeprazole as supporting electrolyte in BR buffer solution (pH 8.0), an analytical method was established. The linearly range of rabeprazole concentration is 1.0×10^{-6} to 2.0×10^{-5}, and the limit of detection was 4.0×10^{-7} M.

Khashaba et al. [25] investigated the effect of adding transition metals to the electrolyte containing proton pump inhibitors, such as rabeprazole sodium (RAB sodium), on the voltammetric response of pencil graphite electrode. Both square wave adsorptive voltammetry (SWAdSV) and cyclic voltammetry (CV) were used to explain and validate the potential complexation of the intermediate metal between RAB sodium and cobalt. The current signal due to the oxidation process was a function of the concentration of RAB sodium, pH of the medium, cobalt concentration, scan rate, frequency, and deposition time at the electrode surface. The oxidation peak current ranged from 0.05–9 to 10^{-9} M and 0.2–8.5 to 10^{-7} M, for SWAdSV and CV, respectively. The detected limits were found to be 0.015×10^{-9} M and 0.06×10^{-7} M for SWAdSV and CV, respectively.

Guo et al. [26] developed a carbon paste electrode (CPE) anodic adsorption voltammetric method for determination of rabeprazole. On the electrode, the solution of Rabeprazole was concentrated and determined by adsorptive voltammetry in supporting solution of 1.8 mol/L H_2SO_4. The rabeprazole anthodic peak potential was at 0.498 V (vs SCE). The linear relationship between the rabeprazole concentration and the peak currents were found in the range of 4.15×10^{-7}–3.73×10^{-6} mol/L, with the detection limit 6.705×10^{-11} mol/L.

5.2.1.2 Polarography

Gartner et al. [27] reported a differential pulse polarography (DPP) method for the electrochemical reduction of proton pump inhibitors (PPIs), namely, lansoprazole and rabeprazole using a dropping mercury electrode (DME). Both experiments were performed with pH values from 3.0 to 11.0 in the Britton-Robinson pump solutions. All PPIs were proved to be thoroughly decomposed, with pH values increasing, containing two principal compounds, one cyclic sulfenamide and another dimer. In this case, pH 8.0 was determined to be stable with lansoprazole and pH 9.0 with rabeprazole. Rabbeprazole was decomposed slightly faster than lansoprazole and hence up to higher pH levels. High currents ranged from 1×10^{-6} M to

7×10^{-5} M at the pH 9.0 linearly with concentrations of the two PPI. Both compounds were displayed similarities in reaction and different variations based on their respective chemical properties and structures.

5.2.2 Spectroscopic methods of analysis

5.2.2.1 Spectrophotometry

El-Gindy et al. [28] proposed three methods for evaluating Rabeprazole in the presence of its degradation products. One of these methods was based on the first derivative of the spectra ratio (^{1}DD), which relied on the measurement of amplitudes at 310.2 nm. The pH-rate profile of rabeprazole degradation in Britton–Robinson buffer solutions within pH range 3–11 was further investigated. In Britton–Robinson buffer solution pH 7, however, the activation energy of rabeprazole degradation was determined. The calibration range of rabeprazole with the ratio first derivative of the spectra ratio was 10–30 μg/mL, the detection limit was 0.019, and the quantitation limit was 0.058.

Patel et al. [29] developed a spectrophotometric method for determination of rabeprazole in pharmaceutical bulk dosage form. The method depended on the ion-pair complexes formation of rabeprazole with four dyes, viz. bromo thymol blue, bromocresol green, bromophenol blue and bromocresol purple in solution of acidic buffer, afterwards their chloroform extraction. The absorption of the organic layer was measured against the corresponding blank reagent at the respective wavelength of the maximum absorbance. The method was evaluated statistically and found to be accurate and precise. Phosphate buffer (pH 2) and bromocresol green dye phosphate buffer gave the maximum absorption of rabeprazole at 454 nm. The Beer law was found within the concentration range of 10–100 μg/mL.

A spectrophotometric derivative method for evaluating Rabeprazole sodium was developed and validated by Garcia et al. [30] in drug formulations. The procedure was performed with water as a diluent (pH 10.0). The derivative spectra of the first order were obtained at $N=5$, $\Delta\lambda=4.0$ nm, and determinations were made at wavelength 304 nm. In the present of drug excipients, the procedure showed a good linearity of 6.0 and 18.0 $\mu g/mL^{-1}$, and the accuracy was 99.15%.

Sabnis et al. [31] developed a spectrophotometric ratio method of spectra derivatives for simultaneous determination of rabeprazole sodium and itopride hydrochlorides in theirs dosage forms. Amplifications were chosen for the first derivatives of the respective at 231 nm (minimum) and 260 nm ratio range. Linearity, accuracy and precision, were validated. The Beer law was found to be within the 4–20 μg/mL concentration range for rabeprazole.

Gul et al. [32] reported a UV spectrophotometric method for determination of rabeprazole in bulk using methanol as solvent. The solution was scanned at 200–400 nm with a maximum absorption at 284 nm. The method was validated according to ICH guidelines. Linearity was in the 12–18 μg/mL range. A recovery was 99.86–100.14%. The %RSD for repeatability was 0.628% and for precision and inter was found at 0.488–0.77%. This method indicated that rabeprazole was interactive with clorazepate dipotassium at pH 7.4.

Garcia et al. [33] developed a UV spectrometric ultraviolet (UV) method for determination of rabeprazole in pharmaceutical formulation and compared the method with a previously accepted capillary electrophoresis (CE) method. The proposed method was applied using water (pH 10.0) as diluted at 291 nm. The method showed a good linearity ($r = 0.9997$) in the concentration range of 6.0–18.0 $\mu g mL^{-1}$.

Gowri et al. [34] developed two new methods for estimation of Rabeprazole in bulk and tablet dosage forms. In the first method, a Folin-Ciocalteu reagent and sodium hydroxide were used. A Folin-Ciocalteu reagent was reduced by Rabeprazole, in the presence of sodium hydroxide, and blue colored chromogen was formed at a maximum absorption of 640 nm. The linearity range was obtained at concentration of 1–8 μg/mL and a correlation coefficient (r^2) was found to be 0.9983. In the second method, the ferric nitrate was reduced by rabeprazole. The product ferrous ions reacted with 2,2,bipyridine, showing maximum at 522 nm. The linearity was obtained at range of concertation of 2–16 μg/mL with a correlation coefficient (r^2) of 0.9994. These two methods were validated according to the ICH guidelines and they can be applied to determination of rabeprazole in bulk and dosage form.

A simultaneous UV spectrophotometric method for determination of Famotidine and Rabeprazole sodium in the laboratory mixture was developed and validated by Dsouza et al. [35]. The proposed method was performed at a maximum absorption of 263 nm for Famotidine and 284 nm for Rabeprazole sodium. The method obeyed Beer's law in the concentration range of 5–30 μg/mL for both famotidine and rabeprazole.

5.2.2.2 Colorimeter

Two direct spectrophotometric methods for determining rabeprazole sodium through complexation of charge transfer reactions were presented by Abdel-Hay et al. [36]. The first method was based on the reaction of the drug with p-chloranilic acid (P-CA) in acetonitrile to produce a red colored product with maximum absorption of 518 nm. The second method was based on the reaction of 7.7,8,8-tetracyanoquinodimethane (TCNQ) with Rabeprazole sodium in acetone. A blue-green complex was formed and measured at 845 nm. The linearity range was obtained at concentration of 20–200 μg/mL for P-CA–rabeprazole and 2–16 μg/mL for TCNQ–rabeprazole.

Validated colorimetric methods were developed by Kanekar et al. [37] for the quantitative estimation of Rabeprazole in bulk and pharmaceutical dosage form. Method A consisted of redox reaction of Rabeprazole in presence of acidic medium reacted with excess amount of Cerric ammonium sulfate. The dye Safranin gave λ_{max} of Rabeprazole at 516 nm. Method B consisted of a redox reaction of Rabeprazole with ferric nitrate and the reduced ferrous ion than reacted with 1,10 Phenanthroline giving maximum absorbance at 511 nm. The methods were validated as per ICH guidelines.

Shaik et al. [38] developed two methods for estimating Rabeprazole in bulk and tablet dosage form. In the first method, Folin-Ciocalteu and hydroxide were used. In the presence of sodium hydroxide, Folin–Ciocalteu reagent was reduced by the Rabeprazole and blue colored chromogen was formed with a maximal absorption at 640 nm and a concentration of 1–8 μg/mL. Linearity was detected with correlation coefficient (r^2) 0.9983. In the second method, rabeprazole reduced ferric nitrate. The reduced ferric ions react with 2,2,bipyidine, exhibiting absorption maximum at 522 and at concentration range 2–16 μg/mL. Linearity was obtained with correlation (r^2) 0.9994. These two method were validated according to ICH guidelines and were suitable for estimation of rebaprazole in bulk and pharmaceutical dosage form.

Two spectrophotometric methods for determination of rabeprazole sodium in commercial dosage forms were developed and validated by Rahman et al. [39]. In method A, Rabeprazole sodium was reacted with 3-methyl-2-benzothiazolinone hydrazone hydrochloride (MBTH) in the presence of ammonium cerium (IV) nitrate in acetic acid medium at room temperature to form red-brown product which absorbs maximally at 470 nm. In method B, Rabeprazole sodium was reacted with 1-chloro-2,4-dinitrobenzene (CDNB) in dimethyl sulfoxide (DMSO) at 45 ± 1 °C to form yellow colored Meisenheimer complex. The band peak of the colored complex was found at 420 nm. The concentration range of rabeprazole sodium was 14–140 μg/mL and 7.5–165 for at method A and B, respectively.

5.2.2.3 Chemiluminescence

Yousef zadeh et al. [40] reported a high-emission efficient chemiluminescence (CL) reaction based on a synergistic improvement effect of silver nanocluster (AgNCs) and graphene quantum points (GQDs). First, AgNCs and GQDs syntheses were carried out simply by reducing $AgNO_3$ chemical material and by thermal treatment of glucose. The useful behavior of the prepared nanomaterial in CL systems was investigated after the characterization phase. A low CL emission was provided by the oxidation reaction of $KMnO_4$-rhodamine B. Nevertheless, the existence of AgNCs and GQDs resulted in a synergetic increase in emissions. Using absorption and fluorescence experiments, a potential mechanism for this effect was studied. In addition, the effect of rabeprazole on the CL emission was relatively selective. The linearity relation between intensity of CL and the Rabeprazole was in concentration range of 4–133 ng/mL, with a detection limit (3Sb/m) of 1.1 ng/mL. The development of the CL method was used to quantify Rabeprazole with acceptable precision and accuracy in biological samples.

5.2.3 Chromatographic methods of analysis

5.2.3.1 Electrophoresis

A capillary electrophoresis method for determining rabeprazole sodium in coated tablets was developed and validated by Garcia et al. [41]. The conditions were as following: a bare fused silica capillary with 48.0 cm length (39.5 cm effective) and 75 μm id; a 10 mM, pH 9.0, sodium tetraborate run buffer; a diode array detector set at 291 nm; hydrodynamic injection (50 mbar/5 s); and a voltage of 20 kV. HP Chemstation CE rev. A.06.03 software was used for system control, data acquisition, and analysis. A linear procedure was found to be within the 5.0–40.0 μg/mL concentration range ($r = 0.9993$), A standard deviation of interday relative was found 0.49. Percentage of mean recovery was 103.1%. The limits of detection and quantitation were 1.29 and 3.91 μg/mL, respectively.

Ma et al. [15] reported an enantioseparation procedure for rabeprazole and omeprazole in acetonitrile—method (60:40, v/v). An ephedrine-based chiral ionic liquid, (+)-*N*,*N*-dimethylephedrinium-bis(trifluoromethanesulfon) imidate ([DMP](+) [Tf(2) N](−)) was used as both a chiral selector and a background electrolyte in nonaqueous capillary electrophoresis. The influences of separation conditions, including the concentration of [DMP](+) [Tf(2) N](−), the electrophoretic media and the buffer, on enantioseparation were evaluated. $[DMP]^+[Tf2N]^-$ concentration in acetonitrile-methanol (60:40, v/v) was investigated between 30 and 70 mM.

Papp et al. [16] developed and validated capillary electrophoresis methods using Cyclodextrins (CDs) for the chiral detection of lansoprazole and rabeprazole. A preliminary analysis was performed on 14 separate neutral and anionic CDs at pH 4 and 7. The most suitable chiral field for both compounds was found to be sulfobutyl—with a number 6.5 and 10 substitution at neutral pH. Specific dual CD systems were also compared and probable enantiomer separation mechanisms were discussed. The most suitable method for separations was a dual selector method comprising sulfobutyl-ether-β-CD substitution 6.5 and native γ-CD. Optimization of the procedure was achieved by an experimental design method with an initial fractional factor screening design and a core composite design to achieve the desired analytical criteria. Optimized procedures (25 mM phosphate buffer, pH 7, 10 mM sulfobutyl-ether-β-CD/20 mM γ-CD, +20 kV voltage; 17 °C temperature; 50 mbar/3 s injection, detection at 210 nm for lansoprazole; 25 mM phosphate buffer, pH 7, 15 mM sulfobutyl-ether-β-CD/30 mM γ-CD, +20 kV voltage; 18 °C temperature; 50 mbar/3 s injection, detection at 210 nm for rabeprazole) provided a baseline separation for the enantiomers with desirable migration order of lansoprazole (Rs = 2.91) and rabeprazole (Rs = 2.53). The developed methods were evaluated in compliance with current guidance.

5.2.3.2 Thin layer chromatography

Raval et al. [17] established a thin layer chromatographic method for the simultaneous estimation of combinations of the ondansetron in solid dose form with omeprazole and rabeprazole. The procedure consisted of separating components with TLC from pre-coated silica gel 60F254 using a dichloromethane mixture:methanol (9:1) as a mobile phase. Spots of ondansetron with omeprazole and rabeprazole were observed at 309 and 294 nm, respectively. The maximum ondansetron and omeprazole retardation factor were found to be 0.42 ± 0.02, 0.54 ± 0.03, respectively. The average factor of retardation was 0.42 ± 0.02, 0.54 ± 0.03 for ondansetrone and omeprazole, respectively. While for ondansetron and rabeprazole was 0.41 ± 0.02 and 0.51 ± 0.02, respectively. The linearity range of three medications was 0.1–0.5 μg/spot.

Pateli et al. [18] was developed high liquid chromatography (HPLC) and thin-layer chromatography (TLC) methods for the quantitative estimation of rabeprazole and mosapride for their combined pharmaceutical formulation.

In the TLC, ethyl acetate-methanol-benzene was used as the mobile phase (2:0.5:2.5, v/v). The retention times of rabeprazole and mosapride were found to be 4.93 ± 0.01 and 9.79 ± 0.02, respectively. The R_f values were 0.42 ± 0.02, and 0.61 ± 0.02, for rabeprazole and mosapride, respectively. In TLC, the detection limit for rabeprazole and mosapride were 132.29 and 98.25 ng/point, respectively. In HPLC, the limit of detection was found to be 97.7 ng/mL for rabeprazole and 97.6 ng/mL for mosapride. The proposed methods can be applied to the determination of rabeprazole and mosapride in pharmaceutical dosage forms.

In order to evaluate Rabeprazole sodium and its degradation components, Osman et al. [19] was established and validated stability indicating spectrofluorometric, TLC-densitometric and HPLC procedures. A TLC procedure was based on acidic medium followed by densitometric estimation of the spot's intact drug at 284 nm. The separation was carried out on Fluka TLC sheets of silica gel 60F254 using isopropyl alcohol–30% ammonia (80 + 2, v/v) mobile phase.

The procedure for a simultaneous measurement of rabeprazole and domperidone was described and validated by Patel et al. [20] using high-performance thin-layer chromatography (HPTLC) in pure powder and in capsule formulations. A HPTLC method was performed using ethyl acetate-methanol-benzene-acetonitrile (30:20:30:20, v/v) as a mobile phase in an aluminum plate of silica gel 60F254 with a UV detection at 287 nm. Quantitation was performed over a 400–1200 and 600–1800 ng/spot concentration range with mean recovery of 99.82 ± 0.74 and 99.43 ± 0.68 for rabeprazole and domperidone, respectively.

A chromatographic high-performance thin layer (HPTLC) approach for simultaneous determination rabeprazol sodium and domperidone in combined tablet dosage was developed and validated by Gandhi et al. [42]. The mobile phase was toluene-acetone-methanol (4.5 + 4.5 + 0.5, v/v/v) with a UV detection at 285 nm. The rabeprazole sodium and domperidone retention factors were 0.53 ± 0.12 and 0.32 ± 0.20. The approach was validated in terms of linearity, accuracy, and robustness. The method was complied with Beer's law in the range of 50–800 ng/band, for both drugs.

5.2.3.3 High-performance liquid chromatography

Elkady et al. [43] developed and validated bio-analytical LC-MS/MS method for the simultaneous extraction and determination of four proton pump inhibitors: esomeprazole, lansoprazole, pantoprazole and rabeprazole in human plasma using escitalopram as an internal standard.

The analytes were extracted from plasma samples using precipitation of proteins with acetonitrile. A mobile phase of 10 mM ammonium formate:acetonitrile:methanol (20:40:40%, v/v) was used for separating at a flow rate 0.8 mL/min. in isocratic mode on a reversed phase C18 INERTSIL ODS-3 (5 μm, 150 × 4.6 mm) and column temperature of 40 °C. Positive mode *electrospray ionization* source was used prior to multiple reaction monitoring (MRM) detection using parent and daughter ions: m/z 346.2 → 198.1 for esomeprazole, m/z 370.1 → 252 for lansoprazole, m/z 384.2 → 200.2 for pantoprazole, m/z 360.1 → 242.1 for rabeprazole and m/z 325.2 → 109 for escitalopram. The curves of calibration were linear in a concentration range of 20–5000 ng/mL. The method was completely validated in accordance with US-FDA and EMA guidelines. Further methods of high-performance liquid chromatography are tabulated in Table 5.

5.3 Determination in body fluids and tissues

5.3.1 Chromatographic methods

Zhang et al. [48] developed and validated a liquid chromatography analytical method, combined with tandem mass detection for estimation of rabeprazole in human plasma using omeprazole as an internal standard. *N*-hexane–dichloromethane–isopropanol (20:10:1, v/v) was used to extract both rabeprazole and omeprazole (internal standard) and was isocratically chromatographed on a column for the Diamonsil C19. The mobile phase was methanol in order to prevent rabeprazole decomposition. The compound was detected using an atmospheric pressure chemical ionization method in the chosen reaction monitoring mode. The cycle was linear in the 2.0–800 ng/mL concentration range. The lower limits were 2.0 ng/mL. The intra- and interday precision, measured using samples from quality control (QC), was less than 9.8%. The accuracy was within ±1.1%. Following oral application of 20 mg of rabeprazole to 18 healthy volunteers the procedure was used in a pharmacokinetic analysis.

Su et al. [69] developed and applied a super-critical fluid chromatography chiral method tandem mass spectrometry to determine (*R*)-rabeprazole in dog plasma and (*S*)-rabeprazole. After the addition of (*R*)-lansoprazole as internal standard, the sample preparation was precipitated by protein with acetonitrile. The simple separation of enantiomers within a time of 4.5 min was performed with an Acquity UPC2 method using a 60 °C column ACQUITY UPC2 Trefoil CEL2. The mobile phase was methanol: CO_2 (30:70, v/v) delivered at flow rate of 2.5 mL/min. The transitions at

m/z 360.0–242.2 (rabeprazole) and 370.3–2252.0 (internal standard) were observed by multiple reaction monitoring methods in the positive ion mode. Inter-day precision in the range of −2.6% to 3.1% was less than 10.0% with accuracy. After the administration of a single oral dose of beagle racemates in 10 mg, the procedure was successfully extended in a pharmacokinetic analysis of rabeprazole enantiomers.

5.3.2 Radioiodination methods

Sanad and Challan [70] used ^{131}I to label Rabeprazole to give ^{131}I-rabeprazole in 98% yield. The optimum conditions were as follows: pH 8, 100 µg of chloramine-T, 30 min, and 0.2 mg of rabeprazole. The labeled Rabeprazole was stable for 24 h. After intravenous injection, biodistribution in mice was analyzed of ^{131}I-rabeprazole and its high absorption in the stomach ulcer, reaching about 38% ID/g at 1 h after injection, was reported. The concentration of the labeled compound was appropriate for ulcer imaging in the stomach ulcer. Easy labeling, high protocol yield and ulcer imaging are the advantages of the method.

6. Stability

6.1 Stability indicating method

In the presence of both impurities and degradation products resulting from decomposition of rabeprazole, Buchireddy et al. [71] developed a stability-indicating reverse-phase high-level liquid chromatographic system for rabeprazole analyzes. As acid hydrolysis, oxidation, photolytic and thermal stress was encountered, degradation, and aqueous hydrolysis were observed. The medication was found to be stable against other stress conditions.

Kuchana et al. [72] developed a stability indicating method for a simultaneous determination of four active pharmaceutical ingredients, namely, Pantaprazole (PAN), Rabeprazole (RAB), Lansaprazole (LAN) and Domperidone (DOM) in their bulk and dosage forms by a RP-HPLC-DAD method. The separation was executed using reverse phase chromatography on X Bridge (3 × 100 mm; 3.9 µm) C18 column with a gradient program. Solvent A was consisted of Triethylamine in phosphate buffer at pH 7.5 adjusted with ortho-phosphoric acid. Solvent B consisted of a mixture of methanol:acetonitrile (85:15). The mobile Phase A was prepared by mixing solvent A with solvent B in the 90:10 (v/v) ratio. The Mobile Phase B was prepared by mixing the Solvent A with Solvent B in the 20:80 (v/v) ratio. The mobile phase-A (0–3 min:70–70; 3–7 min:70–40; 7–20 min:40–40; 20–21 min:40–40; 21–25 min:70–70) with gradient composition and the mobile flow was 0.6 mL/min. The temperature of column oven was maintained at 21–24 °C, run time was 25 min. Quantification was achieved by the PDA detector and the effluents at 283 nm were monitored for four drugs and their combination drug products were exposed to various stress conditions. The correlation coefficient (r^2) of all the calibration curves of four drugs was found to be 0.999. The Limit of detection for Pantaprazole, Rabeprazole, Lansaprazole and Domperidone were found to be 0.782, 0.897, 0.142 and 0.185 μg/mL, respectively. The Limit of quantitation for studied drugs were found to be 2.524, 2.894, 0.459 and 0.599 μg/mL, respectively.

In compliance with ICH Guidelines Q1A (R2), Bhandi et al. [73] studied force degradation products of rabeprazole. A Purospher STAR, C-18 (250–4.5 mm, 5 μ) with 10 mM ammonium acetate (pH −7.0) was used to achieve a chromatographical separation of the compound and its degeneration products. A mixture of 10 degradation products (R1–R10) was identified and characterized by experiments with LC-ESI-MS/MS and precise mass measurements. A study of the fragmentations of rabeprazole $[M+H]^+$ ions and of their degradation products was suggested as the most possibly formation method of the DPs. Semi preparatory HPLCs were isolated using Water's X-bridge Prep C18 (250–10 mm, 5 μ) for major degradation products (R3, R4, and R7). In vitro-toxicity measurement, more than 50% cell inhibition in HepG2 cell and PANC-1 cell lines showed isolated degradation products at concentrations less than 1 μM. Studies of DNA binding using spectroscopic techniques demonstrated that the ligand molecules R3, R4, and R7 binded to the double-stranded DNA surface and stabilized the DNA complex.

6.2 Solid-state stability

Ren et al. [74] reported that Rabeprazole Sodium (RPN) was highly unstable in acid conditions or some acidic pharmaceutical excipients, such as acrylic acid polymers (carbomer 934) with carboxylic acids. Therefore, a Fourier transform infrared (FT-IR) spectroscopy was used to investigate the degradation process of binary rabeprazole blends in a solid state, without the use of solvents in three different ratios (3:1, 1:1, and 1:3). A comparison was also performed with alkalizers (MgO), neutral hydroxypropymethylcellulose (HPMC 4000). The binary compositions were stored under speedy conditions (40 °C/75 percentage relative moisture) for 2 weeks. When the rabeprazole concentration reduced, concentrations of Thioether rabeprazol in binary blends of acidic Carbomer 934 increased. Furthermore, the rabeprazole degradation half-life besides the relative peak area ratios obtained from FTIR spectra of S=O stretching at 1094.1 cm^{-1} decreased consistently as the fraction of carbomer 934 increased due to its sensitivity between the basic benzimidazole nitrogen and carboxylic acid group of carbomer 934. In the presence of carbomer 934, the physical appearance turned to a solid brown color. In comparison, the kinetics of rabeprazole degradation with MgO and HPMC 4000 did not improve substantially in their solid state. This current research indicated that the chromatographic HPLC and FT-IR spectral analysis for solid state compatibility testing may provide a valuable technique to evaluate appropriate pharmaceutical excipients and elucidate the role of degradation in early formulation of drug formulations through RPN.

6.3 Solution-phase stability

Ren et al. [75] investigated the chemical stability of a rabeprazole sodium, which was evaluated in a simulated intestinal fluid (pH 6.8) containing various "Generally Recognized As Safe (GRAS)"-listed excipients, including Brij® 58, Poloxamer 188, Cremophor RH40, Gelucire 44/14 and PEG 6000. The quantities of rabeprazole and the thioether-rabeprazole product were quantified by HPLC after incubation at 37 and 60 °C. The major degradation component was isolated and characterized by LC/MS. Rabeprazole degradation followed kinetics of the first order. The rate constants (k) obtained at 37 and 60 °C was 0.75 and 2.78 h^{-1}, respectively. By comparison, its stability was enhanced by incorporating excipients. Brij®58 had the greatest stabilizing effect of several excipients tested in this experiment. At 37 and 60 °C, for instance, the k values of Brij®58 decreased to 0.22 and 0.53 h^{-1}. In terms of their solubilizing performance and micellar formulation of thioether rabeprazole, the stability mechanisms of these hydrophilic polymer was excellent devices with optimum Hydrophile-lipophile balance (HLB) values explained in detail. In addition, formulations containing sufficient excipients of rabeprazole would improve the stability and bioavailability of the intestinal tract.

7. Uses

Rabeprazole is used for gastric acid suppression, like other proton pump inhibitors such as omeprazole. This effect helps to treat and prevent conditions in which gastric acid directly aggravates symptoms such as duodenal and gastric ulcers. Acid suppression may provide symptomatic relief in the setting of gastroesophageal reflux disease (GERDs), which has long-term gastric exposition in the esophageal because of changes in the stomach or esophageal anatomy such as those due to abdominal obesity. The suppression of acid is also helpful in increasing a gastric acid production including over secretion of gastric acid (excess gastric acid conditions) such as Zollinger-Ellison, several endocrine adenomas and systemic mastocytoses. In addition to antibiotic therapy, Rabeprazole is used in the treatment of *Helicobacter pylori* pathogen, which otherwise thrives in acidic environments. A randomized trial in high risk South Korean people with endoscopically treated early stomach cancer has also demonstrated that *H. pylori* elimination with antibiotics and Rabeprazole prevented developing a second stomach cancer. Rabeprazole is FDA approved for treating symptomatic GERD for adolescents and adults, healing duodenal ulcers in adults and the eradication of *Helicobacter pylori* [76–78].

8. Pharmacology

8.1 Pharmacodynamic properties

In fact, all PPIs that are substituted for benzimidazoles, have the same antisecretory pathways, which is the most effective means for raising the pH in the stomach and hence for achieving the therapeutic level for GERD. They concentrate on the secretory canaliculus in paretic cell. A conversion to an active sulfenamide compound (the rate-restricting phase) is carried out in the protonated molecules which form covalent disulfide-inhibiting bonds with H^+/K^+ ATPases surface-exposed cell cysteines. In terms of acid stability, the PPIs differ as the modified functional replacements in two ring structures provide the highest pK_a (approximately 5.0, the pH at which a drug becomes 50% protonated), and hence the molecule can be activated at higher pH levels much faster than other PPI: Raboprazole took 1.3 min to be half-activated at pH 1.2 (channel room pH after meals), compared to 2.0, 2.8 and 4.6 min for both lansoprazole, omeprazole and pantoprazole, respectively. At pH 5.1 (the pH during fasting), the activation half-life for rabeprazole, lansoprazole, omeprazole and pantoprazole was 7.2, 90, 84, and 282 min, respectively [79].

Rabeprazole has demonstrated its efficient and fast action in an isolated hog vesicle model: the proton pump has been almost completely inhibited within 5 min of Rabeprazole treatment. Lansoprazole and omeprazole achieved the same target after 30 min but only 50% could be blocked by pantoprazole at the end of the 50-min test [80]. Consequently, rabeprazole prevents basal and peptone food-stimulated gastric acid production as a result of dose-related continuous inhibition [81,82].

The antisecretory function of PPIs is well known to predict their success in conditions associated with antacids. The correlation of Double ulcer treatment was found with an intragastric pH > 3 lasting 18–20 h, whereas erosive GERD cure with a pH roughly 24 h > 4 staying time [83,84].

Latest research has established in vivo previous evidence that after first administration, rabeprazole can achieve adequate acid depression and sustain this benefit over the next days of treatment. This results in a mean higher intragastric pH of 24 h and longer spans of pH > 3 and >4 compared to omeprazole [85].

It concluded, in the first 24 h after the dose, the higher pK_a of rabeprazole could justify its significant anti-screening effect compared with the other tested of PPI formulations. Comparative 18H test in a multiple convergence. Adults with pylorio-negative, rabeprazole 20 mg will have a 24-h average gastric pH of 3.4, compared to 2.9 for 30 mg lansoprazole, 2.2 for 40 mg pantoprazole; 1.9 for 20 mg omeprazole; 1.8 for omeprazole 20 mg MUPS, and 1.3 for placebo ($P < 0.05$ vs all other PPI and placebo). Moreover, rabeprazole holds pH > 4 for a longer duration of (8 h) than other agents (7.4, 4.9, 2.9, 3.0 and 0.9 h, respectively, $P \leq 0.04$ for rabeprazole vs other agents) [86].

The acid suppression levels of lansoprazol 15 and 30 mg are greater than those of rabeprazole 10 and 20 mg for hours 0–5 for days 1 and 5, respectively, while rabeprazole is a comparable or a more effective for hours 11–24 on both days. Exactly 72 healthy patients received four PPIs therapies in an open-label, randomized combination trial for 5 days with 2-week washout periods and underwent on-going 24-h intragastric pH, 1 and 5 days, on baseline. The mean pH period >4 was comparable with the mean 24-h pH for both PPIs on days 1 and 5, but rabeprazole 10 mg statistically higher than 15 mg lansoprazole [87].

8.1.1 Mechanism of action

Rabeprazole sodium is in the family of anti-secretory drugs, substituted for benzimidazols, which do not have antagonistic effects with anticholine or H2 histamine but which inhibit the secretions of gastric acid by causing H^+/K^+-ATPase (acid or proton pump) in particular. Dose-connected effect allows both basal and induced acid secretion to be blocked independent of stimulation. Animal tests demonstrate that rabeprazole sodium easily disappears both from plasma and gastric mucosa following administration.

Rabeprazole is consumed easily after any dose as a depleted base and concentrates in the parietal cell acid region. Rabeprazole is converted by protonation into the active sulphenamide form and interacts with the available cysteines on the proton pump afterwards [88].

8.1.2 Clinical efficacy and safety

The anti-secretive effect occurs about 1 h of orally administration of a 20 mg rabeprazole sodium dose, with the full effect-taking place about 2–4 h. The basal-and nutritionally induced secretion of acid is reduced for up to 48 h and is 69–82% 23 h after the first dose of rabeprazole sodium. For repeated once-daily doses, rabeprazole sodium has an inhibitor effect on the secretion of acid, resulting in a steady state inhibition after 3 days. The confidentiality operation normalizes within 2–3 days after the medication is stopped [88].

Reduced stomach acidity by some means, even rabeprazole, reduces the bacteria levels usually present in the gastrointestinal tract. Therapy of PPIs may raise the chance of GI infection including *Salmonella*, *Campylobacter* and *Clostridium difficile*. Therapy may improve. Patients treated with rabeprazole sodium (10 or 20 mg) once daily for up to 43 months in clinical trials leading to rise the levels of serum gastrine during the first 2–8 weeks, indicating the inhibitory effects on acid secretion. The values of gastrine returned to rates of pretreatment normally in 1–2 weeks after therapy stop [88].

Serum gastrine improves in response to the decrease in acid secretion when treated with antisecretory medicines. CgA also improves owing to a reduced acidity in stomach. The higher CgA level will interfere with neuroendocrine tumor studies. Published literature suggests that the antagonists of proton pumps will be stopped between 5 days and 2 weeks before calculating CgA. This is to enable CgA levels to return to the reference range which may be spuriously elevated during PPI treatment. The ECL histology, the degree of gastritis, the effect of gastritis on atrophies, intestinal metaplasia, or distribution of H has not been observed in human gastric biopsies from antrum and fundus in 500 patients under rabeprazole or comparator for up to 8 weeks. There was no substantial improvement in the results present at baseline in over 250 patients who endured 36 months of ongoing therapy. Other impacts: CNS, digestive and respiratory processes have not yet been proven to have a cumulative effect by rabeprazole sodium.

Rabeprazole sodium, administered at an oral intake of 20 mg for 2 weeks, had no effect upon thyroid activity, metabolism of carbohydrates or circulatory levels of parathyroid hormones, cortisol, estrogen, testosterone, prolactin, cholecystokininine, secretin, glucagon, follicular stimulating hormone (FSH).

Studies in healthy voluntaries found that rabeprazole sodium does not interfere with amoxicillin clinically substantially. When co-administrated to extract the upper gastrointestinal H., rabeprazole does not negatively influence the plasma concentrations of amoxaicillin or clarithromycin Infection of the *H. pylori* [88].

8.2 Pharmacokinetic properties

Rabeprazole is an enterically coated formulation since all PPIs are unstable in acidic environments. This is consumed fairly rapidly after oral intake, as the maximum plasma concentration (C_{max}) is reached between 2.8 and 5.1 post-dose [89].

The molecular pharmacokinetics were shown to be linear in the 10–80 mg range, with a total bioavailability of 52%, seen with rabeprazole 20 mg. While C_{max} and AUC are proportional to the dosage taken to the plasma concentration, C_{max} duration and half-life are dosage-independent. The findings suggest that the first-pass metabolism of the rabeprazole is unsaturated and can be absorbed in high doses [89]. However if food sources prolonged the absorption of rabeprazole 20 mg by about 1.7 h and decreased apparent removal halving-life owing to a potentially prolonged gastric emptying, no acid or diet affect the bioavailability of the molecule [89].

The delivery amount of rabeprazole in different tissues, including gastric mucosa, intestine, lung, blood cord, liver, bowel and thyroid have been shown by preclinical studies. It has been shown that bound to plasma proteins are 94.8–97.5% in healthy volunteers [89].

During repetitive administration, no major accumulation occurs as the elimination half-life of 1 h after single administration is 1.5 h after several administrations. A 20 mg dosage of rabeprazole (thioether iron, glucuronide and mercaptopurine metabolites) is excreted in the urine at about 90% and in faces 10% [89–91].

Rabeprazole is unique in its elimination. Whereas other PPIs including omeprazole, lansoprazole, esomeprazole and pantoprazole are mainly metabolized in the liver by CYP2C19, Rabeprazole is mainly metabolized by a non-enzymatic pathway to rabeprazole thioether and, to a lesser extent, by the cytochrome P450 isoenzymes CYP2C19 (demethylated rabeprazole) and CYP3A4 (rabeprazole-sulfone). It is understood that variations in the function of this enzyme are genetically determined. CYP2C19 genotypes have been classified into three groups: rapid extensive metabolism (RM), intermediate metabolism (IM) and poor metabolism (PM). PPIs depend on CYP2C19 genotype status in pharmacokinetics or pharmacodynamics. Plasma PPI concentrations and intragastric pHs are lowest in the RM group during the PPI procedure, the IM group is nearest, and the PM group is highest of the three groups. The variations in CYP2C19 in PPI pharmacokinetics and pharmacodynamics, depending on the genotype, influence the cure levels of gastroesophageal reflux and therapy-based *H. pylori* infection. Dose and dosage schemes of PPIs should be modified depending on the genotype status of CYP2C19 for better PPI-dependent care [92]. The unusual catabolic mechanism suggests that rabeprazole is less sensitive to the effects of genetic polymorphism on CYP2C19, leading to limited pharmacokinetic and pharmacodynamic effects.

While the rabeprazole pharmacokinetic profile is altered in older people (C_{max} increased 60%, and AUC doubled after 7-day rabeprazole treatment by 20 mg), and in patients with mild to moderate hepatic dysfunction (C_{max} increased by 50% and AUC by 20 mg). Nonetheless, no dosage change is required in different populations, despite the evidence that these pharmacokinetic changes were not associated with clinically relevant laboratory parameters or severe adverse overnight effects in patients with renal defects (after a dosage of rabeprazole 20 mg the day after hemodialysis and the second dose after a period of 2 weeks throughout dialysis) [90].

Such results are not the case for antagonists of H2 receptor, where dose changes were reported if the patients have mild to seriously compromised renal function [93] but not associated with those of other PPIs [94].

8.3 Dosing

8.3.1 Dosing: adult

Duodenal ulcer: Tablets: Oral: 20 mg once daily for ≤4 weeks; additional therapy to achieve healing may be required for some patients.

Dyspepsia (off-label use): Oral: 20 mg once daily for up to 8 weeks [95,96].

Gastric ulcer (treatment) (off-label use): Oral: 20 mg once daily [97,98] for generally 8 weeks based on healing and/or ulcer complications [99,100].

8.3.1.1 Gastroesophageal reflux disease (GERD)

Tablets: Oral: Erosive or ulcerative GERD: Treatment: 20 mg once daily for 4–8 weeks; if inadequate response, may repeat up to an additional 8 weeks; maintenance: 20 mg once daily.

Symptomatic GERD (nonerosive): Treatment: 20 mg once daily for ≤4 weeks; if inadequate response, may repeat for an additional 4 weeks.

Helicobacter pylori eradication: Tablets: Oral:

Manufacturer labeling: 20 mg twice daily administered with amoxicillin 1000 mg and clarithromycin 500 mg twice daily for 7 days.

American College of Gastroenterology guidelines ACG [101,102].

Clarithromycin triple regimen: 20–40 mg twice daily in combination with clarithromycin 500 mg twice daily and either amoxicillin 1 g twice daily or metronidazole 500 mg three times per day; continue regimen for 14 days. Note: Avoid use of clarithromycin triple therapy in patients with risk factors for macrolide resistance (e.g., prior macrolide exposure, local clarithromycin resistance rates $\geq$15%, eradication rates with clarithromycin-based regimens $\leq$85%) [101,103].

Bismuth quadruple regimen: 20 mg twice daily in combination with tetracycline 500 mg four times per day, metronidazole 250 mg four times per day or 500 mg three or four times per day, and either bismuth subcitrate 120–300 mg four times per day or bismuth subsalicylate 300 mg four times per day; continue regimen for 10–14 days.

Concomitant regimen: 20 mg twice daily in combination with amoxicillin 1 g twice daily, clarithromycin 500 mg twice daily, and either metronidazole or tinidazole 500 mg twice daily; continue regimen for 10–14 days.

Sequential regimen: 20 mg twice daily plus amoxicillin 1 g twice daily for 5–7 days; then continue rabeprazole along with clarithromycin 500 mg twice daily, and either metronidazole or tinidazole 500 mg twice daily for 5–7 days.

Hybrid regimen: 20 mg twice daily plus amoxicillin 1 g twice daily for 7 days; then continue rabeprazole and amoxicillin along with clarithromycin 500 mg twice daily, and either metronidazole or tinidazole 500 mg twice daily for 7 days.

Levofloxacin triple regimen: 20 mg twice daily in combination with amoxicillin 1 g twice daily and levofloxacin 500 mg once daily; continue regimen for 10–14 days.

Hypersecretory conditions (including Zollinger-Ellison syndrome): Tablets: Oral: Initial: 60 mg once daily; adjust dose to patient needs (some may require divided doses). Doses as high as 100 mg once daily and 60 mg twice daily have been used; continue as long as clinically indicated.

NSAID-induced ulcer prophylaxis (off-label use): Oral: 5–20 mg once daily for 12–24 weeks [104,105].

Prevention of recurrent gastric or duodenal ulcer bleeding post-endoscopy (off-label use): Oral: 20 mg twice daily for 72 h [106].

8.3.2 Dosing: Pediatric

GERD, symptomatic:

Children ≤11 years: Oral: Sprinkle capsules:

<15 kg: 5 mg once daily for ≤12 weeks; if inadequate response may increase to 10 mg once daily.

≥15 kg: 10 mg once daily for ≤12 weeks.

Children ≥12 years and adolescents: Oral: Tablets: 20 mg once daily for up to 8 weeks [107].

8.4 ADME profile

ADME: Oral administration

Rabeprazole's bioavailability is approximately 52%, meaning that 52% of orally administered dose is expected to enter systemic circulation (the bloodstream). The bioavailability of rabeprazole is not affected by either food or ant acids [108].

8.4.1 Absorption

Rabeprazole sodium is an enteric-coated (gastro-resistant) tablet formulation of rabeprazole sodium. This is necessary because rabeprazole is acid-labile. Absorption of rabeprazole therefore begins only after the tablet leaves the stomach. Absorption is rapid, with peak plasma levels of rabeprazole occurring approximately 3.5 h after a 20 mg dose. Peak plasma concentrations (C_{max}) of rabeprazole and AUC are linear over the dose range of 10–40 mg. Absolute bioavailability of an oral 20 mg dose (compared to intravenous administration) is about 52% due in large part to pre-systemic metabolism. Additionally, the bioavailability does not appear to increase with repeat administration. In healthy subjects the plasma half-life is approximately 1 h (range 0.7–1.5 h), and the total body clearance is estimated to be 283 ± 98 mL/min. There was no clinically relevant interaction with food. Neither food nor the time of day of administration of the treatment affect the absorption of rabeprazole sodium [88].

8.4.2 Distribution

Rabeprazole is approximately 97% bound to human plasma proteins [88].

8.4.3 Metabolism and elimination

Rabeprazole sodium, as is the case with other members of the proton pump inhibitor (PPI) class of compounds, is metabolized through the cytochrome P450 (CYP450) hepatic drug metabolizing system. In vitro studies with human liver microsomes indicated that rabeprazole sodium is metabolized by isoenzymes of CYP450 (CYP2C19 and CYP3A4). In these studies, at expected human plasma concentrations rabeprazole neither induces nor inhibits CYP3A4; and although in vitro studies may not always be predictive of in vivo status these findings indicate that no interaction is expected between rabeprazole and cyclosporin. In humans the thioether (M1) and carboxylic acid (M6) are the main plasma metabolites with the sulfone (M2), desmethyl-thioether (M4) and mercapturic acid conjugate (M5) minor metabolites observed at lower levels. Only the desmethyl metabolite (M3) has a small amount of anti-secretory activity, but it is not present in plasma. Following a single 20 mg 14C labeled oral dose of rabeprazole sodium, no unchanged drug was excreted in the urine. Approximately 90% of the dose was eliminated in urine mainly as the two Metabolites: a mercapturic acid conjugate (M5) and a carboxylic acid (M6), plus two unknown metabolites. The remainder of the dose was recovered in faces.

Gender: Adjusted for body mass and height, there are no significant gender differences in pharmacokinetic parameters following a single 20 mg dose of rabeprazole.

Renal dysfunction: In patients with stable, end-stage, renal failure requiring maintenance hemodialysis (creatinine clearance $\leq$5 mL/min/ 1.73 m^2), the disposition of rabeprazole was very similar to that in healthy volunteers. The AUC and the C_{max} in these patients was about 35% lower than the corresponding parameters in healthy volunteers. The mean half-life of rabeprazole was 0.82 h in healthy volunteers, 0.95 h in patients during hemodialysis and 3.6 h post dialysis. The clearance of the drug in patients with renal disease requiring maintenance hemodialysis was approximately twice that in healthy volunteers.

Hepatic dysfunction: Following a single 20 mg dose of rabeprazole to patients with chronic mild to moderate hepatic impairment the AUC doubled and there was a 2–3-fold increase in half-life of rabeprazole compared to the healthy volunteers. However, following a 20 mg dose daily for 7 days the AUC had increased to only 1.5-fold and the C_{max} to only 1.2-fold. The half-life of rabeprazole in patients with hepatic impairment was 12.3 h compared to 2.1 h in healthy volunteers. The pharmacodynamic response (gastric pH control) in the two groups was clinically comparable [88].

8.4.3.1 Elderly

Elimination of rabeprazole was somewhat decreased in the elderly. Following 7 days of daily dosing with 20 mg of rabeprazole sodium, the

AUC approximately doubled, the C_{max} increased by 60% and t½ increased by approximately 30% as compared to young healthy volunteers. However, there was no evidence of rabeprazole accumulation.

CYP2C19 Polymorphism: Following a 20 mg daily dose of rabeprazole for 7 days, CYP2C19 slow metabolizers, had AUC and t½ which were approximately 1.9 and 1.6 times the corresponding parameters in extensive metabolizers while C_{max} had increased by only 40% [88].

8.4.4 Preclinical safety data

The most common metabolites excreted in the urine are the mercapturic acid conjugate and carboxylic acid [109].

Only at exposures which exceeded the maximum human exposure was not observed enough to cause human safety concerns in relation to animal data were non-clinical effects noted. Studies on mutagenicity gave equivocal results. Tests in mouse lymphoma cell line were positive, but in vivo micronucleus and in vivo and in vitro DNA repair tests were negative. Carcinogenicity studies revealed no special hazard for humans. Mutagenicity tests have provided equivocal data. Tests were successful on the cell line of mouse lymphoma, but the in vitro and in situ DNA repair were poor. Trials on mouse lymphoma cells, there was no special threat for humans found in carcinogenicity tests [88].

References

[1] The Metabolomics Innovation Centre (TMIC) [Internet]. Prednisone Information from hmdb.ca/metabolites; [Last Update Date 2020-02-26 21:25:41 UTC, Cited: 22/04/2019]. Available from: https://hmdb.ca/metabolites/HMDB0005026#links.

[2] S. Budavari, The Merck Index, Merck and co. Inc., Whitehouse Station, New Jersey, 2001, 3476

[3] M. Rockville, United States pharmacopeia and national formulary (USP 41-NF 36), 20852 United States Pharmacopeial Convention, 12601 Twinbrook Parkway, Rockville, MD, 2018, 3562.

[4] P.C. Ray, V. Mittapelli, A. Rohatgi, O.D. Tyagi, Efficient synthesis of N-oxide derivatives: substituted 2-(2-(Pyridyl-N-oxide) methylsulphinyl) benzimidazoles, Synth. Commun. 37 (17) (2007) 2861–2868.

[5] ChemSpider Search and share chemistry [Internet]. Prednisone Information from http://www.chemspider.com/; [Version 2020.0.18.0, Cited: 22/04/2019]. Available from: http://www.chemspider.com/Chemical-Structure.4853.html?rid=efee7a9d-a053-423e-bc0d-acf83d5e5dec.

[6] S. Budavari, Raberpazole, The Merck Index, Merck and Co. Inc.: Whitehouse Station, New Jersey, 2001, 3476.

[7] S. Souda, N. Ueda, S. Miyazawa, K. Tagami, S. Nomoto, M. Okita, N. Shimomura, T. Kaneko, M. Fujimoto, M. Murakami, Pyridine derivatives having anti-ulcerative activity, 1991, US5045552A.

[8] Vardanyan, R. and V. Hruby, Chapter 37 - Proton Pump Inhibitors, in Synthesis of Best-Seller Drugs, R. Vardanyan and V. Hruby, Editors. 2016, Academic Press: Boston. p. 765–781.
[9] K. Tagami, S. Chiku, S. Sohda, Synthesis of 14C-labelled sodium pariprazole (E3810), J. Label. Compd. Radiopharm. 33 (9) (1993) 849–852.
[10] https://en.wikipedia.org/wiki/Rabeprazole. 7/06/2020; from.
[11] S. Hilal, Quantitative treatment of solute/solvent interactions, Theoret. Comput. Chem. 1 (1994) 291–353.
[12] Models, EPA Predictive, Tools for Assessing Chemicals under the Toxic Substances Control Act (TSCA), 4/02/2005. http://www.epa.gov/oppt/exposure/pubs/episuitedl.htm, 2005.
[13] A.C. Moffat, M.D. Osselton, B. Widdop, J. Watts, Clarke's analysis of drugs and poisons, Vol. 3, Pharmaceutical press London, London, UK, 2011.
[14] National Center for Biotechnology Information (2020). PubChem Compound Summary for CID 14720269, Rabeprazole sodium. Retrieved July 29, 2020 from https://pubchem.ncbi.nlm.nih.gov/compound/Rabeprazole-sodium.
[15] Z. Ma, L. Zhang, L. Lin, P. Ji, X. Guo, Enantioseparation of rabeprazole and omeprazole by nonaqueous capillary electrophoresis with an ephedrine-based ionic liquid as the chiral selector, Biomed. Chromatogr. 24 (12) (2010) 1332–1337.
[16] L.A. Papp, G. Hancu, Á. Gyéresi, H. Kelemen, Z.I. Szabó, B. Noszál, P. Dubský, G. Tóth, Chiral separation of lansoprazole and rabeprazole by capillary electrophoresis using dual cyclodextrin systems, Electrophoresis 40 (21) (2019) 2799–2805.
[17] P. Raval, M. Puranik, S. Wadher, P. Yeole, A validated HPTLC method for determination of ondansetron in combination with omeprazole or rabeprazole in solid dosage form, Indian J. Pharm. Sci. 70 (3) (2008) 386.
[18] B.H. Patel, B.N. Suhagia, M.M. Patel, J.R. Patel, High-performance liquid chromatography and thin-layer chromatography for the simultaneous quantitation of rabeprazole and mosapride in pharmaceutical products, J. Chromatogr. Sci. 46 (1) (2008) 10–14.
[19] A. Osman, M. Osman, Spectrofluorometry, thin layer chromatography, and column high-performance liquid chromatography determination of rabeprazole sodium in the presence of its acidic and oxidized degradation products, J. AOAC Int. 92 (5) (2009) 1373–1381.
[20] B.H. Patel, B.N. Suhagia, M.M. Patel, J.R. Patel, HPTLC determination of rabeprazole and domperidone in capsules and its validation, J. Chromatogr. Sci. 46 (4) (2008) 304–307.
[21] T. Hishinuma, K. Suzuki, H. Yamaguchi, H. Yamagishi, T. Koike, S. Ohara, T. Shimosegawa, N. Mano, J. Goto, Simple quantification of lansoprazole and rabeprazole concentrations in human serum by liquid chromatography/tandem mass spectrometry, J. Chromatogr. B 870 (1) (2008) 38–45.
[22] United States Pharmacopeial Convention, United States Pharmacopeia and National Formulary (USP 41-NF 36), Rockville, MD., Rabeprazole/Official Monographs 2 (2018) 3562–3563.
[23] The British Pharmacopoeia, The British Pharmacopoeia Commission Secretariat of the Medicines and Healthcare Products Regulatory Agency (MHRA), TSO (The Stationery Office), London, United Kingdom, 2019. 1 January 2019; Available from https://www.tsoshop.co.uk/Medical/MHRA/British-Pharmacopoeia.
[24] A. Radi, N.A. El-Ghany, T. Wahdan, Voltammetric behaviour of rabeprazole at a glassy carbon electrode and its determination in tablet dosage form, Il Farmaco 59 (7) (2004) 515–518.
[25] P.Y. Khashaba, H.R.H. Ali, M.M. El-Wekil, Highly sensitive and selective complexation based voltammetric methods for the analysis of rabeprazole sodium in real samples, RSC Adv. 7 (6) (2017) 3043–3050.

[26] L.-M. Guo, S.-H. Zhong, W. Li, Y.-J. Huang, Determination of rabeprazole at a carbon paste electrode by anodic adsorptive voltammetry, Journal of Pharmaceutical Practice 1 (2008).
[27] P. Gärtner, K. Zschimmer, H. Knoth, Differential pulse polarographic investigation of lansoprazole and rabeprazole using dropping mercury electrode, Die Pharmazie-An International Journal of Pharmaceutical Sciences 64 (6) (2009) 371–375.
[28] A. El-Gindy, F. El-Yazby, M.M. Maher, Spectrophotometric and chromatographic determination of rabeprazole in presence of its degradation products, J. Pharm. Biomed. Anal. 31 (2) (2003) 229–242.
[29] P. Patel, H. Desai, R. Patel, N. Patel, Spectrophotometric method for estimation of rabeprazole, Indian J. Pharm. Sci. 69 (2) (2007) 318–320.
[30] C. Garcia, J. Sippel, M. Steppe, E. Schapoval, Development and validation of derivative spectrophotometric method for determination of rabeprazole sodium in pharmaceutical formulation, Anal. Lett. 39 (2) (2006) 341–348.
[31] S.S. Sabnis, N.D. Dhavale, V.Y. Jadhav, S.V. Gandhi, Spectrophotometric simultaneous determination of rabeprazole sodium and itopride hydrochloride in capsule dosage form, Spectrochim. Acta A Mol. Biomol. Spectrosc. 69 (3) (2008) 849–852.
[32] S. Gul, S. Basheer, UV visible spectrophotometric determination, validation and Invitro interaction of Rabeprazole with Chlorazepate, Saudi J. Med. Pharm. Sci 3 (8) (2017) 813–817.
[33] C.V. Garcia, A.L. Mendez, M. Steppe, E.E. Schapoval, Comparison between UV spectrophotometric and capillary electrophoresis methods for determination of rabeprazole sodium in pharmaceutical formulations, Lat. Am. J. Pharm. 29 (2010) 144–147.
[34] G. Gowri, S. Shaik, D.S.S. Prasanna, New visible spectrophotometric methods for estimation of Rabeprazole in bulk and pharmaceutical formulation, World J. Pharm. Res. 8 (1) (2018) 1057–1066.
[35] L.P. Dsouza, M. Koland, Simultaneous estimation of famotidine and rabeprazole sodium: method development and validation by UV-spectrophotometric method, Int. J. Pharm. Pharm. Sci. 7 (7) (2015) 257–260.
[36] M.H. Abdel-Hay, S.M. Sabry, T.S. Belal, A.A. Mahgoub, Spectrophotometric determination of rabeprazole sodium using two charge transfer complexation reactions, J. Appl. Pharm. Sci. 3 (11) (2013) 128.
[37] P.T. Kanekar, S.S. Kamat, P. Das, E.V.S. Subrahmanyam, A.R. Shabaraya, Development of new methods and its validation for determination of rabeprazole in bulk and markted formulations, Int. J. Pharm. Chem. Biol. Sci. 3 (2) (2013) 411–417.
[38] S. Shaik, T. Hemant Kumar, Y. Srinivasa Rao, K. Vara Prasad Rao, New visible spectrophotometric methods for estimation of Rabeprazole in bulk and pharmaceutical formulation, Der Pharm. Chem. 10 (11) (2018) 34–39.
[39] N. Rahman, Z. Bano, S.N.H. Azmi, Quantitative analysis of rabeprazole sodium in commercial dosage forms by spectrophotometry, Chem. Pharm. Bull. 56 (7) (2008) 995–1001.
[40] A. Yousefzadeh, J. Abolhasani, J. Hassanzadeh, M.H. Somi, A highly efficient chemiluminescence system based on an enhancing effect of Ag nanoclusters/graphene quantum dots mixture for ultrasensitive detection of Rabeprazole, Anal. Sci. 35 (4) (2019) 385–391.
[41] C.V. Garcia, J. Sippel, L.L. Sfair, S.S. Garcia, A. Jablonski, M. Steppe, E.E. Schapoval, Validation of a capillary electrophoresis method for analysis of rabeprazole sodium in a pharmaceutical dosage form, J. AOAC Int. 88 (4) (2005) 1081–1085.
[42] S.V. Gandhi, S.I. Khan, R.T. Jadhav, S.S. Jadhav, G.A. Jadhav, High-performance thin-layer chromatographic determination of rabeprazole sodium and domperidone in combined dosage form, J. AOAC Int. 92 (4) (2009) 1064–1067.

[43] E.F. Elkady, M.A. Fouad, B.M. Jaadan, LC–MS/MS bioassay of four proton pump inhibitors, J. Chromatogr. B 1076 (2018) 61–69.
[44] M. Kim, S.K. Yu, Q.-K. Truong, X.-L. Mai, H.K. Chung, J.-S. Kang, K.H. Kim, Determination of rabeprazole enantiomers in commercial tablets using immobilized cellulose-based stationary phase, Arch. Pharm. Res. 40 (3) (2017) 373–381.
[45] M.S. Tahir, A. Adnan, Q. Syed, Chromatographic resolution of drug analogues: four proton pump inhibitors and two Gastroprokinetic agents, Lat. Am. J. Pharm. 36 (2) (2017) 293–300.
[46] M.K.O. Reddy, C. Bodepudi, P. Shanmugasundaram, Method development and validation of Rabeprazole in bulk and tablet dosage form by RP-HPLC method, Int. J. ChemTechRes 3 (2011) 1580–1588.
[47] N. Ramakrishna, K. Vishwottam, S. Wishu, M. Koteshwara, S.S. Kumar, High-performance liquid chromatography method for the quantification of rabeprazole in human plasma using solid-phase extraction, J. Chromatogr. B 816 (1–2) (2005) 209–214.
[48] Y. Zhang, X. Chen, Q. Gu, D. Zhong, Quantification of rabeprazole in human plasma by liquid chromatography–tandem mass spectrometry, Anal. Chim. Acta 523 (2) (2004) 171–175.
[49] S. Elumalai, K. Aher, G. Bhavar, S. Gupta, Development and validation of RP-HPLC method for determination of content uniformity of rabeprazole sodium in its tablets dosage form, J. Appl. Pharm. Sci. 1 (6) (2011) 165.
[50] T.G. Barot, V. Prajapati, P. Pate, D. Bhoi, L. Patel, A method development and validation of Rabeprazole sodium (API) by high-performance liquid chromatography, Int. J. ChemTech Res. 4 (2009) 1345–1353.
[51] R. Kalirajan, K. Anandarajagopal, S.M. Mathew, B. Gowramma, S. Jubie, B. Suresh, Simultaneous determination of Rabeprazole and Domperidone in dosage forms by RP-HPLC, Rasayan J. Chem 1 (2) (2008) 232–235.
[52] B.H. Patel, M.M. Patel, J.R. Patel, B.N. Suhagia, HPLC analysis for simultaneous determination of rabeprazole and domperidone in pharmaceutical formulation, J. Liq. Chromatogr. Relat. Technol. 30 (3) (2007) 439–445.
[53] P.R. Battu, M. Reddy, Development and validation of RP-HPLC for the rabeprazole sodium in pharmaceutical formulations and human plasma, Asian J. Res. Chem. 2 (1) (2009) 49–51.
[54] R. Sharma, G.P. Mishra, S.C. Chaturvedi, Development and validation of RP-HPLC method for the simultaneous determination of rabeprazole sodium and itopride hydrochloride in solid dosage form, J. Chem. 7 (3) (2010) 947–952.
[55] A.L. Rao, B. Kumar, G.G. Sankar, Development of RP-HPLC method for the estimation of rabeprazole in pure and tablet dosage form, J. Chem. 5 (S2) (2008) 1149–1153.
[56] A. Heda, D. Gadade, J. Kathiriya, P. Puranik, HPLC method development and validation for simultaneous analysis of diclofenac sodium and rabeprazole sodium, J. Chem. 7 (S1) (2010) S386–S390.
[57] S.L. Khokra, B. Choudhary, H. Mehta, RP-HPLC analysis for the simultaneous estimation of rabeprazole sodium and aceclofenac in a combined dosage form, Int. Curr. Pharm. J. 1 (12) (2012) 410–413.
[58] V. Kulkarni, P. Mahulikar, Estimation of Rabeprazole Sodium in Tablet Dosage Form by Rapid Isocratic Reversed Phase High Performance Liquid Chromatography Using Volatile Buffer Additives, CSIR, 2006.
[59] S.S. Perumal, S.P. Ekambaram, S. Raja, Analytical method development and validation of simultaneous estimation of rabeprazole, pantoprazole, and itopride by reverse-phase high-performance liquid chromatography, J. Food Drug Anal. 22 (4) (2014) 520–526.

[60] K. Karunakaran, G. Navaneethan, K. Elango, Development of a new RP-UPLC method for the determination of rabeprazole sodium in pharmaceutical formulation and application in dissolution studies, Trop. J. Pharm. Res. 10 (5) (2011) 655–661.
[61] S.M. Sharna, M.H. Shuvo, S. Ferdaus, S. Chowdhury, Development and validation of RP-HPLC method for analysis of Rabeprazole sodium, Dhaka Univ. J. Pharm. Sci. 18 (1) (2019) 125–128.
[62] U. Khan, S.V. Luhar, S.B. Narkhede, RP-HPLC method development and validation for simultaneous estimation of clidinium bromide, chlordiazepoxide, dicyclomine HCl and rabeprazole sodium in pharmaceutical dosage form, Eur. J. Biomed. Pharm. Sci. 5 (5) (2018) 628–634.
[63] M.F. Khan, S.S. Zuthi, M.S. Kayser, M.S. Islam, S. Asad, M.A. Rashid, A simple RP-HPLC method development and validation for the simultaneous estimation of naproxen and Rabeprazole, J. Appl. Pharm. Sci. 6 (11) (2016) 147–152.
[64] S.A. Kumar, M. Debnath, V. Padala, An isocratic method development and validation for simultaneous estimation of Rabeprazole and Mosapride in tablet dosage forms by using RP-HPLC, PharmaTutor 4 (7) (2016) 41–51.
[65] T.N. Rao, A new analytical method development and validation for related substances of rabeprazole in active pharma ingredient by HPLC-PDA detector, Int. J. ChemTech Res. 10 (5) (2017) 241–251.
[66] V.A. Ranjani, J.P. Kumar, K.S.B. Kumar, A new method development and validation of rabeprazole sodium in formulation by using HPLC, Int. J. Pharm. Pharm. Res. 6 (4) (2016) 241–254.
[67] C.-W. Park, Y.-S. Rhee, B.-W. Go, S.-H. Kam, K.-H. Lee, H.-S. Lee, E.-S. Park, High performance liquid chromatographic analysis of rabeprazole in human plasma and its pharmacokinetic application, Arch. Pharm. Res. 31 (9) (2008) 1195–1199.
[68] V. Asha Ranjani, Analytical study of Rabeprazole sodium in formulation by using HPLC, EC Pharm. Sci. 1 (2015) 86–94.
[69] C. Su, H. Yang, X. Meng, J.P. Fawcett, J. Cao, Y. Yang, J. Gu, Determination of rabeprazole enantiomers in dog plasma by supercritical fluid chromatography tandem mass spectrometry and its application to a pharmacokinetic study, J. Sep. Sci. 40 (4) (2017) 1010–1016.
[70] M. Sanad, S. Challan, Radioiodination and biological evaluation of rabeprazole as a peptic ulcer localization radiotracer, Radiochemistry 59 (3) (2017) 307–312.
[71] R. Buchireddy, K. Mukkanti, P. Srinivasulu, K.S. Srinivas, A validated, stability-indicating, LC method for rabeprazole sodium, Chromatographia 68 (3–4) (2008) 275–280.
[72] V. Kuchana, B. Shamili, C.N. Kumar, Development of spectrum index based stability indicating analytical method studies for determination of various proton pump inhibitors in its bulk and multicomponent pharmaceutical dosage forms, J. Sci. Res. Pharm. 6 (12) (2017) 157–168.
[73] M.M. Bhandi, R.M. Borkar, G. Shankar, S. Raut, N. Nagesh, R. Srinivas, Identification and characterization of stressed degradation products of rabeprazole using LC-ESI/MS/MS and 1 H-NMR experiments: in vitro toxicity evaluation of major degradation products, RSC Adv. 6 (13) (2016) 10719–10735.
[74] S. Ren, T.T.-D. Tran, P.H.-L. Tran, Y.-S. Rhee, B.-J. Lee, Investigation of degradation mechanism of Rabeprazole with solid state pharmaceutical excipients, J. Pharm. Investig. 40 (6) (2010) 367–372.
[75] S. Ren, M.-J. Park, H. Sah, B.-J. Lee, Effect of pharmaceutical excipients on aqueous stability of rabeprazole sodium, Int. J. Pharm. 350 (1–2) (2008) 197–204.
[76] displayed similarities in reaction and different variations b Drugs.com [Internet]. Prednisone Information from Drugs.com; c1996-2018 [Updated: 13 February 2018, Cited: 22/04/2019]. Available from: https://www.drugs.com/pro/rabeprazole.html.

[77] I.J. Choi, M.-C. Kook, Y.-I. Kim, S.-J. Cho, J.Y. Lee, C.G. Kim, et al., Helicobacter pylori Therapy for the Prevention of Metachronous Gastric Cancer, N. Engl. J. Med. 378 (2018) 1085–1095. https://www.nejm.org/doi/10.1056/NEJMoa1708423, 22/04/2019.
[78] Alia Dadabhai, and Frank K Friedenberg. Rabeprazole: a pharmacologic and clinical review for acid-related disorders. Expert Opin. Drug Saf. 8, 2009. 119-126. https://www.tandfonline.com/doi/abs/10.1517/14740330802622892?journalCode=ieds20, 22/04/2019.
[79] W. Kromer, U. Krüger, R. Huber, M. Hartmann, V. Steinijans, Differences in pH-dependent activation rates of substituted benzimidazoles and biological in vitro correlates, Pharmacology 56 (2) (1998) 57–70.
[80] M. Besancon, A. Simon, G. Sachs, J.M. Shin, Sites of reaction of the gastric H, K-ATPase with extracytoplasmic thiol reagents, J. Biol. Chem. 272 (36) (1997) 22438–22446.
[81] E. Lew, R. Barbuti, T. Kovacs, B. Sytnic, T. Humphries, J. Walsh, An ascending single-dose safety and tolerance study of an oral formulation of rabeprazole (E3810), Aliment. Pharmacol. Ther. 12 (7) (1998) 667–672.
[82] G. Ohning, J. Walsh, J. Pisegna, A. Murthy, J. Barth, T. Kovacs, Rabeprazole is superior to omeprazole for the inhibition of peptone meal-stimulated gastric acid secretion in helicobacter pylori-negative subjects, Aliment. Pharmacol. Ther. 17 (9) (2003) 1109–1114.
[83] N. Bell, D. Burget, C. Howden, J. Wilkinson, R. Hunt, Appropriate acid suppression for the management of gastro-oesophageal reflux disease, Digestion 51 (Suppl. 1) (1992) 59–67.
[84] D.W. Burget, S.G. Chiverton, R.H. Hunt, Is there an optimal degree of acid suppression for healing of duodenal ulcers?: a model of the relationship between ulcer healing and acid suppression, Gastroenterology 99 (2) (1990) 345–351.
[85] M. Williams, J. Sercombe, M. Hamilton, R. Pounder, A placebo-controlled trial to assess the effects of 8 days of dosing with rabeprazole versus omeprazole on 24-h intragastric acidity and plasma gastrin concentrations in young healthy male subjects, Aliment. Pharmacol. Ther. 12 (11) (1998) 1079–1089.
[86] D. Pantoflickova, G. Dorta, M. Ravic, P. Jornod, A. Blum, Acid inhibition on the first day of dosing: comparison of four proton pump inhibitors, Aliment. Pharmacol. Ther. 17 (12) (2003) 1507–1514.
[87] K.G. Tolman, J. Täubel, S. Warrington, Y.-L. Chiu, B.L. Pilmer, W.-J. Pan, Comparison of the effects of single and repeated oral doses of lansoprazole and rabeprazole on ambulatory 24-hour intragastric pH in healthy volunteers, Clin. Drug Investig. 26 (1) (2006) 21–28.
[88] Prednisone Information from medicines.org.uk; c1996-2018 [Last updated on emc: 16 Apr 2020, Cited: 22/04/2019]. Available from: https://www.medicines.org.uk/emc/product/2842/smpc. [Internet].
[89] S.K. Swan, A.M. Hoyumpa, G.J. Merritt, Review article: the pharmacokinetics of rabeprazole in health and diseases, Aliment. Pharmacol. Ther. 13 (1999) 11–17.
[90] U. Fuhr, A. Jetter, Rabeprazole: pharmacokinetics and pharmacokinetic drug interactions, Pharmazie 57 (9) (2002) 595–601.
[91] B. Thjodleifsson, Review of rabeprazole in the treatment of gastro-oesophageal reflux disease, Expert Opin. Pharmacother. 5 (1) (2004) 137–149.
[92] T. Furuta, N. Shirai, M. Sugimoto, A. Nakamura, A. Hishida, T. Ishizaki, Influence of CYP2C19 pharmacogenetic polymorphism on proton pump inhibitor-based therapies, Drug Metab. Pharmacokinet. 20 (3) (2005) 153–167.
[93] M. Feldman, M.E. Burton, Histamine2-receptor antagonists: standard therapy for acid-peptic diseases, N. Eng. J. Med. 323 (24) (1990) 1672–1680.

[94] F. Pace, S. Pallotta, S. Casalini, G.B. Porro, A review of rabeprazole in the treatment of acid-related diseases, Ther. Clin. Risk Manag. 3 (3) (2007) 363.
[95] B. Dewan, N. Philipose, Lafutidine 10 mg versus Rabeprazole 20 mg in the treatment of patients with heartburn-dominant Uninvestigated dyspepsia: a randomized, Multicentric Trial, Gastroenterol. Res. Pract. 640685 (10) (2011) 5.
[96] M.I. Pinto-Sanchez, Y. Yuan, A. Hassan, P. Bercik, P. Moayyedi, Proton pump inhibitors for functional dyspepsia, Cochrane Database Syst. Rev. (2017) 21(11).
[97] M.L. Cloud, N. Enas, T.J. Humphries, S. Bassion, Rabeprazole in treatment of acid peptic diseases: results of three placebo-controlled dose-response clinical trials in duodenal ulcer, gastric ulcer, and gastroesophageal reflux disease (GERD). The Rabeprazole study group, Dig. Dis. Sci. 43 (5) (1998) 993–1000.
[98] C.P. Dekkers, J.A. Beker, B. Thjodleifsson, A. Gabryelewicz, N.E. Bell, T.J. Humphries, Comparison of rabeprazole 20 mg vs. omeprazole 20 mg in the treatment of active gastric ulcer–a European multicentre study. The European Rabeprazole study group, Aliment. Pharmacol. Ther. 12 (8) (1998) 789–795.
[99] V. Boparai, J. Rajagopalan, G. Triadafilopoulos, Guide to the use of proton pump inhibitors in adult patients, Drugs 68 (7) (2008) 925–947.
[100] M.M. Wolfe, G. Sachs, Acid suppression: optimizing therapy for gastroduodenal ulcer healing, gastroesophageal reflux disease, and stress-related erosive syndrome, Gastroenterology 118 (2 Suppl 1) (2000) S9–31.
[101] W.D. Chey, G.I. Leontiadis, C.W. Howden, S.F. Moss, ACG clinical guideline: treatment of helicobacter pylori infection, Am. J. Gastroenterol. 112 (2) (2017) 212.
[102] W.D. Chey, B.C. Wong, American College of Gastroenterology guideline on the management of helicobacter pylori infection, Am. J. Gastroenterol. 102 (8) (2007) 1808.
[103] C.A. Fallone, N. Chiba, S.V. van Zanten, L. Fischbach, J.P. Gisbert, R.H. Hunt, N.L. Jones, C. Render, G.I. Leontiadis, P. Moayyedi, The Toronto consensus for the treatment of Helicobacter pylori infection in adults, Gastroenterology 151 (1) (2016) 51–69. e14.
[104] R. Iwakiri, K. Higuchi, M. Kato, M. Fujishiro, Y. Kinoshita, T. Watanabe, T. Takeuchi, M. Yamauchi, M. Sanomura, H. Nakagawa, Randomised clinical trial: prevention of recurrence of peptic ulcers by rabeprazole in patients taking low-dose aspirin, Aliment. Pharmacol. Ther. 40 (7) (2014) 780–795.
[105] T. Sanuki, T. Fujita, H. Kutsumi, T. Hayakumo, S.-i. Yoshida, H. Inokuchi, M. Murakami, Y. Matsubara, H. Kuwayama, T. Kawai, Rabeprazole reduces the recurrence risk of peptic ulcers associated with low-dose aspirin in patients with cardiovascular or cerebrovascular disease: a prospective randomized active-controlled trial, J. Gastroenterol. 47 (11) (2012) 1186–1197.
[106] H.-K. Kim, J.-S. Kim, T.-H. Kim, C.-W. Kim, Y.-S. Cho, S.-S. Kim, H.-S. Chae, S.-W. Han, Y.-W. Park, H.-S. Son, Effect of high-dose oral rabeprazole on recurrent bleeding after endoscopic treatment of bleeding peptic ulcers, Gastroenterol. Res. Pract. 2012 (2012) 317125.
[107] Wolfe, M.M. Proton pump inhibitors: overview of use and adverse effects in the treatment of acid related disorders. UpToDate. Waltham, MA: UpToDate (2018); from: https://www.uptodate.com/contents/rabeprazole-drug-information?sectionName=Adult&topicId=9519&search=rabeprazole&usage_type=panel&anchor=F216713&source=panel_search_result&selectedTitle=1~53&kp_tab=drug_general&display_rank=1#F6189304, 22/04/2019.
[108] S. Yasuda, S. Higashi, M. Murakami, Y. Tomono, M. Kawaguchi, Antacids have no influence on the pharmacokinetics of rabeprazole, a new proton pump inhibitor, in healthy volunteers, Int. J. Clin. Pharmacol. Ther. 37 (5) (1999) 249–253.
[109] T. Setoyama, A. Laurent, T. Humphries, J. Hasegawa, Pharmacokinetics of rabeprazole following single intravenous and oral administration to healthy subjects, Int. J. Clin. Pharmacol. Ther. 43 (1) (2005) 37–42.

CHAPTER FIVE

Irbesartan (a comprehensive profile)

Ibrahim A. Darwish[a,b,*], Hany W. Darwish[a,c], Ahmed H. Bakheit[a,d], Hamad M. Al-Kahtani[a], and Zahi Alanazi[a]

[a]Department of Pharmaceutical Chemistry, College of Pharmacy, King Saud University, Riyadh, Kingdom of Saudi Arabia
[b]Department of Pharmaceutical Analytical Chemistry, Faculty of Pharmacy, Assiut University, Assiut, Egypt
[c]Department of Analytical Chemistry, Faculty of Pharmacy, Cairo University, Cairo, Egypt
[d]Department of Chemistry, Faculty of Science and Technology, Al-Neelain University, Khartoum, Sudan
*Corresponding author: e-mail address: idarwish@ksu.edu.sa

Contents

Profiles of Drug Substances, Excipients, and Related Methodology, Volume 46
ISSN 1871-5125
https://doi.org/10.1016/bs.podrm.2020.07.004

1. Description

1.1 Nomenclature

1.1.1 Chemical names

1.1.1.1 IUPAC name

2-Butyl-3-[[4-[2-(2*H*-tetrazol-5-yl)phenyl]phenyl]methyl]-1,3-diazaspiro[4.4]non-1-en-4-one.

1.1.1.2 Synonym names

2-Butyl-3-(*p*-(o-1H-tetrazol-5-ylphenyl)benzyl)-1,3-diazaspiro[4.4]non-1-en-4-one.

2-Butyl-3-[[2′-(2H-tetrazol-5-yl)[1,1′-biphenyl]-4-yl]methyl]-1,3-diazaspiro[4.4]non-1-en-4-one.

2-Butyl-3-{[2′-(1H-tetrazol-5-yl)-4-biphenylyl]methyl}-1,3-diazaspiro[4.4]non-1-en-4-on [German] [ACD/IUPAC Name].
2-Butyl-3-{[2′-(1H-tétrazol-5-yl)-4-biphénylyl]méthyl}-1,3-diazaspiro[4.4]non-1-én-4-one [French] [ACD/IUPAC Name] [1–3].

1.1.2 Generic names

Irbesartaani, Irbésartan, Irbesartán, Irbesartanum, Irbesartan D4, Irbesartan-D4, Irbesartan-d7, Irbesartin, BMS-186295; SR-47436, DB01029 [1–4].

1.1.3 Proprietary names

Aprovel, Avapro, Karvea, Avalide [1–6].

1.2 Chemical structure

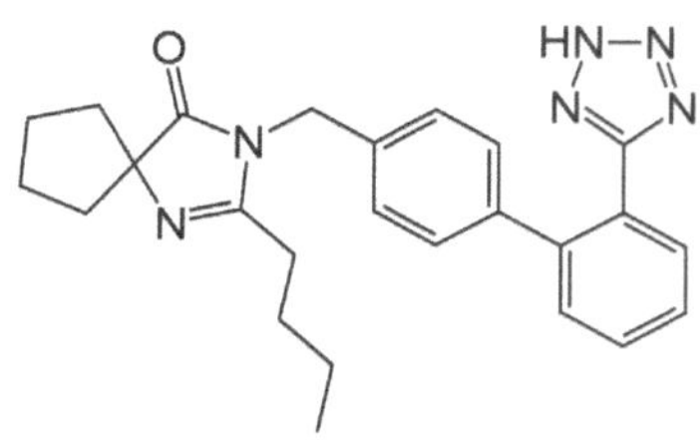

1.3 Chemical formula and molecular weight

$C_{25}H_{28}N_6O$ 428.5

1.4 Elemental analysis

Carbon: 70.07%, Hydrogen: 6.59%, Nitrogen: 19.61%, Oxygen: 3.73% [5].

1.5 Registry numbers

Chemical Abstracts Service (CAS): 138402-11-6.
European Community (EC) Number: 604-078-2.
Unique Ingredient Identifier (UNII): J0E2756Z7N.

1.6 International chemical identifier key

International Chemical Identifier (InChI) key computed from the chemical structure by InChI 1.0.5 (PubChem release 2019.06.18): YOSHYTLCDANDAN-UHFFFAOYSA-N.

1.7 Canonical SMILES

Simplified Molecular-Input Line-Entry System (SMILES) computed from the chemical structure by OEChem 2.1.5 (PubChem release 2019.06.18): CCCCC1=NC2(CCCC2)C(=O)N1CC3=CC=C(C=C3)C4=CC=CC=C4C5=NNN=N5.

2. Methods of synthesis

Many methods were reported for synthesis of irbesartan [7–17] and the relevant methods are described as the following:

2.1 Method 1

The first reported synthetic method for the preparation of irbesartan 1 is shown in Scheme 1. It involved the condensation of 4′-bromomethyl-biphenyl-2-carbonitrile 2 with 2-butyl-1,3-diaza-spiro[4,4]non-1-en-4-one hydrochloride 3 using sodium hydride (NaH) as a base and dimethyl formamide (DMF) as solvent medium. The second step involved the tetrazole formation by treating tributyltinazide in xylene followed by trityl protection to obtain trityl irbesartan. The trityl group was cleaved by treating with aqueous hydrochloride in methanol and tetrahydrofuran mixture [7,8].

This method suffers from the disadvantages such as (a): NaH is used as base in the first step of the method for which utmost care should be taken to perform the reaction under inert atmosphere as it is an exothermic

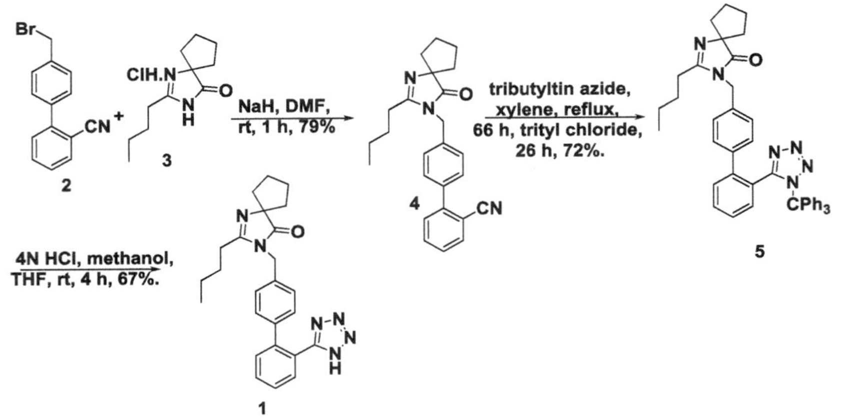

Scheme 1 Preparation of irbesartan by condensation of 4′-bromomethyl-biphenyl-2-carbonitrile with 2-butyl-1,3-diaza-spiro[4,4]non-1-en-4-one hydrochloride, formation of trityl tetrazole derivative, followed by cleavage of trityl group.

reaction, highly sensitive toward moisture, it is unsafe and not suitable for production on commercial scale; (b): the purification of the compound 4 and compound 5 was done through column chromatography which is not commercially viable; (c): tri-*n*-butyltinazide was used as reagent for the tetrazole formation in the step 2. Toxicity of tin in humans is most frequently reported as the exposure to tin compounds will cause memory loss and insomnia as well as other symptoms like death. Neurotoxicity and liver damage has been reported with tin compounds exposure in humans; (d): the temperature for the tetrazole formation step is high and the time cycle is more, i.e., reflux in xylene (~150 °C) for 66 h which is not an economic and ecofriendly process; (e): the tetrazole group was protected by tritylation using trityl chloride. As the deprotection of the trityl group in the final step will give trityl alcohol as byproduct which is the potential impurity in this process and it is not removed easily, and (f): the over yield of the process is 38% only.

2.2 Method 2

Rao and Babu [9] described an efficient inexpensive method for large-scale preparation of irbesartan 1, as shown in Scheme 2. This five-stage process involved preparation of the key intermediate, 1-benzyl-5-(4′-bromomethyl-biphenyl-2-yl)-1H-tetrazole 5 starting from 4′ methyl-biphenyl-2-carboxylic

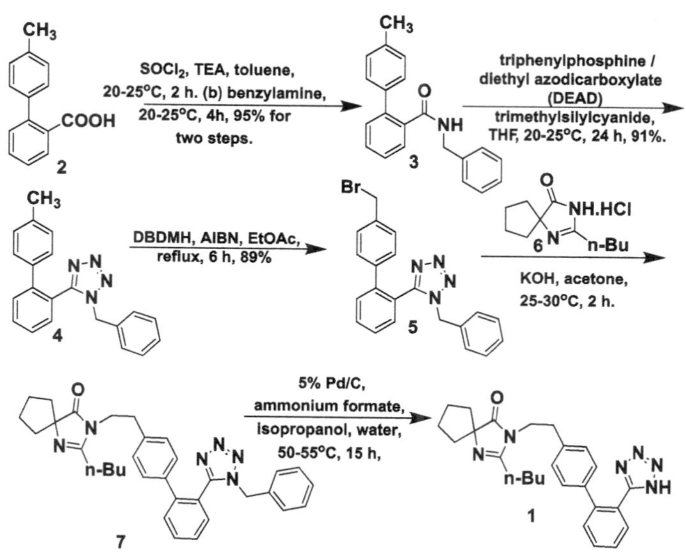

Scheme 2 Preparation of irbesartan starting from 4′-methyl-biphenyl-2-carboxylic acid.

acid 2 and the subsequent condensation with 2-butyl-1,3-diaza-spiro[4.4]non-1-en-4-one hydrochloride 6 and *in situ* debenzylation of the intermediate compound 7 to provide irbesartan 1.

This method has the advantages like (a): toxic metal compounds such as tri-*n*-butyl tin azide are not used in the formation of tetrazole ring; (b): hazardous reagents like sodium hydride are avoided for the condensation reaction; (c): column chromatographic purifications are not required in any stage; (d): the use of protecting groups which will give to the byproducts that can arise as impurities in the finished product are avoided; and (e): the overall yield of the process (66%) is higher that of the other reported synthesis [7,8].

2.3 Method 3

Singh et al. [10] described an improved green method for synthesis of irbesartan, based on the bromination by green method and Suzuki coupling reaction; the synthetic method is given in Scheme 3. The synthetic approach started with 4-bromo toluene 2. The bromination of 4-bromo toluene using *N*-bromosuccinamide (NBS) and ultrasonicator to give 4-bromo benzyl bromide 3. 4-Bromobenzyl bromide was then condensated with 2-butyl-1,3-diaza-Spiro [4.4] non-1-en-4-one 4 and then the key intermediate efficiently catalyzed the homogeneous catalyst by Suzuki-Miyaura cross-coupling conditions to give the phenyl-phenyl bond by using 2-cyano phenylboronic acid 6 to give 2-*n*-butyl-3-[(2-cyano biphenyl-4-yl) methyl] 1,3-diazo Spiro [4,4] non-1-en-4-one 7. The final steps were formation of tetrazole, formed from the organic nitriles and azide, to give irbesartan 1.

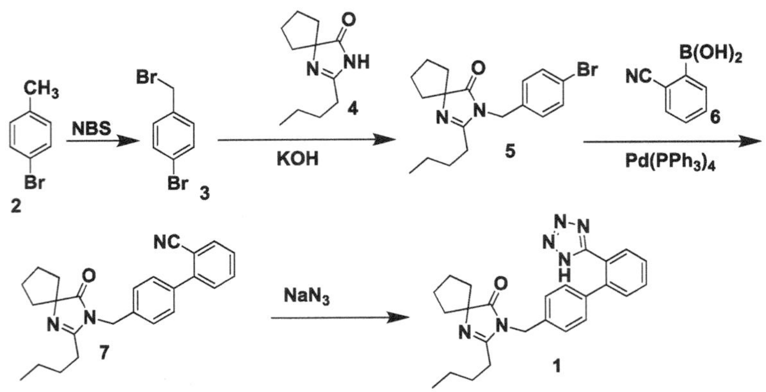

Scheme 3 Preparation of irbesartan based on the bromination of 4-bromo toluene by green method and Suzuki coupling reaction.

2.4 Method 4

Rao et al. [11] described an improved methods for the synthesis of irbesartan (1) as illustrated in Scheme 4. The first part of the procedures involved the preparation of 1-pentanamidocyclopentane-1-carboxylic acid by reaction between 1-aminocyclopentane-1-carbonitrile hydrochloride and pentanoyl chloride in CH_2Cl_2 in presence of triethylamine (TEA) which yielded *N*-(1-cyanocyclopentyl)pentanamide. This is followed by acidic hydrolysis of the latter compound to produce 1-pentanamidocyclopentane-1-carboxylic acid in excellent yields (90%). The second part involves preparation of 4′-(aminomethyl)-[1,1′-biphenyl]-2-carbonitrile hydrochloride by treating 4′-(bromomethyl)-[1,1′-biphenyl]-2-carbonitrile with NaN_3 in DMF and the resulting azide derivative is reduced by triphenylphosphine (PPh3) to yield the title compound in 92% yield. Coupling reaction between 4′-(aminomethyl)-[1,1′-biphenyl]-2-carbonitrile hydrochloride and 1-pentanamidocyclopentane-1-carboxylic in presence of DCC and HOBt gave *N*-((2′-cyano-[1,1′-biphenyl]-4-yl)methyl)-1-pentanamido-cyclopentane-1-carboxamide (93%). The latter was converted to irbesartan in 70–75% yield upon treatment with Bu_3SnCl and NaN_3 in DMF.

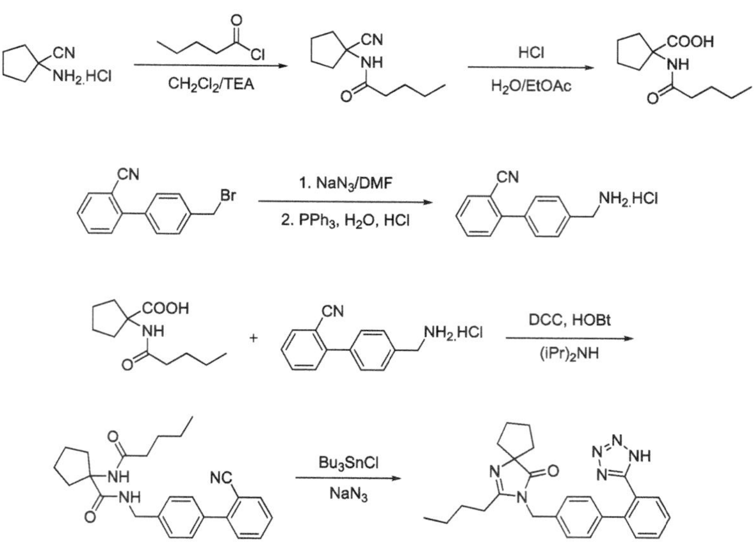

1

Scheme 4 Synthesis of irbesartan starting from the reaction 1-aminocyclopentane-1-carbonitrile hydrochloride and pentanoyl chloride.

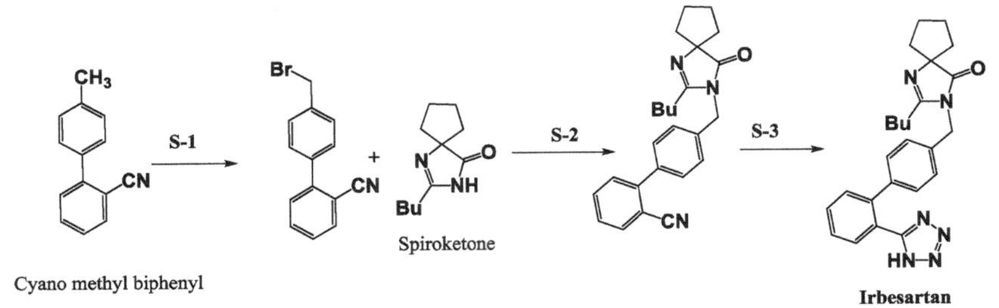

Scheme 5 Preparation of irbesartan based on the alkyl bromination of cyano methyl biphenyl and spirolactone followed by spiroketone substitution and cyclization with a reduction.

2.5 Method 5

Rao et al. [12] described a commercial scalable 3-steps process for the preparation of irbesartan as illustrated in Scheme 5. The step 1 (S-1) corresponds to alkyl bromination of cyano methyl biphenyl by reaction with 4′-bromomethyl-biphenyl-2-carbonitrile and 2-butyl-1,3-diaza-spiro[4,4] non-1-en-4-one (spirolactone) followed by spiroketone substitution in step 2 (S-2) and cyclization with a reduction to form tetrazole ring in step 3 (S-3). The purity of the compound was found to be 99.21% with five trace level impurities.

2.6 Method 6

DiPoto and Wu [13] described the synthesis of irbesartan starting from cyclopentanecarbonyl chloride as illustrated in Scheme 6. This method focuses on synthesis of the key intermediate 2-butyl-1,3-diazaspiro[4.4] non-1-en-4-one. It involves bromination of cyclopentanecarbonyl chloride and the resulting product was converted to formation of *N*-(benzyloxy)-1-bromocyclopentane-1-carboxamide (77%). annulation reaction between α-bromocarboxamide and valeronitrile gave 3-(benzyloxy)-2-butyl-1, 3-diazaspiro[4.4]non-1-en-4-one (59%). This was followed by benzyloxy group with SmI2 to give the key intermediate 2-butyl-1,3-diazaspiro[4.4] non-1-en-4-one (50%) which was used by Bernhart to prepare irbesartan.

3. Physical characteristics

3.1 Appearance

White or almost white crystalline powder [6].

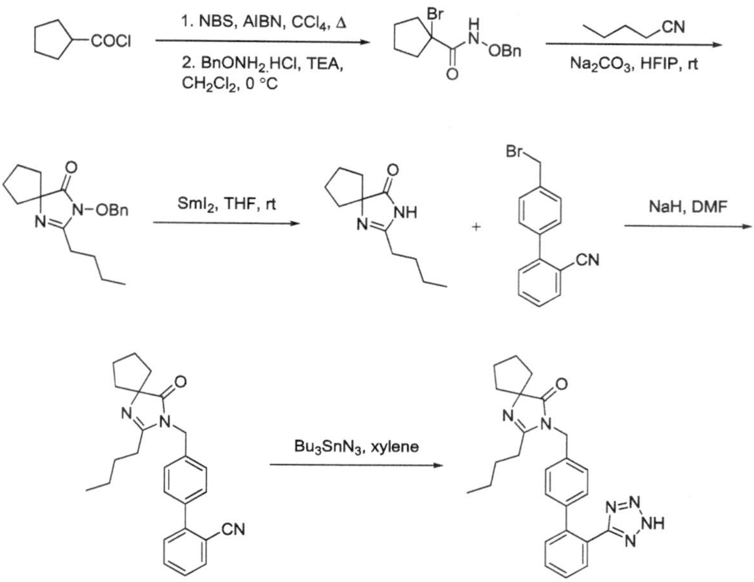

Scheme 6 synthesis of irbesartan starting from cyclopentanecarbonyl chloride.

3.2 Melting and boiling points

Melting point is 180–181 °C and boiling point is 848.6 °C [3].

3.3 Solubility

Practically insoluble in water 5.9×10^{-2} mg/L at 25 °C; slightly soluble in ethanol (8.84×10^{-3} g/L), dimethylsulfoxide (59 g/L) and methylene chloride [2,3,6].

3.4 Ionization constant

pKa1 = 4.08 (amidazole); pKa2 = 4.29 (tetrazol) [3].

3.5 Partition coefficient

Log P (octanol/water) = 4.564 [2].
C log P (calculated) = 5.31 [3].

3.6 Refractivity and polarizability

Refractivity = $136.72\,m^3 \times mol^{-1}$ [18].
Polarizability = $47.59\,Å^3$ [18].

3.7 Crystallographic properties

The X-ray powder diffraction for characterization of irbesartan powder was performed by high resolution X-ray diffraction using Rigaku Ultima IV diffractometer (Rigaku Corporation, Tokyo, Japan) with a scintillation detector and a reflection mode, Cu K{acute over (α)} radiation (1.5406 Å), scanning range: 3–60°2θ, step size: 0.02°2θ and time per step: 1 s monochromatized with a graphite crystal. The pattern was collected at 40 kV of tube voltage and 40 mA of tube current in step scan mode. The obtained peaks (reflections) of irbesartan powder are given in Fig. 1 and Table 1 [19,20].

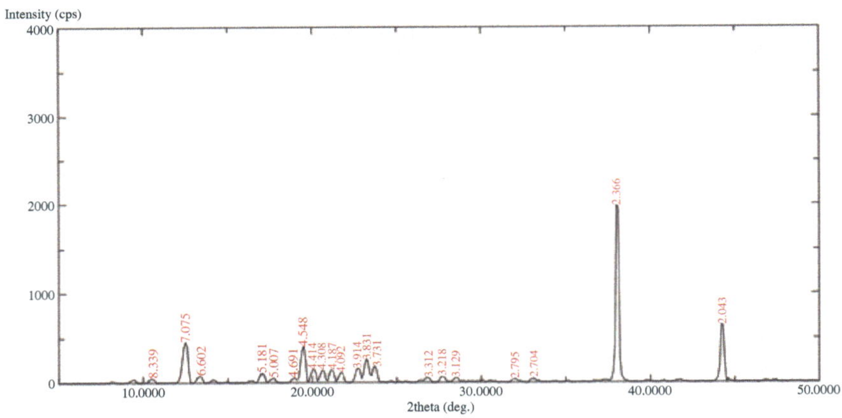

Fig. 1 Powder-X-ray diffraction pattern (reflections) of irbesartan powder.

Table 1 The peaks (reflections) of irbesartan powder.

Peak no.	2theta	Flex width	d-value	Intensity	I/I_0
1	10.600	0.353	8.3390	50	3
2	12.500	0.471	7.0754	462	24
3	13.400	0.471	6.6022	79	4
4	17.100	0.471	5.1811	105	6
5	17.700	0.353	5.0068	51	3
6	18.900	0.471	4.6915	52	3
7	19.500	0.471	4.5485	412	21
8	20.100	0.353	4.4140	153	8
9	20.600	0.353	4.3080	139	7

Table 1 The peaks (reflections) of irbesartan powder.—cont'd

Peak no.	2theta	Flex width	d-value	Intensity	I/I_0
10	21.200	0.353	4.1874	147	8
11	21.700	0.353	4.0920	108	6
12	22.700	0.353	3.9140	165	9
13	23.200	0.353	3.8308	269	14
14	23.700	0.353	3.7511	187	10
15	26.900	0.471	3.3117	52	3
16	27.700	0.353	3.2178	74	4
17	28.500	0.353	3.1293	61	4
18	32.000	0.353	2.7945	42	3
19	33.100	0.118	2.7041	49	3
20	38.000	0.235	2.3660	1982	100
21	44.300	0.235	2.0430	642	33

4. Thermal analysis

4.1 Melting behavior

Melting point of irbesartan was determined using a Büchi, B-545 melting point apparatus (Büchi Laboratories AG, Postfach, Switzerland). The melting point was in the range of 185–186 °C, which is nearby to that of described in the literature, 180–181 °C [3].

4.2 Differential scanning calorimetry

Differential scanning calorimetry data of irbesartan was obtained from a PerkinElmer DSC8000 differential scanning calorimeter with a TAC 71DX thermal analysis controller (PerkinElmer Inc., Norwalk, CT, USA). Samples were accurately weighed (3.8 mg) into aluminum pans and thermograms obtained at a heating rate of 10 °C/min over a temperature range of 40–300 °C. Pyris software 2.04 (PerkinElmer Inc., Norwalk, CT, USA) was used for analysis. Differential scanning calorimetric curve of irbesartan (Fig. 2) showed an endothermic process around 186 °C. An exothermic process follows immediately accompanied by loss of mass due to decomposition (at 198–201 °C).

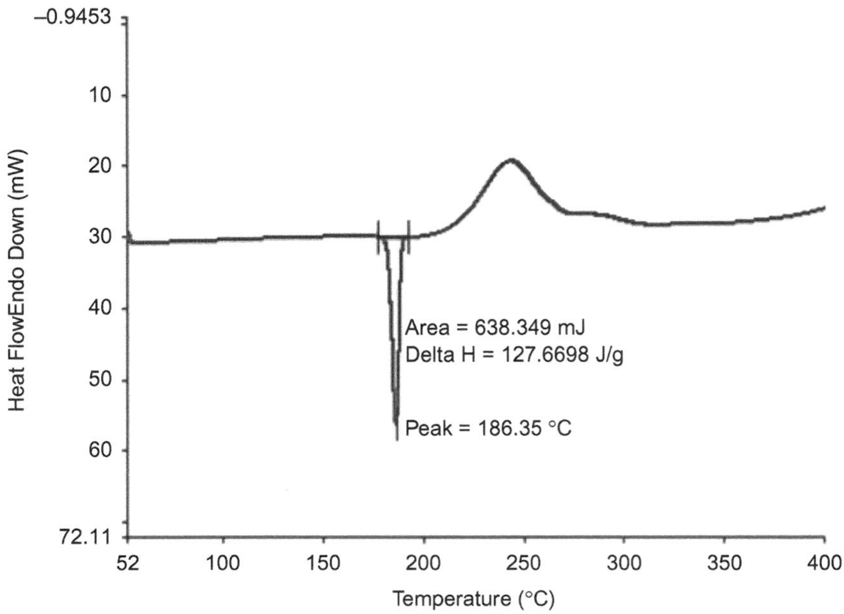

Fig. 2 Differential scanning calorimetric curve of irbesartan.

5. Spectroscopic analysis

5.1 Ultraviolet spectroscopy

The ultraviolet (UV) absorption spectrum of methanolic solution of irbesartan (20 μg/mL) in methanol was recorded using a Shimadzu UV-spectrophotometer, model UV-1800 (Shimadzu Corporation, Tokyo, Japan) with 1 cm matched quartz cells. The absorption spectrum was recorded over the range of 200–400 nm. The spectrum shows that irbesartan exhibited two maxima at 257.4 and 202.6 nm (Fig. 3).

5.2 Fluorescence spectroscopy

Fluorescence spectra of irbesartan were recorded on a JASCO FP-8200 Fluorescence Spectrometer (JASCO Corporation, Osaka, Japan) equipped with a 150 W xenon lamp and 1 cm quartz cells. The slit widths for both the excitation and emission monochromators were set at 5.0 nm. The calibration and linearity of the instrument were checked with standard quinine sulfate (0.01 μg/mL). Wavelength calibration was performed by measuring λ_{ex} at 254.5 nm and λ_{em} at 364.5 nm; no variation in the wavelength was

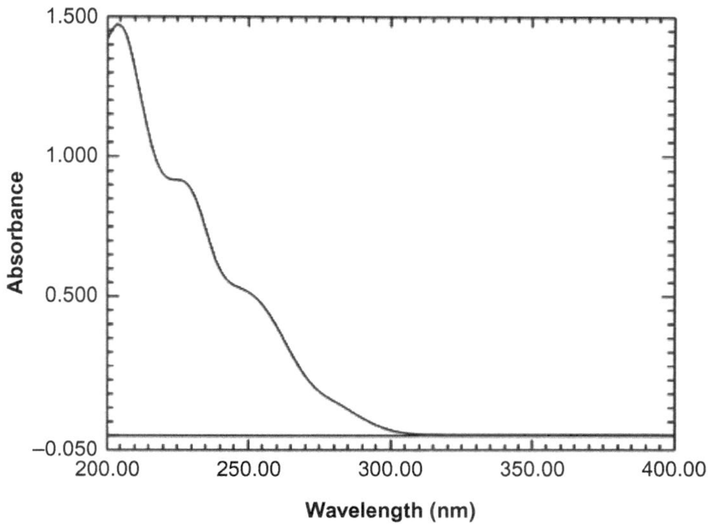

Fig. 3 Ultraviolet spectrum of irbesartan solution (20 μg/mL, in methanol).

observed. The recorded spectral data was converted to ASCII format by Spectra Manager® software (JASCO Corporation, Osaka, Japan). The fluorescence spectrum (Fig. 4) shows that irbesartan exhibited maxima excitation at 229; 254.5 and maxima emission at 364.5 nm.

5.3 Infrared spectroscopy

Fourier-transform infrared (FT-IR) absorption spectrum of irbesartan (as KBr disk) was recorded using the PerkinElmer FT-IR Spectrum BX apparatus (PerkinElmer, Norwalk, CT, USA); the FT-IR spectrum is given in Fig. 5. Characteristic peaks of irbesartan at 3436.51 cm^{-1} (N—H stretching), 2959.52 cm^{-1} (C—H stretching), 1732.90 cm^{-1} (C=O stretching), 1615.90 cm^{-1} (N—H bending) and 1485.77 cm^{-1} (C=C stretching) were observed.

5.4 Nuclear magnetic resonance spectrometry

5.4.1 ^{1}H NMR spectrum

The ^{1}H NMR spectrum of irbesartan was scanned in DMSO-d_6 on a Brucker NMR spectrometer (Bruker Co., Cremlingen, Germany) operating at 500 MHz. Chemical shifts are expressed in δ-values (ppm) relative to tetramethylsilane (TMS) as an internal standard. Coupling constants (*J*) are expressed in Hz (Fig. 6 and Table 2).

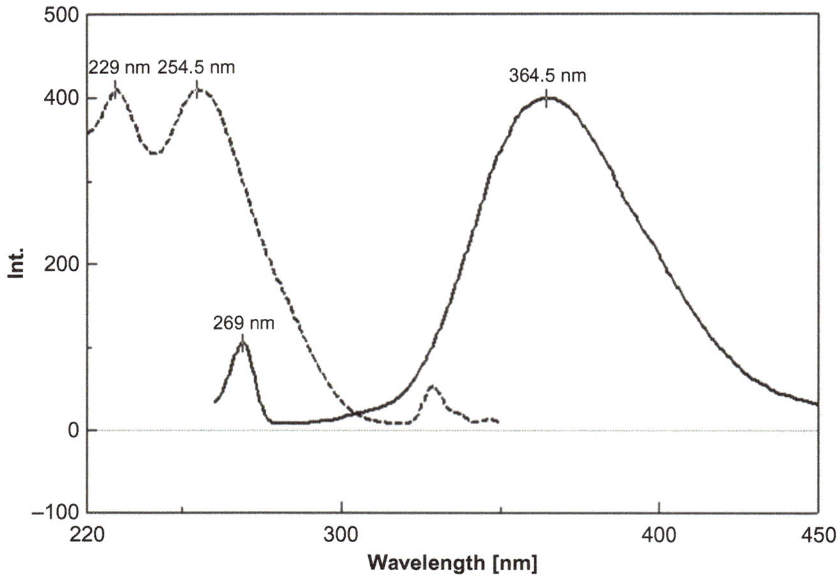

Fig. 4 Fluorescence spectrum of irbesartan solution (400 ng/mL, in methanol).

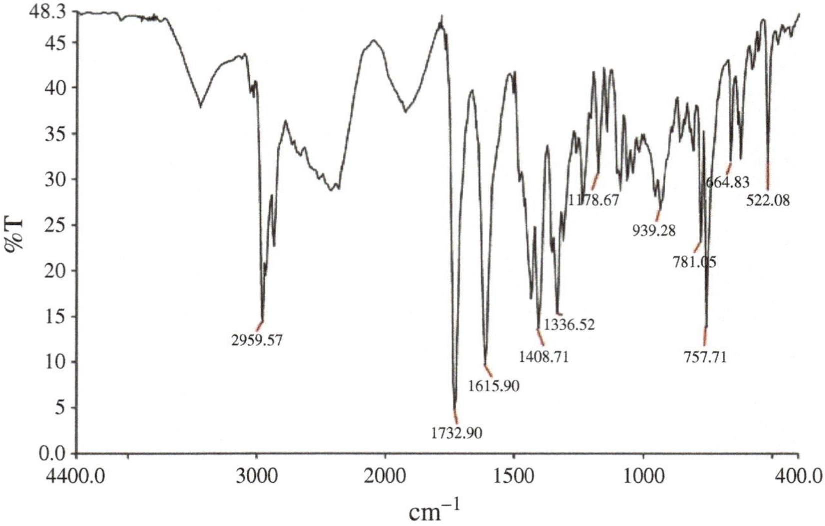

Fig. 5 Fourier-transform infrared (FT-IR) absorption spectrum of irbesartan.

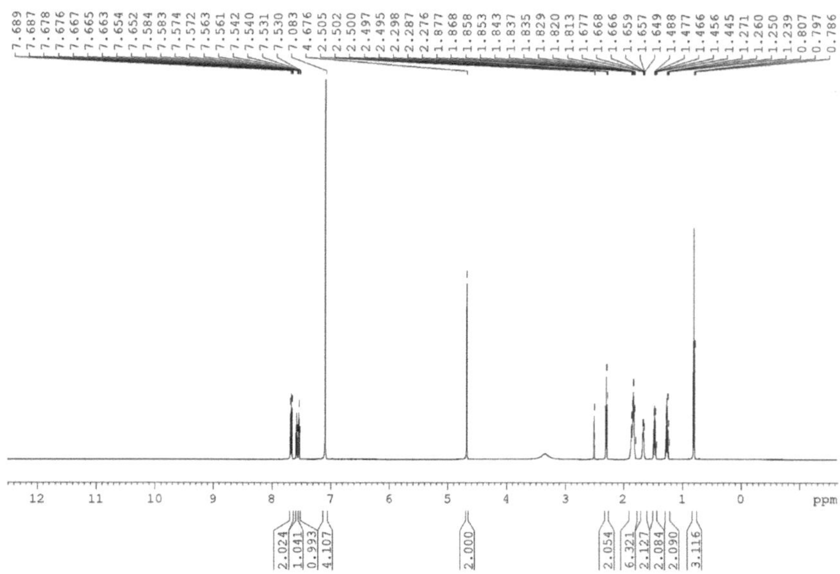

Fig. 6 ^{1}H NMR Spectrum of irbesartan.

Table 2 ^{1}H NMR of irbesartan (DMSO-d_6).

Signal	Chemical shift	Splitting	Integration	Assignment
1	0.80	t (J=7.34 Hz)	3	$CH_2CH_2CH_2CH3$
2	1.26	Sextet (J=7.34 Hz)	2	$CH_2CH_2CH_2CH_3$
3	1.47	Quintet (J=7.34 Hz)	2	$CH_2CH_2CH_2CH_3$
4	1.61–1.71	m	2	Cyclopentane
5	1.78–1.91	m	6	Cyclopentane
6	2.29	t (J=7.49 and 7.63 Hz)	2	$CH_2CH_2CH_2CH_3$
7	4.68	s	2	$PhCH_2$
8	7.08	s	4	ArH
9	7.52–7.55	m	1	ArH
10	7.56–7.60	td	1	ArH
11	7.64–7.70	m	2	ArH

5.5 ^{13}C NMR spectrum

^{13}C NMR spectrum of irbesartan was scanned in DMSO-d_6 on a Brucker NMR spectrometer (Bruker Co., Cremlingen, Germany) operating at 125 MHz. Chemical shifts are expressed in δ-values (ppm) relative to TMS as an internal standard (Fig. 7 and Table 3).

5.6 Mass spectrometry

The mass spectrum of irbesartan ($C_{25}H_{28}N_6O$, 428) was obtained using an Agilent 6320 Ion trap mass spectrometer (Agilent technologies, Santa Clara, CA, USA) equipped with an electrospray ionization interface (ESI) interface. A connector was used instead of column. Mobile phase composed of a mixture of acetonitrile:water (50:50, v/v). Irbesartan stock solution (1 mg/mL) was prepared by dissolving the solid material in methanol and then diluted with the mobile phase. Irbesartan test solutions were prepared by diluting the stock solutions with the mobile phase in the range of 10–30 µg/mL, according to the ion's intensities. A flow rate was 0.4 mL/min and the run time was 5 min. The scan was in an ultra-scan mode. MS2 scans were performed in the mass range of m/z 50–600. The ESI was operated in positive mode. The source temperature was set to 350 °C nebulizer gas pressure of 55.00 psi. Fig. 8A shows the peak of molecular ion at $m/z = 429$ $[M+1]^+$. Fragments are also described. The ESI negative ion mode is also reported; the important fragments (m/z) are 207, 385 and 180 as shown in Fig. 8B.

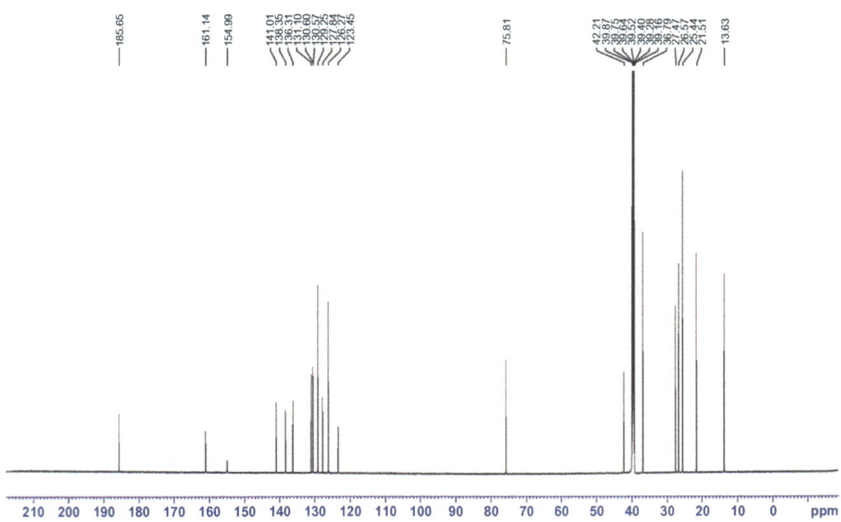

Fig. 7 ^{13}C NMR spectrum of irbesartan.

Table 3 ^{13}C NMR of irbesartan (DMSO-d_6).

Signal	Chemical shift	Assignment
1	13.63	$CH_2CH_2CH_2CH_3$
2	21.51	$CH_2CH_2CH_2CH_3$
3	25.44	$CH_2CH_2CH_2CH_3$
4	26.57	Cyclopentane
5	27.47	Cyclopentane
6	36.79	$CH_2CH_2CH_2CH_3$
7	42.21	$PhCH_2$
8	75.80	Cyclopentane (spiro carbon)
9	123.45	ArC (benzene)
10	126.27	ArC (benzyl) X2
11	127.84	ArC (benzene)
12	129.25	ArC (benzyl) X2
13	130.57	ArC (benzene)
14	130.60	ArC (benzene)
15	131.10	ArC (benzene)
16	136.31	ArC (benzyl)
17	138.35	ArC (benzene)
18	141.01	ArC (benzyl)
19	154.99	Tetrazole carbon
20	161.14	Imidazolone carbon
21	185.65	C=O of imidazolone

6. Methods of analysis

Numerous methods were reported for determination of irbesartan and its combinations in pharmaceutical dosage forms and biological matrices. Nalini and Nivedhitha [21] and Yogeesh et al. [22] reviewed many ultraviolet (UV) and high performance liquid chromatographic methods. The methods used for analysis of irbesartan are classified and described in the following sections.

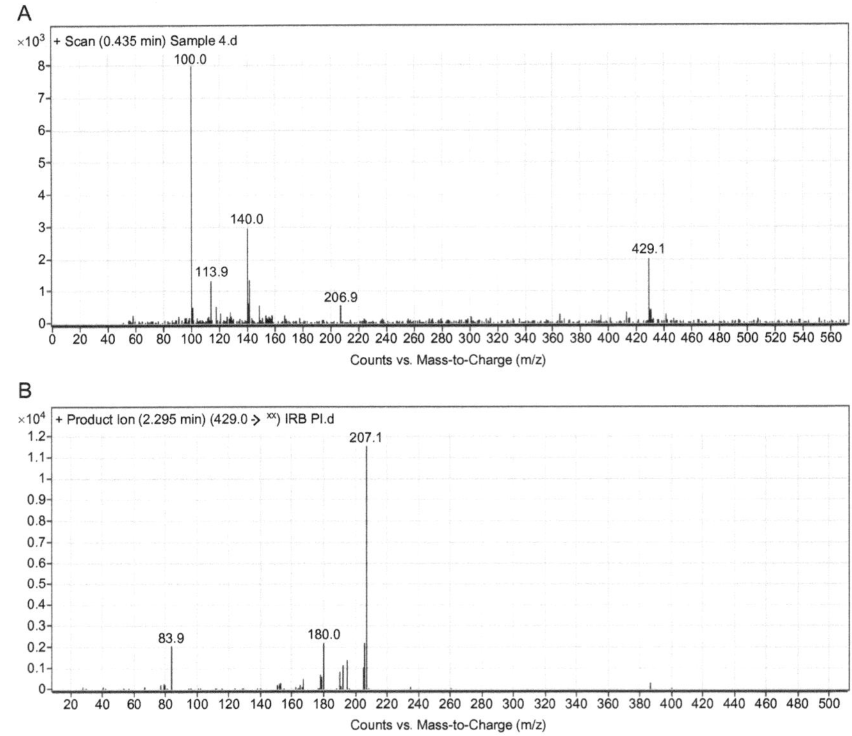

Fig. 8 (A) Positive ion spray mass spectra corresponding to irbesartan, parent ion at *m/z* of 429. (B) Positive ion spray mass spectra corresponding to irbesartan, daughter ion at *m/z* of 429.

6.1 Compendial methods

6.1.1 United States pharmacopoeia

6.1.1.1 Identification

The United States Pharmacopeia 42-National Formulary 37 (USP42-NF37) [23] uses infrared absorption (IR) spectroscopic, ultraviolet absorption (UV) and chromatographic methods, for identification of irbesartan bulk drug substance. The IR absorption spectrum of as substance, compared with that obtained concomitantly for the corresponding USP Reference Standard, provides perhaps the most conclusive evidence of the identity of the substance that can realized from any other single test. The UV absorption spectrum, on the other hand, does not exhibit a high degree of specificity. Conformance with both IR absorption and UV test specificities, as called for in a large proportion of compendial monographs, leaves little doubt, if any, regarding the identity of the specimen under examination.

6.1.1.2 Infrared absorption

Six methods are indicated for preparation of previously dried test specimen and Reference Standards for analysis. The reference ⟨197K⟩ in a monograph signifies that the substance under examination is mixed intimately with potassium bromide. The reference ⟨197M⟩ in a monograph signifies that the substance under examination is finely ground and dispersed in mineral oil. The reference ⟨197F⟩ in a monograph signifies that the substance under examination is suspended neat between suitable (for example, sodium chloride and potassium bromide) plates. The reference ⟨197S⟩ in a monograph signifies that a solution of designated concentration is prepared in the solvent specified in the individual monograph, and the solution is examined in 0.1-mm cells unless a different cell path length is specified in the individual monograph. The reference ⟨197A⟩ in a monograph signifies that a substance under examination is mixed intimately in contact with an internal reflection element for attenuated total reflectance (ATR) analysis. The reference ⟨197E⟩ signifies that the substance under examination is pressed as a thin sample against a suitable plate for IR spectroscopic analysis. The ATR ⟨197A⟩ and the ⟨197E⟩ techniques can be used as alternative methods for ⟨197K⟩, ⟨197M⟩, ⟨197F⟩, and ⟨197S⟩ where testing is performed qualitatively, and the Reference Standard spectra are similarly obtained.

Record the spectra of the test specimen and the corresponding USP Reference Standard over the range from about 2.6 to 15 μm ($3800–650\,cm^{-1}$) unless otherwise specified in the individual monograph. The IR absorption spectrum of the preparation of the test specimen, previously dried under conditions specified for the corresponding Reference Standard unless otherwise specified, or unless the Reference Standard is to be used without drying, exhibits maxima only at the same wavelengths as that of a similar preparation of the corresponding USP Reference Standard. Differences that may be observed in the spectra so obtained sometimes are attributed to the presence of polymorphs, which are not always acceptable. Unless otherwise directed in the individual monograph, therefore, continue as follows. If a difference appears in the IR spectra of the analyte and the standard, dissolve equal portions of the test specimen and the Reference Standard in equal volumes of a suitable solvent, evaporate the solution to dryness in similar containers under identical conditions, and repeat the test on the residues.

6.1.1.3 Ultraviolet absorption

The reference ⟨197U⟩ in a monograph signifies that a test solution and a Standard solution are examined spectrophotometrically, in 1-cm cells, over the spectral range from 200 to 400 nm unless otherwise specified in the individual monograph.

Dissolve a portion of the substance under examination in the designated *Medium* to obtain a test solution having the concentration specified in the monograph for *Solution*. Similarly prepare a Standard solution containing the corresponding USP Reference Standard. Record and compare the spectra concomitantly obtained for the test solution and the Standard solution. Calculate absorptivities and/or absorbance ratios where these criteria are included in an individual monograph. Unless otherwise specified, absorbances indicated for these calculations are those measured at the maximum absorbance at about the wavelength specified in the individual monograph. Where the absorbance is to be measured at about the specified wavelength other than that of maximum absorbance, the abbreviations (min) and (sh) are used to indicate a minimum and shoulder, respectively, in an absorption spectrum. The requirements are met if the UV absorption spectra of the test solution and the Standard solution exhibit maxima and minima at the same wavelengths and absorptivities and/or absorbance ratios are within specified limits.

6.1.1.4 Retention time

The retention time of the major peak in the chromatogram of the assay preparation corresponds to that in the chromatogram of the standard preparation, as obtained in the *Assay*.

6.1.1.5 Water content

Water content of irbesartan is determined by Method I ⟨921⟩: not more than 5%.

6.1.1.6 Heavy metals content

Heavy metals content of irbesartan is determined by Method II ⟨231⟩: 0.002%.

6.1.1.7 Limit of azide

Mobile phase—Prepare a filtered and degassed 0.1 N sodium hydroxide solution.

Standard solution—Transfer about 25 mg of sodium azide, accurately weighed, to a 100-mL volumetric flask, dissolve in and dilute with mobile phase to volume, and mix. Pipet 250 μL of this solution into a 200-mL volumetric flask, dilute with mobile phase to volume, and mix. This solution contains about 0.312 µg of sodium azide per mL.

Test solution—Transfer about 100 mg of irbesartan, accurately weighed, to a 5-mL volumetric flask, dissolve in and dilute with *Mobile phase* to volume, and mix.

Chromatographic system—The liquid chromatograph is equipped with a conductimetric detector and a 4.0-mm × 25-cm column that contains packing L31. The flow rate is about 1.0 mL per minute. Chromatograph the standard solution, and record the peak responses as directed for *Procedure*: the signal-to-noise ratio for the azide peak is not less than 10.

Procedure—Separately inject equal volumes (about 200 µL) of the standard solution and the test solution into the chromatograph, record the chromatograms, and measure the peak areas for azide. Calculate the amount of azide, in ppm, in the portion of irbesartan taken by the formula:

$$1000(C_S/C_T)(42.02/65.01)(r_U/r_S)$$

in which C_S is the concentration, in µg per mL, of sodium azide in the standard solution; C_T is the concentration, in mg per mL, of irbesartan in the test solution; r_U is the peak area for azide obtained from the test solution; and r_S is the peak area for azide obtained from the standard solution: not more than 10 ppm of azide is found.

6.1.1.8 Related compounds

USP irbesartan related compound A RS: [1-pentanoylamino-cyclopentanecarboxylic acid [2′-(*1H*-tetrazol-5-yl)-byphenyl-4-ylmethyl]-amide] ($C_{25}H_{30}N_6O_2$: 446.54).

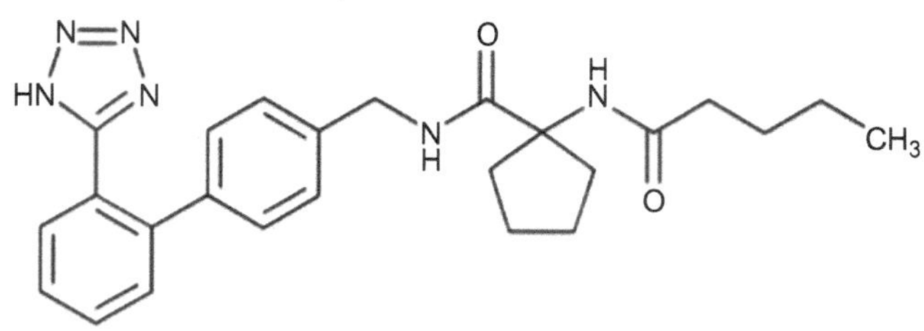

Mobile phase—pH 3.2 phosphate buffer.

Standard solution—Prepare as directed for the system suitability solution in the assay.

Test solution—Dissolve an accurately weighed quantity of irbesartan in methanol to obtain a solution having a known concentration of about 1 mg per mL.

Chromatographic system—Proceed as directed in the assay. Chromatograph the standard solution and record the peak responses as directed for procedure: the relative standard deviation for replicate injections is not more than 2.0%.

Procedure—Separately inject equal volumes (about 10 μL) of the standard solution and the test solution into the chromatograph, record the chromatograms, and measure the area for the irbesartan related compound A peak. Calculate the percentage of irbesartan related compound A in the portion of irbesartan taken by the formula:

$$100(C_S/C_T)(r_U/r_S)$$

in which C_S is the concentration, in mg per mL, of USP irbesartan related compound A RS in the standard solution; C_T is the concentration, in mg per mL, of irbesartan in the test solution; r_U is the peak response for irbesartan related compound A obtained from the Test solution; and r_S is the peak response for irbesartan related compound A obtained from the standard solution.

Calculate the percentage of other impurities in the portion of irbesartan taken by the formula:

$$100(C_S/C_T)(r_U/r_S)$$

in which C_S is the concentration, in mg per mL, of USP irbesartan RS in the standard solution; C_T is the concentration, in mg per mL, of irbesartan in the Test solution; and r_U and r_S are the peak responses for each of the other impurities and USP irbesartan RS obtained from the test solution and the standard solution, respectively: not more than 0.2% of irbesartan related compound A is found; not more than 0.1% of any other impurity is found; and not more than 0.5% of total impurities is found.

6.1.1.9 Assay

Mobile phase—pH 3.2 phosphate buffer: Mix 5.5 mL of phosphoric acid with about 950 mL of water, and adjust pH to 3.2 with triethylamine.

Mobile phase—Prepare a filtered and degassed mixture of pH 3.2 phosphate buffer and acetonitrile (67:33). Make adjustments if necessary.

System suitability solution—Dissolve accurately weighed quantities of USP irbesartan RS and USP irbesartan related compound A RS in methanol to obtain a solution having a known concentration of about 0.05 mg per mL of each USP Reference Standard.

Standard preparation—Dissolve an accurately weighed quantity of USP irbesartan RS in methanol to obtain a solution having a known concentration of about 0.5 mg per mL.

Assay preparation—Transfer about 50 mg of irbesartan, accurately weighed, to a 100-mL volumetric flask, dissolve in and dilute with methanol to volume, and mix.

Chromatographic system—The liquid chromatograph is equipped with a 220-nm detector and a 4.0-mm × 25-cm column that contains packing L1. The flow rate is about 1.0 mL per minute. Chromatograph the system suitability solution, and record the peak responses as directed for procedure: the relative retention times are about 0.8 for irbesartan related compound A and 1.0 for irbesartan; the resolution, R, between irbesartan and irbesartan related compound A is not less than 2.0. Chromatograph the standard preparation, and record the peak response as directed for Procedure: the standard deviation for replicate injections is not more than 1.0%.

Procedure—Separately inject equal volumes (about 10 μL) of the standard preparation and the assay preparation into the chromatograph, record the chromatograms, and measure the responses for all the peaks. Calculate the quantity, in mg, of $C_{25}H_{28}N_6O$ in the portion of irbesartan taken by the formula:

$$100C(r_U/r_S)$$

in which *C* is the concentration, in mg per mL, of USP irbesartan RS in the standard preparation; and r_U and r_S are the peak responses obtained from the Assay preparation and the standard preparation, respectively.

6.1.2 European pharmacopoeia

European Pharmacopoeia [24] uses infrared absorption (IR) spectroscopic method for identification and potentiometric titration for assay of irbesartan bulk drug substance. IR spectroscopy is performed with solid state irbesartan CTR. For assay, the solvent used is 0.1 M perchloric acid and the end point is determined potentiometrically.

6.2 Voltammetric methods

Gupta et al. [25] developed a sensitive voltammetric method for the determination of irbesartan in pharmaceutical formulations. In a Britton-Robinson buffer medium, irbesartan exhibited a well-defined cathodic peak over the entire pH range from 2.0 to 12.0. The mechanism of reduction was postulated based on controlled potential electrolysis, coulometry, and spectral analysis. Under optimal conditions, a linear response of irbesartan was obtained in the range from 3.0×10^{-5} to 5.7×10^{-3} mol/L and with a limit of detection of 5.33×10^{-7} mol/L. The effect of cationic surfactant on the voltammetric reduction peak of irbesartan in Britton-Robinson buffer was also described.

Bozal et al. [26] developed a differential pulse and square wave voltammetric techniques for the determination of irbesartan. The electrochemical behavior of irbesartan was investigated by cyclic voltammetry, differential pulse voltammetry, and square wave voltammetry at the hanging mercury drop electrode. Different parameters were tested to optimize the conditions of the determination. It was found that in the range of $8 \times 10^{-6} \times 10^{-4}$ M, the currents measured by both of methods presented a good linear property as a function of the concentration of irbesartan. In addition, validation parameters, such as reproducibility, sensitivity, and recovery were evaluated as well. The slope of the log Ip-log ν linear plot was 0.58 indicating the diffusion control for 0.5 M sulfuric acid without the need for separation or complex sample preparation, since there was no interference from the excipients and endogenous substances. The methods were successfully applied to the analysis of irbesartan in the pharmaceutical tablet formulations and in human serum samples.

El-Desoky et al. [27] developed and validated a square wave adsorptive cathodic stripping voltammetric method for quantification of irbesartan in the bulk form, pharmaceutical formulations and spiked human serum. Cyclic voltammograms of irbesartan at the hanging mercury dropping electrode in the Britton-Robinson buffer of pH values lower than 4.5 exhibited a single 2-electron irreversible cathodic peak corresponding to the reduction-saturation of the C=N double bond of its tetrazolyl moiety. Over the pH range 4.5–5.5, the voltammograms exhibited two irreversible cathodic peaks (first and second peaks) of lower peak current but of relatively equal heights which were attributed to reduction of the C=N group of the acidic and basic forms of irbesartan, respectively. The developed stripping voltammetric method was successfully applied for quantitation of irbesartan in pharmaceutical formulations and spiked human serum, without the necessity for samples pretreatment and/or time-consuming extraction steps prior to the analysis. Insignificant interferences from its active ingredient "hydrochlorothiazide," excipients, common metal ions and co-administrated drugs were obtained during the analysis. Limits of detection of 9.0×10^{-10} and 2.1×10^{-9} mol/L and limits of quantitation of

3.0×10^{-9} and 7.0×10^{-9} mol/L irbesartan in the bulk form and in spiked human serum were achieved, respectively, by the developed stripping voltammetric method. Moreover, pharmacokinetic parameters of irbesartan in plasma of healthy volunteers following an oral administration of a single dose of Aprovel® tablets were also estimated.

Habib et al. [28] described an adsorptive stripping voltammetric method for the determination of irbesartan using a hanging mercury drop electrode. The method was based on adsorptive accumulation of the species at the hanging mercury drop electrode, followed by a first-harmonic alternating current stripping sweep at pH 6. The behavior of the adsorptive stripping response was studied under various experimental conditions, e.g., type of supporting electrolyte, pH, accumulation time, scan rate, and mode of sweep. In Britton-Robinson buffer solution, a quasi-reversible reduction process involving transfer of two electrons and two protons takes place. The response was linear over the concentration range of 0.16–1.60 μg/mL with regression coefficient 0.997 and detection limit 0.11 μg/mL. The average of determinations of the cited compound in oral dosages with its standard deviation was 99.39–4.33%. The result obtained by the proposed method was compared with that obtained by the UV-spectrophotometric technique. Furthermore, the proposed method was also successfully applied as a stability-indicating method.

6.3 Spectroscopic methods

6.3.1 Spectrophotometry

Shah et al. [29] reviewed some spectrophotometric methods reported for irbesartan before 2015. The other methods and those reported thereafter are described in the following paragraphs.

Asati et al. [30] developed a UV spectroscopy method for quantitative analysis of irbesartan in bulk drug. Irbesartan showed a maximum absorbance at 246 nm. Beer's law was obeyed in the concentration range of 5–30 μg/mL. The method was successfully validated in accordance to USP pharmacopeia and ICH guideline for accuracy, precision, rang and linearity. The linear regression analysis data for calibration plots showed a good linear relationship and the correlation factor was greater than 0.999 for irbesartan. The % recovery/accuracy was within the range between 98% and 102%. The percentage relative standard deviation for precision was found to be less than 2%. Thus, method was successfully applied for routine analysis of irbesartan in bulk samples.

Pradhan et al. [31] developed and validated a stability-indicating UV-method for estimation of irbesartan. Many experiments were conducted by taking combination of drugs and by single drug but no suitable validation with stability studies were reported. Irbesartan had the absorbance maxima at 246 nm. Experiments involved development and validation showed sharp peak at 246 nm. Methanol was utilized as a solvent for all methods. Linearity for the detector response was observed in the concentration range of 5–45 μg/mL. Validation experiments were performed to demonstrate system suitability, specificity, precision, linearity, accuracy precision, robustness, ruggedness. Furthermore, stability studies of irbesartan were carried out under acidic, alkali, neutral, oxidation & photolytic degradation as per stability-indicating assay methods. The results of analysis were validated. Recovery studies were in the range (99.335–100.583%) by adding specific drug amount (80%, 100%, and 120%).

Nissankararao et al. [32] developed a hydrotropic agent assisted spectrophotometric method using 1 M sodium bicarbonate and 2 M urea (50:50%, v/v), as hydrotropic agent; to increase the solubility of poorly water-soluble irbesartan. The method was developed for the estimation of irbesartan in its bulk and pharmaceutical dosage forms. An absorption maximum was found to be at 246.4 nm where sodium bicarbonate, urea and other excipients did not show any absorbance above 228 nm and thus no interference in the estimation. Irbesartan was obeyed Beer's law in the concentration range from 10 to 35 μg/mL. The method was validated according to the International Conference of Harmonization (ICH) guidelines and values of accuracy, precision and other statistical analysis were found to be in good accordance with the prescribed values with correlation coefficient of 0.9998. The percentage recovery of irbesartan ranged from 99.4% to 101.3% in pharmaceutical dosage form. Results of the analysis for accuracy, precision, limit of detection, limit of quantitation and were found to be satisfactory.

Biyani and Yadav [33] developed and validated an UV-spectrophotometric method for estimation of irbesartan by the hydrotrophy technique. In this method, various hydrotropes were screened for solubilization of irbesartan. It was found that 1 M sodium acetate enhanced the solubilization by 33 times. After checking the effect of molarity from 0.5 to 4 M, 1 M sodium acetate showed characteristic absorption peaks and obeyed the Beer-Lambert law for 10–20 mg/mL solution of irbesartan hydrotrope mixture. The calibration curve equation was found to be $y = 0.040x - 0.013$ with a correlation coefficient of 0.998. The LOD and LOQ values were found to be 0.3 and 1. The method was validated for accuracy and precision at three levels of concentrations (10, 14, and 18 mg/mL, for which the RSD was found to be 0.09%, 0.07%, and 0.1%, respectively). The method was robust and specific in the presence of other excipients and showed RSD values of 0.4%, 0.5%, and 0.6%, which were within acceptable limits.

Virani et al. [34] developed an UV spectroscopic method for simultaneous determination of irbesartan and atorvastatin by simultaneoufs equation spectroscopic method. In this method, 226.00 and 246.00 nm wavelengths were selected for measurement of absorptivity. Both the drugs show linearity in a concentration range of 05–30 μg/mL at their respective λ_{max}. Accuracy, precision and recovery studies were done by quality control samples covering lower, medium and high concentrations of the linearity range. The relative standard deviation for accuracy, precision studies were found to be within the acceptance range (<2%). The limit of detection was 0.033and 0.125 μg/mL for irbesartan and atorvastatin, respectively. The limit of quantification was 0.1008 and 0.3792 μg/mL for irbesartan and atorvastatin, respectively. Recovery of irbesartan and atorvastatin were found to be 99.75% and 99.52%, respectively. Confirming the accuracy of the proposed method. The proposed method was recommended for routine analysis since they were rapid, simple, accurate and also sensitive and specific by no heating and no organic solvent extraction.

Fayez [35] described the simultaneous determination of some antihypertensive drugs including irbesartan in their binary mixture by manipulating their ratio spectra. These methods were the ratio subtraction coupled with constant multiplication, ratio difference and constant center methods. The accuracy, precision and linearity ranges of the proposed methods were determined, and the methods were validated, and the specificity was assessed by analyzing synthetic mixtures containing the cited drugs. The three methods were applied for the determination of the cited drugs in tablets and the obtained results were statistically compared with each other and with those of official methods. The comparison showed that there is no significant difference between the proposed methods and the official methods regarding both accuracy and precision.

Rahman et al. [36] developed a kinetic spectrophotometric method for the determination of irbesartan in pharmaceutical formulations. The method was based on the reaction of carboxylic acid group of the oxidized irbesartan with a mixture of potassium iodate (KIO_3) and iodide (KI) to form yellow colored triiodide ions in aqueous medium at 30 ± 1 °C. The reaction was followed spectrophotometrically by measuring the rate of change of absorbance at 352 nm. The initial rate and fixed time (ΔA) methods were adopted for constructing the calibration curves, which were found to be linear over the concentration ranges of 10.0–60.0 and 7.5–60.0 μg/mL, respectively.

The regression analysis of calibration data yielded the linear equations: rate = $-2.138 \times 10^{-6} + 1.058 \times 10^{-4}$ C and $\Delta A = -3.75 \times 10^{-3} + 3.25 \times 10^{-3}$ C for initial rate and fixed time (ΔA) methods, respectively. The limit of detection for initial rate and fixed time methods were 0.21 and 2.40 μg/mL, respectively. The various activation parameters such as E_a ΔH, ΔS and ΔG were also calculated for the reaction and found to be 70.95 ± 0.43, 68.48 ± 0.21, 16.54 ± 0.24 and -4.94 ± 0.07 kJ/mol, respectively. The proposed methods were optimized and validated as per the guidelines of International Conference on Harmonization. The point and interval hypothesis tests were performed which indicate that there was no significant difference between the proposed methods and the reference method. The methods were successfully applied to the determination of irbesartan in commercial dosage forms.

Divya et al. [37] developed UV spectroscopy method for simultaneous estimation of irbesartan and hydrochlorothiazide in combined tablet dosage form utilizing a concept of standard addition. The method is based upon determination of irbesartan at 215 nm and hydrochlorothiazide at 230 nm, in methanol. Different analytical performance parameters such as linearity, precision, accuracy and ruggedness were determined according to ICH guidelines. Irbesartan and hydrochlorothiazide at their respective λ_{max} showed linearity in the concentration range of 2–16 μg/mL. The method was validated statistically. The results of analysis formulation given as percentage of label claim were found to be 106.6 and 101.6 for irbesartan and hydrochlorothiazide respectively. Therefore, the proposed methods could be used for the routine analysis of both drugs in quality control laboratories.

Chabukswar et al. [38] described ratio spectra derivative and absorption correction spectrophotometry method for the simultaneous determination of irbesartan and hydrochlorothiazide in combined tablet dosage form the amplitudes at 252.6 nm and 230 nm in the first derivative of the ratio spectra method were selected to determine irbesartan and hydrochlorothiazide, respectively. The second method was based on the absorption correction method in which irbesartan and hydrochlorothiazide exhibited λ_{max} at 244 and 275.02 nm, respectively, in 0.1 N HCL. Hydrochlorothiazide had some interference due to irbesartan at 275.02 nm, while hydrochlorothiazide did not show any absorption at 244 nm. The % assay for commercial formulations was found to be in the range 98.29–99.80% for irbesartan and 98.91–100.10% for hydrochlorothiazide by the methods, for both the formulations. Recovery was found to be in the range of 98.11–100.08% for irbesartan and 98.54–99.29% for hydrochlorothiazide by ratio derivative method and 98.59–99.56% for irbesartan and 99.33–100.12% for hydrochlorothiazide by absorbance correction method for both the Formulations. The methods were validated, as per ICH guidelines. Absorbance correction method was applied for dissolution studies and drug release pattern was evaluated.

Albero et al. [39] developed a first derivative spectrophotometric method for the determination of irbesartan alone and in the presence of hydrochlorothiazide. Measurements were made at the zero-crossing wavelength at 263.0 nm for irbesartan. The calibration graph was linear over the range 1.0–12.0 mg/L of irbesartan, and the detection limit was 0.15 mg/L. Hydrochlorothiazide, in the presence of irbesartan, was determined by a direct spectrophotometric method at 317 nm. The calibration graph was linear over the range 2.0–50.0 mg/L of hydrochlorothiazide and the detection limit was 0.25 mg/L. The proposed methods were successfully applied to the determinations of irbesartan and hydrochlorothiazide in combined tablets.

El-Bagary et al. [40] developed spectrophotometric, spectroflourimetric, and liquid chromatographic methods for the determination of irbesartan. The spectrophotometric method represented a stability-indicating assay for the determination of irbesartan in presence of its alkaline degradation product in the range of 2.5–30 mg/mL. This method was based on measuring the first derivative of ratio spectra at 236.5 nm. The results were statistically compared using one-way analysis of variance. The developed method was satisfactorily applied to the analysis of the pharmaceutical formulations and proved to be suitable for the quality control of the cited drugs in pharmaceutical dosage forms.

Fawzy et al. [41] developed two simple and accurate spectrophotometric methods for the determination of binary mixtures of each of eprosartan and irbesartan with hydrochlorothiazide in dosage forms. Concerning the mixture of irbesartan/hydrochlorothiazide, irbesartan was determined by the first derivative zero-crossing technique after extraction with chloroform, while hydrochlorothiazide was determined by a specific colorimetric method using chromotropic acid. The method was proved to be able to cope with the strong spectral interference of the binary mixture components, without any interference from the other component or from the excipients. The proposed method could be used for the routine analysis of the mixture in quality control laboratories.

Erk [42] developed three analytical methods for the simultaneous determination of hydrochlorothiazide and irbesartan in a binary mixture without previous separation. The first method, based on compensation technique, was presented for the derivative spectrophotometric determination of binary mixtures with overlapping spectra. By using ratios of the derivative maxima or the derivative minimum, the exact contribution of either component in the binary mixture could be measured and the amounts were quantified. The second method was the first derivative of the ratio spectra. The ratio spectra were obtained by dividing the absorption spectra of the binary mixture by that of one of the components. For determination of hydrochlorothiazide and irbesartan in binary mixtures, the amplitudes in the first derivative of the ratio spectra at 231, 266, 279, 238 and 248 nm were selected. The concentrations of the other components were then determined from their respective calibration graphs treated similarly. In the third method, the absorbance ratio method, the determination of hydrochlorothiazide and irbesartan was performed using the absorbances read at 272, 241 and 263 nm in the zero-order spectra of their mixture. The absorbance ratio was also developed as a comparison method. The three methods required no preliminary separation steps and could, therefore, be used for routine analysis of both drugs in quality control laboratories.

Joseph-Charles et al. [43] developed a second-derivative spectrophotometric method for the simultaneous determination of irbesartan and hydrochlorothiazide in tablets. Measurements were made at the zero-crossing wavelengths at 230.1 nm for irbesartan and 232.7 nm for hydrochlorothiazide. The calibration graphs were linear in the range of 14.4–33.6 mg/L for irbesartan and 1.2–2.8 mg/L for hydrochlorothiazide. The limits of quantitation were 5.0 and 1.1 mg/L, respectively. The proposed method was successfully applied to the simultaneous determination of irbesartan and hydrochlorothiazide in commercial tablets.

Farouk et al. [44] developed and validated a spectrophotometric method for determination of irbesartan, torasemide and olmesartan medoxomil. Irbesartan was determined simultaneously in a binary mixture with hydrochlorothiazide by a ratio subtraction technique. The proposed method was successfully applied to determine the mentioned studied drugs in pure form, in laboratory prepared mixtures and in pharmaceutical preparations. The obtained results were statistically compared to the reference methods of analysis and no significant difference was found.

Vetuschi et al. [45] developed a fourth-order UV derivative spectrophotometric method for the determination of hydrochlorothiazide and irbesartan in pharmaceuticals. The alcoholic sample was used for the fourth-order derivative spectrum. Hydrochlorothiazide was determined by a specific peak-trough, of low intensity, at 330–340 nm. For irbesartan, a peak-trough around 250–310 nm was used, common to both products, whose amplitude increased linearly only for low concentration values, while it decreased at higher values. Both signals were evaluated simultaneously. The best results were achieved by using a linear regression for hydrochlorothiazide and a regression plane for irbesartan.

Sivasubramanian and Lakshmi [46] developed partial least squares and principal component regression methods for the spectrophotometric analysis of a ternary mixture consisting of irbesartan, hydrochlorothiazide and ramipril without prior separation. In these chemometric techniques, the measurements of absorbance values were realized in the spectral range from 200 to 350 nm in the intervals of $\Delta\lambda = 1$ nm at the 151 wavelengths in the zero-, first- and second-order spectra of the different ternary mixtures of these ingredients in 0.1 M NaOH. The prepared calibrations of both techniques using the absorbance data and concentration matrix data sets were used to predict the concentrations of the unknown concentrations of irbesartan, hydrochlorothiazide and ramipril in their ternary mixture. The techniques were successfully applied to pharmaceutical formulations.

Shakya et al. [47] developed a UV-spectrophotometric method for the simultaneous determination of irbesartan and hydrochlorothiazide, involving measuring area under the curve. Methanol was selected as a solvent. Irbesartan showed maximum absorbance at 224 nm and hydrochlorothiazide showed maximum absorbance at 271 nm. For the area under curve method, the wavelength ranges between 225–230 and 258–265 nm, respectively were selected with reference to the absorbance curves plotted between the wavelengths of 200–400 nm. This method allowed rapid analysis of two drug combinations. The result of analysis was validated statistically following the International Conference on Harmonization (ICH) method validation guideline. Tablet containing both drugs was assayed using the method developed, showing a good accuracy and precision.

Pattanaik and Annapurna [48] developed and validated two spectrophotometric methods for the simultaneous determination of irbesartan and hydrochlorothiazide in pharmaceutical formulations using hydrochloric acid. Both irbesartan and hydrochlorothiazide showed linearity over the concentration range 1.0–35.0 and 0.1–30 μg/mL in both simultaneous equation method as well as absorbance ratio method (Q-analysis).

Kumar and Annapurna [49] developed and validated two spectrophotometric methods for the simultaneous determination of irbesartan and hydrochlorothiazide in pharmaceutical formulations (tablets). These methods were the simultaneous equation and absorbance ratio method (Q-analysis) in phosphate buffer (pH 7.5). Irbesartan and hydrochlorothiazide showed linearity over the concentration range 5–35 μg/mL and 0.2–40 μg/mL in both methods.

Syahputra et al. [50] developed a spectrophotometric method by dual wavelength measurements for the simultaneous analysis of irbesartan and hydrochlorothiazide in tablet preparations without separation. The measurements were carried out at λ 263.4 and 281 nm for irbesartan and at λ 243.4 and 247.6 nm for hydrochlorothiazide. The results obtained by irbesartan and hydrochlorothiazide in tablets were 108.04 ± 2.696 and 94.28 ± 4.48%, respectively, with good precision and accuracy. The method successfully applied for the determination of the concentration of irbesartan and hydrochlorothiazide in tablets.

Sivasubramanian and Lakshmi [51] described an UV-visible spectrophotometric method for the simultaneous determination of irbesartan, hydrochlorothiazide and telmisartan in tablets using absorbance correction H-point standard addition method without any prior separation of samples. The linear range was 0.5–3 μg/mL for irbesartan, hydrocholrothiazide and telmisartan. The relative standard deviation for the simultaneous determination of 1 μg/mL of irbesartan and 2 μg/mL of hydrochorothiazide and telmisartan was 0.819, 0.388 and 1.365, respectively. The method could be claimed as green analytical chemistry as it did not involve the use of organic solvents and hence could be utilized for the routine analysis of these drugs from tablets.

Abdellatef [52] developed an extractive spectrophotometric method for determination of disopyramide and irbesartan in their pharmaceutical formulation. Picric acid, bromocresol green, bromothymol blue, cobalt thiocyanate and molybdenum(V) thiocyanate were tested as spectrophotometric reagents for the determination of both disopyramide and irbesartan. Reaction conditions were optimized to obtain colored complexes of higher sensitivity and longer stability. The absorbance of ion-pair complexes formed were found to increase linearity with increase in concentrations of disopyramide and irbesartan which were corroborated by correction coefficient values. The developed methods were successfully applied for the determination of disopyramide and irbesartan in bulk drugs and pharmaceutical formulations. The common excipients and additives did not interfere in their determination. The results obtained by the proposed methods were statistically compared by means of student *t*-test and by the variance ratio *F*-test. The validity was assessed by applying the standard addition technique. The results were compared statistically with the official or reference methods showing a good agreement.

Omar et al. [53] described a non-extractive spectrophotometric method for determination of irbesartan and other angiotensin receptor blockers, namely losartan l telmisartan and valsartan in their pure forms and pharmaceutical tablets. The method was based on the interaction of the drugs with sulfophthalein dyes to form stable, yellow colored, ion-pair complexes peaking at 413–419 nm. The dyes used were bromothymol blue, bromophenol blue and bromocresol green. The proposed method was successfully applied for determination of these drugs in their tablets and in presence of their degradation products with good recoveries. The method was recommended for quality control and routine analysis where time and cost effectiveness are of great importance.

El-Sutohy et al. [54] developed two non-extractive spectrophotometric methods for determination of irbesartan and losartan potassium in bulk forms as well as in their pharmaceutical dosage forms. The methods were based on using chloroformic solutions of either bromcresol purple or cresol red for determination of the cited drugs giving highly colored complexes that measured at 415 or 435 nm for both methods, respectively. Different variables affecting the reactions were studied and optimized. Under the optimum reaction conditions, linear relationships with good correlation coefficients (0.9960–0.9999) were obtained over the concentration ranges of 2–50 μg/mL. Good accuracy and precision were successfully obtained for the analysis of tablets containing each drug alone or combined with other diuretic drug hydrochlorothiazide. No interferences could be observed from the co-formulated hydrochlorothiazide as well as the commonly encountered excipients present in the tablets.

Fathima et al. [55] developed a spectrophotometric method for the estimation of irbesartan, paroxetine, famotidine, quetiapine, dextromethorphan, levofloaxacin, and itraconazole. The method was based on formation of charge-transfer complexes with 2,3 dichloro 5,6 dicyano-*p*-benzoquinone reagent. Acetonitrile was found to be suitable solvent for the analysis. The methods were validated in terms of ICH guidelines. The complexes were found to have 1:1 ratio. For irbesartan, the colored complex was measured at 545 nm and Beer's law was obeyed in the range of 13.4–121 μg/mL. The limits of detection and quantitation were 1.37 and 4.12 μg/mL, respectively.

Darwish et al. [56] developed and validated two novel 96-microwell-based spectrophotometric assays for determination of irbesartan in its tablets.

The two methods were based on the formation of colored charge-transfer complexes between irbesartan as a n-electron donor and each of 2,3-dichloro-5,6-dicyano-1,4-benzoquinone and *p*-chloranilic acid as π-electron acceptors. The proposed assays were carried out in 96-microwell plates, and the absorbances were measured by microwell-plate reader. Under the established optimum conditions, Beer's law was obeyed in irbesartan concentration ranges of 1–160 and 0.5–180 μg/mL, for the methods used 2,3-dichloro-5,6-dicyano-1,4-benzoquinone and *p*-chloranilic acid methods, respectively. No interference was observed from the additives that are present in the tablets or from hydrochlorothiazide that is co-formulated with irbesartan in some tablets. The assays were successfully applied to the determination of irbesartan in tablets with good accuracy and precision. The two assays had high throughput properties, consume minimum volumes of organic solvents thus reduce the exposures of the analysts to their toxic effects, and they reduce the analysis cost by 50-folds. Although the proposed assays were validated for irbesartan, however, the same methodology could be used for any electron-donating analyte for which a charge-transfer reaction can be performed.

Abdelmonem et al. [57] developed a spectrophotometric method for the determination of irbesartan and bisoprolol hemifumarate in their tablets dosage form. The method was based on the reaction between the studied drugs and 7-chloro-4-nitrobenzen-2-oxa-1,3-diazole (NBD-Cl). The developed orange products were measured at their absorption maxima at 476 and 479 nm for irbesartan and bisoprolol hemifumarate, respectively. The absorbance-concentration plots were rectilinear over the range of 20–90 μg/mL for irbesartan and 40–160 μg/mL for bisoprolol hemifumarate with lower detection limits of 1.37 and 3.98 μg/mL and lower quantification limits of 4.17 and 12.06 μg/mL for irbesartan and bisoprolol hemifumarate respectively. The method was successfully applied to the analysis of the two selected drugs. The study showed that the results were in good agreement with the reference methods.

Baraka et al. [58] developed two spectrophotometric methods for determination of irbesartan, losartan, hydrochorothiazide and atenolol in bulk and tablet forms. The first method based on a formation of the colored chromogen by a condensation reaction between irbesartan, losartan and atenolol and vanillin in acidic conditions and the products were measured at λ_{max} 546, 552 and 560 nm for irbesartan, losartan and atenolol, respectively. Under the indicated conditions, this method was linear over the concentration range of 40–240, 80–240 and 40–200 μg/mL for irbesartan, losartan and atenolol, respectively. In the second method, 1,2-naphthoquinone-4-sulfonate sodium reacted with irbesartan, losartan and hydrochlorothiazide through nucleophilic substitution reaction producing orange colored products in alkaline medium showing maximum absorption peaks at λ_{max} 465 nm for irbesartan, losartan and hydrochlorothiazide. The method was linear over the concentration range of 1–6, 0.2–1 and 0.25–1.25 μg/mL for irbesartan, losartan and hydrochlorothiazide, respectively. The methods were statistically applied for the determination of drugs in both bulk and tablet forms. Results were compared with reference methods and no statistically difference was obtained.

Ibrahim et al. [59] developed a spectrophotometric method for evaluating the reducing properties of irbesartan and other antihypertensive (valsartan and lisinopril) and antibacterial drugs (cefixime and cefprozil). The method was based on the reaction of the cited drugs with iron (III) ions as an oxidizing agent. The resulted iron (II) ions reacted with O-phenanthroline forming the well-known highly stable orange-red colored chelate complex, which exhibited an absorption maximum at 510 nm. Different conditions were thoroughly studied and optimized. The reducing properties of the selected drugs were compared to each other (under the applied experimental conditions) and results showed that these drugs act as reducing agents. The method was applied to investigate the stability of these drugs in air and results showed that cefprozil, cefixime and irbesartan are very labile to autoxidation.

Ashour and Bayram [60] developed a kinetic spectrophotometric method for determination of irbesartan in pure or pharmaceutical formulations. The method based on oxidation of 3-methyl-2-benzothiazolinone hydrazone hydrochloride monohydrate with Ce(IV) in 2% sulfuric acid medium, followed by coupling the produced electrophilic intermediate (diazonium salt of the reagent) with irbesartan to give greenish-blue colored product (1:1, stoichiometry) having maximum absorption at 629 nm and the colored species were stable for more than 1 h. The initial rate and fixed time (at 35 min) methods were adopted for determination of irbesartan concentration. The linearity was in the ranges of 5.0–40.0 and 2.0–45.0 μg/mL and the limit of detection is 0.46 and 0.40 μg/mL for initial rate and fixed time methods, respectively. Molar absorptivity for the method was found to be 1.50 × 104 L/mol/cm. The validated kinetic methods could be successfully applied to the analysis of irbesartan in bulk and tablet dosage form and in the routine quality control analysis. The percentage recoveries were above 100% for both methods. The excipients did not interfere in the analysis.

Tekkeli and Önal [61] developed a spectrophotometric method for the determination of irbesartan among five angiotensin II type 1 receptor antagonists in tablets and plasma. The spectrophotometric method involved the addition of a measured excess of bromate-bromide in HCl medium and subsequent estimation of the residual bromine by reacting with a fixed amount of methyl orange. The proposed method was successfully applied to the determination of the investigated drugs in their pure forms and pharmaceutical preparations.

Ashour and Bayram [62] developed a spectrophotometric method for the determination of irbesartan and fexofenadine hydrochloride in bulk and pharmaceutical preparations. The method is based on the reaction of the above cited drugs with naphthol blue black dye in solutions containing Britton buffer to form ion-pair complexes extractable with chloroform and subsequently measured spectrophotometrically at 625 nm. All the reaction conditions for the proposed methods were studied. The reactions were extremely rapid at room temperature and the absorbance values remained unchanged for at least 24 h. The linear range of the method for irbesartan was 10–244 μg/mL with a detection limit of 0.75 μg/mL. The proposed method was applied successfully for the determination of irbesartan in pharmaceutical formulations. Interferences of the other ingredients and excipients were not observed. The results obtained were compared statistically with those obtained by the official method and showed no significant differences regarding accuracy and precision.

Rani et al. [63] developed a spectrophotometric method for the estimation of irbesartan and telmisartan in bulk and pharmaceutical formulations. The method was based on the formation of ion association complex of the drug with eriochrome black-T in acidic buffer of pH 3.5 followed by extraction into chloroform. For irbesartan, the measurements were carried out 510 nm. The linearity range of irbesartan was found to be 50–250 μg/mL with correlation coefficient of 0.9931 and molar absorptivity of 2.090×10^5 mL/mol^{-1} cm^{-1}. The developed method was validated and successfully applied for analysis of dosage formulations.

Attia et al. [64] developed an analytical method for determination of some angiotensin receptor antagonists including irbesartan in pure forms, pharmaceutical tablets, urine, and blood plasma samples. The method was based on formation of an ion associate complex between studied drugs and eosin-Y dye in aqueous buffered medium. Under the optimum conditions, the binary complexes showed absorption maxima at 547 nm. Calibration curve of irbesartan was linear in the range of 2–8 μg/mL with

LOD of 0.082 μg/mL. The reaction mechanism was studied. The method had the advantage of being applicable for the determination of the studied drugs without prior extraction. The method was recommended for quality control and routine analysi where time, cost effectiveness and high specificity of analysis techniques were of great importance.

6.3.2 Spectrofluorimetry

El-Shaboury et al. [65] developed a spectrofluorimetric method for determination of some angiotensin II receptor antagonists including irbesartan: losartan potassium, valsartan and candesartan cilexetil in pure forms as well as in their pharmaceutical dosage forms. All the variables affecting the relative fluorescence intensity were studied and optimized. Under the optimum conditions, linear relationships with good correlation coefficients (0.9982–0.9991) were obtained over the concentration range from 0.006 to 1.7 μg/mL. Good accuracy and precision were successfully obtained for the analysis of tablets containing each drug alone or combined with hydrochlorothiazide without interferences from the co-formulated hydrochlorothiazide or the additives commonly present in tablets.

Abdelmonem et al. [57] developed a spectrofluorimetric method for the determination of irbesartan and bisoprolol hemifumarate in their tablets dosage form. The method was based on the reaction between the studied drugs and 7-chloro-4-nitrobenzen-2-oxa-1,3-diazole. The developed orange products were measured in the appropriate solvent fluorometrically at 534 and 538 nm after excitation at 465 and 470 nm for irbesartan and bisoprolol hemifumarate, respectively. The fluorescence-concentration plots were rectilinear over the range of 2.5–8 μg/mL for irbesartan and 6–16 μg/mL for bisoprolol hemifumarate with lower detection limits of 0.18 and 0.39 μg/mL and a lower quantification limit of 0.55 and 1.17 μg/mL for irbesartan and bisoprolol hemifumarate, respectively. The method was successfully applied to the analysis of the two selected drugs. The results were in good agreement with the reference methods.

El-Bagary et al. [40] developed a spectroflourimetric method based on measuring the native fluorescence of irbesartan in the range of 1–6 μg/mL (λ_{em} at 785 nm upon excitation at 250 nm) in 0.1 N H_2SO_4. The method was satisfactorily applied to the analysis of the pharmaceutical formulations and the results proved to be specific and accurate for the quality control of the cited drugs in pharmaceutical dosage forms.

Tekkeli and Önal [61] developed and optimized a spectrofluorimetric method for the determination of five Angiotensin II type 1 receptor antagonists, including irbesartan, in tablets and plasma. The method based on the oxidation of the drugs with cerium(IV) and subsequent monitoring of the fluorescence of the induced cerium(III) at 365 nm with excitation at 255 nm. The method was successfully applied to the determination of the investigated drugs in their pure forms and pharmaceutical preparations. Besides, the method was applied to the determination of irbesartan and telmisartan in biological fluids with good accuracy and precision.

Farouk et al. [44] developed and validated a spectrofluorimetric method for determination of irbesartan, torasemide and olmesartan medoxomil. The proposed methods were validated according to International Conference of Harmonization (ICH) guidelines. For irbesartan, the method was linear in the range of 300–2300 ng/mL with a correlation coefficient of 0.9998. The accuracy of the method was $100.86 \pm 1.24\%$ and its repeatability and intermediate precisions were 99.8 ± 0.66 and 99.6 ± 0.78, respectively. The method was successfully applied to determine irbesartan in its pharmaceutical preparations (Approvel® tablets) and the label claim was found to be $98.5 \pm 0.47\%$. The obtained results were statistically compared to the reference method of analysis and no significant difference was found.

Omar et al. [53] described a spectrofluorimetric method for determination of certain angiotensin receptor blockers, namely: irbesartan, losartan, telmisartan and valsartan in their bulk forms and pharmaceutical tablets. The method was based on the formation of non-extractive binary complex between eosin and the drugs without using any surfactant. The proposed methods were successfully applied for determination of these drugs in their tablets and in presence of their degradation products with good recoveries. They were recommended for quality control and routine analysis where time and cost effectiveness are of great importance.

Fu et al. [66] described the simultaneous determination of irbesartan and repaglinide in biological plasmas using micellar enhanced excitation-emission matrix fluorescence coupled with ATLD method. The method was assisted with the use of second-order calibration method based on alternating trilinear decomposition (ATLD) method. The weak fluorescence of irbesartan was enhanced by changing the microenvironment in samples and improving the fluorescence quantum yield by using an appropriate micellar enhanced surfactant. The enhanced excitation-emission matrix fluorescence of irbesartan could be accurately resolved and simultaneously attained the optimal concentration even in the presence of a potentially strong intrinsic fluorescence from complex biological matrixes, such as rat and human plasmas, by using the ATLD method, which completely exploited the "second-order advantage." The average recovery of irbesartan obtained from ATLD with the factor no. of 3 ($N = 3$) was 99.1% ± 4.1% for rat plasma and 97.1% ± 3.6% for human plasma. Several statistical methods, including Student's *t*-test, figures of merit, and elliptical joint confidence region, were utilized to evaluate the accuracy of the proposed method. Results showed that the developed method can maintain second-order advantage in simultaneous determinations of the weak fluorescent analytes of interest in different biological plasma matrixes.

6.3.3 *Fourier-transform infrared spectroscopy*

Khalid et al. [67] developed a non-destructive analytical method for the quantification of irbesartan in tablets using Fourier Transform Infrared (FT-IR) spectroscopy. The FT-IR quantification of irbesartan in tablets was performed using two chemometric approaches, partial least squares and principal component regression methods. To compare the predictive ability of these models, the standard errors of prediction were calculated. Standard error prediction error values in the range of 0.941–1.62% for calibration and validation data sets were obtained for the two procedures applied. A successful quantification of irbesartan in tablets containing 150 mg active ingredient/tablet was performed using the partial least squares model, with good recovery. The proposed procedure could be used for quantification of irbesartan in commercial tablets.

6.4 Chromatographic methods

6.4.1 *Capillary electrophoresis*

Dal and Koyutürk [68] developed and validated a capillary electrophoretic method for the simultaneous determination of irbesartan and hydrochlorothiazide in tablets. A 40 cm effective length (48.5 cm total, 50 μm i.d.) fused silica capillary was used to separate the compounds. The run buffer was used 10 mM borate buffer containing 10% methanol (pH 9.0). The samples were injected for 10 s under 50 mbar pressure and the applied potential was 25 kV. Under the optimum conditions, compounds were separated with less than 7 min. The limit of quantification values were 3.370×10^{-7} M (0.144 μg/mL) and 3.140×10^{-7} M (0.094 μg/mL) for irbesartan and hydrochlorothiazide, respectively. The method was applied successfully to both irbesartan and irbesartan/hydrochlorothiazide combined tablets with good recoveries almost 100%.

Hillaert and Bossche [69] developed capillary zone electrophoretic and micellar electrokinetic capillary chromatographic methods to simultaneously separate hydrochlorothiazide and six angiotensin-II-receptor antagonists, namely: candesartan, eprosartan mesylate, irbesartan, losartan potassium, telmisartan, and valsartan. These two methods were suitable for the qualitative and quantitative determination of combined drugs in pharmaceutical formulations. Depending on the angiotensin-II-receptor antagonists, at least one of the two methods can be used for each combination. The two methods were validated in terms of their linearity of response, reproducibility, and accuracy.

Fayez and Hegazy [70] developed a micellar electrokinetic chromatographic method for the determination of hydrochlorothiazide in the presence of irbesartan, bisoprolol and hydrochlorothiazide main impurity. Four factors affecting the separation, namely: sodium dodecyl sulfate concentration, buffer concentration, temperature and voltage—were studied. Optimization studies were performed with the aid of a central composite design with six central points. The optimal separation conditions were obtained by applying Derringer's desirability function, and the conditions were borate buffer 17 mM (pH 9), sodium dodecyl sulfate 5.2 mM, temperature 25 °C and voltage 12 kV. Hydrochlorothiazide, irbesartan and bisoprolol were successfully determined in their pure form and pharmaceutical formulations with separation in less than 8 min. Calibration curves ($R > 0.999$) were prepared and complete method validation was performed according to ICH guidelines. The results obtained were statistically compared with that of the official methods.

Zhang et al. [71] developed a novel polymer monolith microextraction (PMME) using a poly(methacrylic acid-ethylene glycol dimethacrylate) (poly(MAA-EGDMA)) monolith in conjunction with capillary zone electrophoresis for the determination of several angiotensin II receptor antagonists in human urine. The extraction device consisted of a regular plastic syringe (1 mL), a poly(MAA-EGDMA) monolithic capillary (2 cm × 530 μm I.D.) and a plastic pinhead connecting the former two components seamlessly. The extraction was achieved by driving the sample solution through the monolithic capillary tube using a syringe infusion pump, and for the desorption step, an aliquot of organic solvent was injected via the monolithic capillary and collected into a vial for subsequent analysis by capillary zone electrophoresis. The best separation was realized at 25 kV using a buffer that consisted of 50% acetonitrile and 50% buffer solution (v/v) containing 10 mM disodium hydrogen phosphate (adjusted to pH 2.3 with 1 M hydrochloric acid). The method was successfully applied to the determination of telmisartan, irbesartan and losartan in urine samples with candesartan as internal standard, yielding the detection limit of 15–20 ng/mL. Close correlation coefficients ($R > 0.999$) and excellent method reproducibility were obtained for all the analytes over a linear range of 0.08–3 μg/mL.

González et al. [72] optimized a capillary zone electrophoretic method for the separation of five angiotensin II receptor antagonists (losartan, irbesartan, valsartan, telmisartan and eprosartan) and two of their metabolites (EXP 3174 and Candesartan M1) by means of experimental design methodologies. The effects of the buffer (pH, concentration and composition), the organic modifier and voltage were studied. Critical factors were identified in a screening design (fractional factorial design) and sequentially an optimization design (central composite design) was used to choose optimal conditions for separation. The most favorable electrophoretic conditions were found by setting the resolution at a threshold value ($Rs \leq 1.5$) and minimizing, if possible, analysis time. Successful results were obtained with a 50 mM potassium dihydrogen phosphate:boric acid (25:75 v/v) buffer at pH 5.5 in the presence of 5% methanol and application of a 25 kV voltage. Analysis time was 8 min in a conventional fused silica capillary (50 cm effective length) in a normal cationic mode (anode at the inlet and cathode at the outlet) after hydrostatical sample injection for 30 s.

6.4.2 Thin layer chromatography

Shah et al. [73] developed and validated a high performance thin layer chromatographic (HPTLC) method for the simultaneous estimation of irbesartan and hydrochlorothiazide in combined dosage forms. The stationary phase used was pre-coated silica gel 60F254. The mobile phase used was a mixture of acetonitrile:chloroform:glacial acetic acid (7:3:0.1, v/v/v). The detection of spots was carried out at 260 nm. The method was validated in terms of linearity, accuracy, precision and specificity. The calibration curve was found to be linear between 100 to 700 ng/spot for irbesartan and 100 to 350 ng/spot for hydrochlorothiazide. The limit of detection and the limit of quantification for the irbesartan were found to be 30 and 100 ng/spot, respectively, and for hydrochlorothiazide 25 and 100 ng/spot respectively. The proposed method can be successfully used to determine the drug content of marketed formulation.

Hussein et al. [74] described a high performance thin layer chromatographic (HPTLC) method for simultaneous quantification of four angiotensin II receptor antagonists, namely, losartan, irbesartan, valsartan, and olmesartan. The method was based on HPTLC separation of the four drugs using a reflectance-absorbance mode at 255 nm. Separation was carried out on pre-coated silica gel HPTLC plates, using chloroform:glacial acetic acid (7.5:2.5, v/v) as mobile phase. The four drugs were satisfactorily resolved with R_F values of 0.23 ± 0.01, 0.36 ± 0.01, 0.45 ± 0.01, and 0.78 ± 0.01 for olmesartan, losartan, irbesartan, and valsartan, respectively. The method was validated according to the International Conference on Harmonization (ICH) guidelines. The linear regression analysis data for the calibration plots of all studied drugs produced a good linear relationship with correlation coefficients ranging from 0.9988 to 0.9996 and coefficients of determination ranging from 0.9976 to 0.9990 over the concentration range of 90–750 ng per band. The limits of detection and quantification for all studied drugs ranged from 14.85 to 30.42 and from 45.00 to 92.18 ng per band, respectively. The developed method was applied successfully for the determination of the studied drugs in pharmaceutical formulations and spiked human plasma with good precision and accuracy.

Khodke et al. [75] develop and validated a stability-indicating high performance thin layer chromatographic (HPTLC) method for simultaneous estimation of irbesartan and hydrochlorothiazide using TLC plates precoated with silica gel 60F254 and the mobile phase comprising acetonitrile:chloroform in the ratio of 5:6 (v/v). Irbesartan and hydrochlorothiazide were well resolved with Rf values of 0.27 ± 0.03 and 0.45 ± 0.03, respectively. Wavelength selected for the quantization was 270 nm. Inherent stability of these drugs was studied by exposing both drugs to various stress conditions as per the guidelines of the International Conference on Harmonization (ICH). Dry heat, oxidative, photolysis (ultraviolet and cool white fluorescent light) and hydrolytic conditions under different pH values. The results revealed that both the drugs were not degraded under dry heat and photolytic conditions but showed degradation under hydrolytic condition. The degraded products of irbesartan and hydrochlorothiazide were well resolved from the individual bulk drug response. The study concluded that the method was simple, specific, precise and stability indicating. The specificity of the method was confirmed by peak purity profile of the resolved peaks.

Shah et al. [76] developed a simple, sensitive, precise high performance thin layer chromatographic (HPTLC) method for estimating irbesartan and amlodipine besylate in combination. The method involved TLC aluminium plates pre-coated with silica gel 60F254 as the stationary phase and a solvent system of chloroform:toluene:methanol:acetic acid (6:2.5:1.5:0.5, v/v/v/v). The Rf values were 0.57 ± 0.02 for irbesartan and 0.30 ± 0.02 for amlodipine besylate. The separated spots were analyzed densitometrically in absorbance mode at 244 nm. The results were linear in the range 50–500 ng per band for irbesartan and 400–900 ng per band for amlodipine besylate, with limits of detection of 14.4 ng per band and 12.7 ng per band, respectively. The limits of quantification were 47.4 ng per band for irbesartan and 42.0 ng per band for amlodipine. After irbesartan and amlodipine stock solutions were subjected to acid and alkaline hydrolysis, chemical oxidation, dry heat degradation and photodegradation, the peaks for degraded product were well resolved from that for pure drug, with significant differences in their Rf values when assayed by the HPTLC method. The method was validated with respect to linearity, accuracy, precision and robustness and was used successfully to estimate irbesartan and amlodipine besylate in a synthetic mixture.

Sivasubramanian and Lakshmi [77] developed and validated a high performance thin layer chromatographic (HPLC) method for estimation of irbesartan, telmisartan, hydrochlorothiazide and ramipril. The method employed TLC aluminium plates pre-coated with silica gel 60F254 as the stationary phase. The solvent system consisted of acetonitrile:toluene:methanol:formic acid (8:10:2:0.6, v/v/v/v). Densitometric analysis of drugs was carried out in the absorbance mode at 210 nm. The method was validated with respect to linearity, specificity, precision, limit of detection, limit of quantification and recovery. The linear regression analysis data for the calibration plots showed a good linear relationship with R^2 of 0.999 for all the drugs in the concentration range of 250–1250 ng/spot for telmisartan and ramipril and 500–2500 ng/spot for irbesartan and hydrochlorothiazide, respectively. Recovery of all drugs was achieved in the range of 98.85–101.31% by the method. Statistical analysis proved that the method was repeatable and specific for the estimation of all four drugs. The method was successfully applied for estimation of all four drugs in their combined dosage form.

Rosangluaia et al. [78] developed and validated a high performance thin layer chromatographic method for the simultaneous estimation of irbesartan and hydrochlorothiazide in combined dosage forms. The stationary phase used was pre-coated silica gel 60F254. The mobile phase was a mixture of acetonitrile:ethyl acetate (8:2, v/v). The detection of spots was carried out at 260 nm. The method was validated in terms of linearity, accuracy, precision and specificity. The calibration curve was found to be linear between 100 to 600 ng/spot for irbesartan and 50 to 250 ng/spot for hydrochlorothiazide. The limit of detection and the limit of quantification for the irbesartan were found to be 22.19 and 73.98 ng/spot, respectively, and for hydrochlorothiazide 12.42 and 41.394 ng/spot, respectively. The proposed method can be successfully used to determine the drug content of marketed formulation.

Mehta and Morge [79] developed and validated a high performance thin layer chromatographic (HPTLC) method for determination of irbesartan and hydrochlorothiazide in pharmaceutical dosage forms. Chromatography was performed on silica gel 60GF254 TLC plates using acetone:chloroform:ethyl acetate:methanol (3:3:3:0.5, v/v/v/v) as the mobile phase. The detection was carried out at 250 nm. The Rf values were found to be 0.27 for irbesartan and 0.37 for hydrochlorothiazide. The regression data curve showed good linear relationship in the concentration range 300.4–1802.4 μg for irbesartan, 25.0–150.4 μg for hydrochlorothiazide. The percent drugs estimated from marketed formulations were found to be in the range 99.42% and 99.58% for irbesartan and hydrochlorothiazide, respectively. The percent recoveries of drugs were found to be 99.2% and 98.5% for irbesartan and hydrochlorothiazide respectively. The method was validated for accuracy, precision, specificity, linearity and detector response. The proposed method provided a quality control tool for routine analysis of irbesartan and hydrochlorothiazide in formulation.

Kalaiselvi and Lalitha [80] developed a high performance thin layer chromatography (HPTLC)-densitometry method for separation and determination of chlorthalidone and irbesartan in pharmaceutical dosage forms. The compounds were separated on silica gel 60 GF254 HPTLC plates using mobile phase of toluene:ethyl acetate:acetonitrile: methanol:ammonia solution-25% (5:2:2:1:0.2, v/v/v/v) as compact spots at Rf of 0.57 for chlorthalidone and Rf of 0.36 for irbesartan. Densitometric detection was performed at 254 nm. The method was validated in terms of linearity, precision, accuracy, limit of detection, and limit of quantification. The calibration curve of irbesartan was linear in the range of 150–900 ng/spot. Recovery of irbesartan varied in range of 99.76–100.40%. The limits of detection and quantitation of irbesartan were 11.34 and 14.37 ng/spot, respectively. It was observed that the HPTLC method could be used for efficient analysis of both chlorthalidone and irbesartan in combined tablet dosage forms.

Sane et al. [81] developed a stability-indicating high performance thin layer chromatographic (HPTLC) method for the analysis of irbesartan in tablet dosage form. Irbesartan and the internal standard (losartan potassium) was separated on the plate coated with silica gel 60F254. Chromatograms were developed using a mixture of ethylacetate:toluene:acetic acid (6:4:1, v/v/v). Quantification was carried out by the use of densitometer absorbance mode at 247 nm. These HPTLC-system was quantitatively evaluated in terms of stability, accuracy and calibration proving the utility in the analysis of its tablet dosage form.

Zeng et al. [82] performed the transfer of thin layer chromatography (TLC) Global Pharma Health Fund E.V. Minilab Manual and U.S. Food and Drug Administration (FDA) Compendium protocols for detecting falsified or substandard drugs in pharmaceutical products to quantitative high performance TLC (HPTLC)-densitometry methods suitable for support of regulatory authorities following a previously published model process for moxfloxacin HCl, ofloxacin, and amoxicillin trihydrate. In addition, the guidelines of the process were followed to develop and validate HPTLC-densitometry methods for pharmaceutical products containing acetylsalicylic acid + acetaminophen + caffeine, nimesulide, irbesartan, and pantoprazole that are not included in the Minilab Manual or FDA Compendium. The new methods involved silica gel 60 F_{254} plates; semiautomated bandwise sample and standard solution application; ascending mobile phase development of plates in a glass chamber; and automated slit scanning densitometry for detection, identification, and quantification. Data for standard and sample solution preparation, calibration curves, assay of pharmaceutical products compared to label declarations, standard addition validation, and peak purity and peak identity are provided. Supplemental FDA Compendium screening methods corresponding to the HPTLC-densitometry methods for caffeine, nimesulide, irbesartan, pantoprazole, and acetylsalicylic acid co-formulated with acetaminophen and caffeine were developed and published.

Abd El-Rahman et al. [83] developed two HPTLC methods for the determination of irbesartan suitable for quality control laboratories and using reflectance-fluorescence mode for detection increased the sensitivity accompanied with the selectivity of HPTLC methods for the determination of irbesartan in plasma. The methods used TLC plates pre-coated with silica gel 60F254 and a mobile phase composed of acetonitrile-chloroform-glacial acetic acid (6:3.8:0.2 v/v/v). The methods provided high sensitivity (LOQ = 6.02 ng/band), such high sensitivity inherent to fluorometric based detection. The recoveries of irbesartan from plasma ranged from 95.8% to 97.4% with a standard deviation did not exceed 2.96. These results indicated that the method was suitable to quantify the target irbesartan in human plasma without interference from various plasma components.

Abd El-Rahman et al. [84] developed HPTLC method with reflectance/fluorescence densitometry was developed to separate and quantify some angiotensin II receptor blockers, namely, losartan potassium, irbesartan, valsartan, and olmesartan medoxomil, in tablets and plasma. Separation of the target analytes was performed on pre-coated silica gel HPTLC plates developed using chloroform:glacial acetic acid (7.5:2.5, v/v) mobile phase. The drugs were adequately resolved. After the exposure of the chromatograms to concentrated hydrochloric acid vapor for 10 min, the scanner was set in the reflectance/fluorescence mode with 255 nm excitation wavelength and >400 nm emission filter. The linear regression analysis data for the calibration curves of all studied drugs produced a good linear relationship with correlation coefficients ranging from 0.9991 to 0.9999 over the concentration range of 2–20 ng/band. LOD and LOQ values of all studied drugs ranged from 0.6 to 0.8 and from 1.7 to 2.4 ng/band, respectively. The developed method was successfully applied for analysis of the target drugs in tablets and spiked human plasma samples with good precision and accuracy.

A summary for the chromatographic conditions for determination of irbesartan by high performance thin layer chromatography (HPTLC) is give in Table 4.

6.4.3 Liquid chromatography

The high performance liquid chromatographic (HPLC) methods for determination of irbesartan in pharmaceuticals and biological matrix (as a single component or in combinations), published before 2014 [85–122], have been reviewed and described by Virani et al. [123]. Also, Nalini and Nivedhitha [21] and Yogeesh et al. [22] reviewed many UV and HPLC methods for irebseartan and its combinations in its pharmaceutical dosage forms. These methods and others are described herein in the following paragraphs, and summarized in Tables 5 and 6.

Water Corporation [155] transferred the USP HPLC method for irbesartan from the original HPLC method to the ultra-performance liquid chromatography (UPLC) platform. The transfer was facilitated by use of the ACQUITY UPLC columns calculator and the Waters reversed-phase column selectivity chart. System suitability requirements for the USP method transferred from the Alliance HPLC system to the ACQUITY UPLC system were met. The ACQUITY UPLC HSS T3, 1.8 μm column demonstrated robust and reliable performance at low pH (3.2) conditions for over 3000 injections. Other benefits realized by transferring this methodology were an 85% reduction in run time, 91% reduction in solvent consumption and 86% reduction in sample consumption. The ease of transfer using the simplistic workflow design 3 helped in the seamless adoption of UPLC technology with conspicuous gains in overall laboratory efficiency and productivity for running USP monographs.

Table 4 Chromatographic conditions for determination of irbesartan by HPTLC.

Stationary phase	Mobile phase	Detection (nm)	Rf value	Reference
TLC plates pre-coated with silica gel 60F254 (10 × 10 cm, Merck)	Acetonitrile:chloroform:glacial acetic acid (7:3:1, v/v/v)	UV: 260	0.6 ± 0.04	[73]
HPTLC plates pre-coated with silica gel 60F254 (20 × 10 cm, Merck)	Chloroform:glacial acetic acid (7.5:2.5, v/v)	UV: 255	0.45 ± 0.01	[74]
TLC plates pre-coated with silica gel 60F254 (10 × 10 cm, Merck)	Acetonitrile:chloroform (5:6, v/v)	UV: 270	0.27 ± 0.03	[75]
TLC plates pre-coated with silica gel 60F254 (10 × 10 cm, Merck)	Chloroform:toluene:methanol:acetic acid (6:2.5:1.5:0.5, v/v/v/v)	UV: 244	0.57 ± 0.02	[76]
Plates pre-coated with silica gel 60F254 (10 × 10 cm, Merck)	Acetonitrile:toluene:methanol:formic acid (8:10:2:0.6, v/v/v/v)	UV: 210	0.6	[77]
TLC plates pre-coated with silica gel 60F254 (10 × 10 cm, Merck)	Acetonitrile:ethyl acetate (8:2, v/v)	UV: 260	0.27	[78]
TLC plates pre-coated with silica gel 60GF254	Acetone:chloroform:ethyl acetate:methanol (3:3:3:0.5, v/v/v/v)	UV: 250	0.27	[79]
HPTLC pre-coated with silica gel 60GF254	Toluene:ethyl acetate:acetonitrile: methanol: ammonia solution-25% (5:2:2:1:0.2, v/v/v/v)	UV: 254	0.36	[80]
TLC plates pre-coated with silica gel 60F254	Ethylacetate:toluene:acetic acid (6:4:1, v/v/v)	UV: 247	–	[81]
TLC plates pre-coated with silica gel 60F254	Ethyl acetate-acetone-glacial acetic acid (18:4:0.1, v/v/v)		0.4	[82]
TLC plates pre-coated with silica gel 60F254	Acetonitrile:chloroform:glacial acetic acid (6:3.8:0.2, v/v/v)	Fluorescence: 255 nm excitation and > 400 nm emission	0.4 ± 0.1	[83]
TLC plates pre-coated with silica gel 60F254	Chloroform:glacial acetic acid (7.5:2.5, v/v)	Fluorescence: 255 nm excitation and > 400 nm emission	0.45	[84]

Table 5 Liquid chromatographic methods reported after 2014 for determination of irbesartan in pharmaceutical samples.[a]

Analyte	Sample	Separation conditions	Detection (nm)	Validation parameters	Reference
Irbesartan Atorvastatin Simvastatin Telmisartan	Bulk dosage form	Column: Symmetry C18 (75 × 4.6 mm; 3.5 μm); Mobile phase: ammonium acetate buffer: acetonitrile (40:60, v/v); Flow rate: 1 mL/min for 3.5 min and then changed to 2 mL/min	UV: 220	Range: 1–16 μg/mL; LOD: 0.189 μg/mL; LOQ: 0.630 μg/mL; Recovery: ≤100 ± 2%	[124]
Irbesartan	Bulk dosage form	Column: Phenomenex C18 column (150 × 4.6 mm, 5 μm); Mobile phase: 0.1% formic acid buffer:aetonitrile (60:40, v/v); Flow rate: 1.0 mL/min	UV: 220	Range: 1.5–120 μg/mL; Recovery: 97.68 ± 0.146%	[125]
Irbesartan	Bulk dosage form	Column: Inert ODS C18 (250 × 4.6 mm, 5 μm); Mobile phase: methanol:water (90:10, v/v); Flow rate: 1 mL/min	UV: 246	LOD: 6.51 μg/mL; LOQ: 1.973 μg/mL; Recovery 99.94–99.97%	[126]
Irbesartan related impurities	Pharmaceutical ingredient	Column: Kromasil C8 (3.5 μm, 150 × 4.6 mm) short column with 0.1% (v/v) orthophosphoric acid and acetonitrile	UV: 220	LOD: 0.019 μg/mL; Recovery: 99.6–100.7%	[127]
Irbesartan Hydrochlorthiazide	Dosage form	Column: Intersil-BDS C18 (250 × 4.6 mm, 5 μm); Mobile phase: methanol:0.05% orthophosphoric acid (90:10, v/v)	UV: 226	Range: 36–216 μg/mL; LOD: 0.65; LOQ: 1.97 μg/mL	[128]

Irbesartan Hydrochlorothiazide	Dosage form	Column: X Bridge shield C18 (150 × 3.0 mm, 3.5 μm); Mobile phase: 0.1% v/v formic acid: acetonitrile (40:60, v/v); Flow rate: 0.8 mL/min	UV: 235	LOD: 0.349 μg/mL; LOQ: 1.056 μg/mL	[129]
Irbesartan with 8 sartans and statins	Pure and dosage forms	Column: Kinetex C18 (100 × 4.6 mm, 2.6 μm); Mobile phase: KH_2PO4:acetonitrile; Flow rate: 1 mL/min	DAD: 280	Range: 11.46–233.52 mM; LOD: 2.01 mM; LOQ: 6.11; Recover: 99.64%	[130]
Irbesartan Chlorthalidone Cilnidipine	Dosage form	Column: C18 (250 × 4.6 mm, 5 μm); Mobile phase: potassium dihydrogen phosphate buffer: acetonitrile:triethylamine (80:20:0.1%, v/v/v); Flow rate: 1.0 mL/min	UV: 222	Range: 30–90 μg/mL; LOD; 0.862 μg/mL; LOQ: 2.612 μg/mL; Recovery: 99.27–99.81%	[131]
Irbesartan Atorvastatin	Synthetic mixture	Column: Thermo C18 (250 × 4.6 mm, 5 μm); Mobile phase: acetonitrile: methanol:0.1% formic acid (50:10:40, v/v/v); Flow rate: 1 mL/min	UV: 271	Range: 300–1500 ng; LOD: 10.94 ng; LOQ: 33.14 ng; Recovery: 99.84–99.98%	[132]
Irbesartan Atorvastatin	Synthetic mixture	Column: Inertsil C18 (150 × 4.6 mm, 5 μm); Mobile phase: acetonitrile:0.1% formic acid (40 60, v/v); Flow rate: 1.0 mL/min	UV: 262	Range: 400–800 μg/mL; Recovery: 99.88%	[133]

Continued

Table 5 Liquid chromatographic methods reported after 2014 for determination of irbesartan in pharmaceutical samples.[a]—cont'd

Analyte	Sample	Separation conditions	Detection (nm)	Validation parameters	Reference
Irbesartan Candesartan Hydrochlorothiazide	Dosage form	Column: Sunfire C18 (150 × 4.6 mm; 5 μm); Mobile phase: 0.1% acetic acid in water: acetonitrile (33:77, v/v); Flow rate: 0.7351 mL/min	UV: 225	Range: 0.05–120 μg/mL; Recovery: 98.88–99.79%	[134]
Irbesartan Cilnidipine	Dosage form	Column: Hypersil BDS C18 (25 × 0.46 cm); Mobile phase phosphate buffer:acetonitrile (40:60, v/v); Flow rate: 1 mL/min	UV: 240	Range: 30–90 μg/mL; LOD: 1.428 μg/mL; LOQ: 4.327 μg/mL; Recovery: 100.68 ± 0.80%	[135]
Irbesartan Cilnidipine	Dosage form	Column: Hypersil C18 BDS (25 × 0.46 cm); Mobile phase: phosphate buffer:methanol (85:15, v/v); Flow rate: 1 mL/min	UV: 248	Range: 15–45 μg/mL; LOD: 3.538 μg/mL; LOQ: 0.139 μg/mL	[136]
Irbesartan and 6 angiotensin II receptor blockers	Bulk forms	Column: C18 (250 × 4.6 mm, 5 μm); Elution: gradient; Mobile phase: methanol:sodium phosphate monobasic buffer (0.01 M, pH 6.5)	UV: 254	Precision RSD: < 1; Linearity: $r^2 \geq 0.999$); Accuracy: 98–102%; LOD and LOQ: 40 and 160 ng/mL, respectively	[137]
Irbesartan, Amlodipine besylate	Dosage forms	Column: Enable C18G (250 × 4.6 mm; 5 μm); elution mode: isocratic; Mobile phase: sodium acetate buffer (pH 4.0): acetonitrile (30:70, v/v); Flow rate: 0.6 mL/min; Retention time: 6.321 min	UV: 238	Range: 10–70 μg/mL; intr- and inter-day precision RSD: 0.435 and 0.30, respectively	[138]

Irbesartan	Dosage forms	Column: An Inertsil ODS C18 (250 × 4.6 mm, 5 μm); Elution: isocratic; Mobile phase: acetonitrile,water (40:60, v/v); Flow rate: 1.5 mL/min; retention time: 4.5 min	UV: 260	LOD and LOQ: 10 and 30 ng, respectively; Recovery: 100.61%	[139]
Irbesartan	Dosage forms	Column: A Microsorb-MV 100-5 C18 (250 × 4.6 mm, 5 mm); Mobile phase: methanol:0.02% formic acid (70:30, v/v); Flow rate: 1.0 mL/min: Retention time: 8.18 min	UV: 234	Range: 5–30 μg/mL; LOD and LOQ: 0.0010 and 0.0033 μg/mL, respectively; Recovery: 99.36–107.06%; Precision RSD: 0.56–1.36%	[140]
Irbesartan	Dosage forms	Column: Odyssil C18 (4.6 × 250 mm, 5 μm); Mobile phase: SDS-butyl alcohol: cyclohexane:water (3.0:6.0:0.8:92.2, v/v); Flow rate: 1.0 mL/min	UV: 230	Range: 50.01–175.04 μg/mL ($r = 0.9992$, $n = 6$); Recovery: 100.4% (RSD = 1.0%, $n = 9$); LOD: 0.02 μg/mL	[141]
Irbesartan, valsartan hydrochlorothiazide, losartan potassium	Bulk and dosage forms	Column: Intersil ODS-3 C18 (4.6 × 250 mm, 5 mm); Mobile phase: acetonitrile:50 mM KH2PO4 (pH 4.5) (45:55, v/v); Flow rate: 1 mL/min	UV: 210	Range: 10–100 mg/mL LOD: 0.01 mg/mL; Recovery: 97–103%; Reproducibility: within 1.67%	[142]

Continued

Table 5 Liquid chromatographic methods reported after 2014 for determination of irbesartan in pharmaceutical samples.[a]—cont'd

Analyte	Sample	Separation conditions	Detection (nm)	Validation parameters	Reference
Irbesartan	Dosage forms	Column: Lithosphere 100 RP-18e (250 × 4.0 mm, 5 μm); Mobile phase: phosphate buffer: acetonitrile (50:50, v/v); Flow rate: 0.8 mL/min	UV: 220	Range: 20–60 mg/mL.; Recovery: 99.0–101.6%; Repeatability and intermediate RSD: 0.68% and 1.35%, respectively	[143]
Irbesartan, related impurities	Dosage forms	Column: Phenomenex Luna C18 (250 × 4.6 mm, 5 μm); Elution: isocratic; Mobile phase: methanol: acetonitrile:5 mM KH_2PO_4 buffer (40:30:30, v/v); Flow rate: 0.5 mL/min	UV: 260	Range: 25–250 μg/mL; LOD: 18.51 mg/mL; Intra- and Inter-assay precision RSD: 0.5054 and 0.9327, respectively; Recovery: 101.3–103.5%	[98]

[a]Abbreviations are: UV: ultraviolet, PDA; photodiode array; SDS: sodium dodecylsulfate; RSD: relative standard deviation, LOD: limit of detection, LOQ: limit of quantitation.

Table 6 Liquid chromatographic methods reported after 2014 for determination of irbesartan in biological matrices.[a]

Analyte	Matrix	Separation conditions	Detection	Validation parameters	Reference
Irbesartan, Candesartan Gliquidone, Pioglitazone	Human serum	Column: Purospher® C18 (25 cm × 0.46 cm, 5 μm); Mobile phase: methanol:water (90:10, v/v); flow rate: 1 mL/min	UV: 230 nm	Range: 5–25 μg/mL; LOD: 0.83 μg/mL; LOQ: 2.78 μg/mL	[144]
Irbesartan Hydrochlorothiazide	Human plasma	Column: C_4; mobile phase: acetonitrile:phosphate buffer (pH 3.6); elution: gradient; flow rate: 1 mL/min	PDA	Range: 20–4000 ng/mL; LOQ: 20 ng/mL	[145]
Irbesartan, Valsartan Telmisartan	Bulk and dosage forms	Column: LiChrospher C18 column (250 × 4 mm, 5 mm); Elution: isocratic; Mobile phase: 10 mM phosphate buffer:acetonitrile (65:35, v/v); Flow rate: 1.0 mL/min	UV: 225	Range: 5–40 mg/mL; Recovery: 96.5–103.1%; LOD and LOQ: 0.06 and 0.16 mg/mL, respectively; Precision RSD: 0.83–2.9%; Recovery: 98.09%	[146]
Irbesartan	Human plasma	Column: Acquity UPLC BEH C18 (50 × 2.1 mm, 1.7 μm); Mobile phase: acetonitrile:methanol:10 mM ammonium acetate (70:15:15, v/v/v); Flow rate: 0.4 mL/min	MS/MS: MRM *m/z* 427.2 → 193.08	Range: 2–500 ng/mL; LLOQ: 2 ng/mL; Precision RSD: ≤ 9.91%	[147]
Irbesartan Hydrochlorothiazide	Human plasma	Column: Ace 5 C18 (100 × 4.6 mm, 5 μm); Mobile phase: methanol:0.1% formic acid in water (70:30, v/v); Flow rate: 1.0 mL/min	MS/MS: MRM *m/z* 427.1 → 193.0	Range: 10–5000 ng/mL; Precision RSD <15%; Accuracy: >85%	[148]
Irbesartan Hydrochlorothiazide	Human plasma	Column: C18 (50 × 4 mm, 3 m); Mobile phase: 2.5% formic acid: methanol:acetonitrile (40:45:15, v/v/v); Flow rate: 0.70 mL/min	MS/MS: MRM *m/z* 427 → 175	LOD: 0.01 μg/mL; LLOQ: 0.06 μg/mL; Recovery: 48.5–60.7%	[149]

Continued

Table 6 Liquid chromatographic methods reported after 2014 for determination of irbesartan in biological matrices.[a]—cont'd

Analyte	Matrix	Separation conditions	Detection	Validation parameters	Reference
Irbesartan Sinensetin	Rat plasma	Column: C_{18} Acquity® (100 × 2.1 mm, 1.7 μm); Mobile phase: acetonitrile: 0.1% formic acid (40:60, v/v); Flow rate: 0.3 mL/min	TQ-ESI: MRM *m/z* 429.1 → 205.9	LLOQ: 25 ng/mL; Precision RSD: 9.3–5.25% and 1.52–5.47%; Accuracy: >80%	[150]
Irbesartan, Candesartan	Rabbit eye tissue	Column: C18; elution: gradient	MS/MS: ESI, MRM: *m/z* 433 → 211	Range: 5–5000 ng/g; within-run and between-run accuracies (% bias): −8.0–10.0; RSD: 0.6–7.3%	[151]
Irbesartan	Human plasma	Column: 4.0 × 25 cm column having L1 packing from plasma; Mobile phase: phosphate buffer:acetonitrile (67:33. v/v); Flow rate: 1 mL/min	UV: 220	Range: 0.1–6 mg/mL; Accuracy: 89.33–96.37%; Intra-day, inter-day precision: 0.02 and 2.15, respectively; Recovery: 97.28%	[152]
Irbesartan	Human plasma	Column: Diamonsil C18; Mobile phase: water:acetonitrile (57:43, v/v); Flow rate: 1.0 mL/min; Retention time: 4.3 min	FD: 254 nm, emission and 371 nm excitation	Range: 0.05–5 mg/mL ($r = 0.9996$); LOD: 0.023 mg/mL. Recovery: 98.45–102.22%; Precision RSD: <10%	[153]
Irbesartan, Rosuvastatin	Rat plasma	Column: Agilent Eclipse Plus ODS (4.6 × 100 mm, 3.5 mm); Elution: gradient; Mobile phase: 6 mM ammonium Formate:0.1% formic acid:acetonitrile; Flow rate 0.4 mL/min	MS/MS: MRM: positive; *m/z* 492.1 → 206.9	Range: 1–10,000 ng/mL; LOD: 0.05 ng/mL	[154]

[a]Abbreviations are: UV: ultraviolet, PDA; photodiode array; FD: fluorescence detection; ESI: electrospray ionization, MS: mass detector, MRM: multiple reaction monitoring, RSD: relative standard deviation, TQ: triple quadrupole, *m/z*: mass/charge ratio, LOD: limit of detection, LOQ: limit of quantitation, LLOQ: lower limit of quantification.

Alhazmi et al. [124] developed and validated a reversed-phase HPLC method for simultaneous determination of irbesartan and other cardiovascular agents (atorvastatin, simvastatin and telmisartan) in bulk drugs and tablet oral dosage forms. The chromatographic separation was accomplished by using Symmetry C18 column (75 × 4.6 mm; 3.5 μm) with a mobile phase consisting of ammonium acetate buffer (10 mM; pH 4.0) and acetonitrile in a ratio 40:60 (v/v). Flow rate was maintained at 1 mL/min up to 3.5 min, and then suddenly changed to 2 mL/min till the end of the run (7.5 min). The data was acquired using ultraviolet detector monitored at 220 nm. The method was validated for linearity, precision, accuracy and specificity. The developed method showed an excellent linearity ($r^2 > 0.999$) over the concentration range of 1–16 μg/mL. The limits of detection and quantification of irbesartan were 0.189 and 0.630 μg/mL, respectively. Inter-day and intra-day accuracy and precision data were recorded in the acceptable limits. The method was successfully applied for quantification of all drugs in their tablet dosage forms with percent recovery within 100 ± 2%.

Mondal et al. [125] developed a stability-indicating RP-HPLC method for the irbesartan in bulk as well as tablet dosage form. Chromatographic separation was performed on the Phenomenex C18 column (150 × 4.6 mm, 5 μm), using 0.1% formic acid buffer (pH 3.2):acetonitrile (60:40, v/v) as the mobile phase at a flow rate of 1.0 mL/min and the detection was set at 220 nm. The linear range of the method was 1.5–120 μg/mL with a correlation coefficient of 0.999. The retention time of irbesartan was found 2.226 min using optimized condition. In force degradation study, the acidic and alkaline conditions decomposed the irbesartan more in compared to other stressed conditions. The method was found suitable to assay the marketed tablet dosage form with an average recovery value of 97.68 ± 0.146. All the other validation parameters were found within the limit that has been conducted as per the ICH guidelines.

Pradhan [126] developed a stability-indicating RP-HPLC method for the determination of irbesartan in bulk and dosage forms using telmisartan (10 μg/mL) as the internal standard. The method used Inert ODS C18 column (250 × 4.6 mm, 5 μm) and isocratic mode for a mobile phase consisted of methanol:water (90:10, v/v) and the pH was adjusted to 3 with 1% glacial acetic acid. The flow rate was 1 mL/min and effluents were monitored at the wavelength of 246 nm. The retention time for irbesartan was 2.3 min. The method was validated as per ICH guidelines for linearity, accuracy, precision, specificity, limit of detection, limit of quantification and robustness. Limit of detection and quantification were found 6.51 and 1.973 μg/mL, respectively and recovery from bulk and dosage forms was found from 99.94% to 99.97%. As the separation of the degradants using this mobile phase was quite good, isolation of the degradants with preparative techniques could also be achieved using this mobile phase. Irbesartan was prone to degrade more in acidic, alkaline, oxidative and thermal conditions. The method could be economically very useful in both research and industrial aspect.

Rao et al. [127] developed and validated a stability-indicating reverse phase rapid resolution liquid chromatographic method for simultaneous estimation of irbesartan and six related impurities in active pharmaceutical ingredient samples. The chromatographic separation was achieved on Kromasil C8 short column (150 × 4.6 mm, 3.5 μm) with 0.1% (v/v) orthophosphoric acid and acetonitrile as mobile phase using gradient elution. The method showed good resolution between irbesartan and its six related impurities and were eluted within 15 min run time of chromatogram. Regression analyses indicated correlation coefficient value greater than 0.999 for irbesartan and its six related impurities. The limit of detection for irbesartan and the known related impurities were observed at a level below 0.004% (0.019 μg/mL) and the method showed good recoveries for irbesartan (99.6–100.7%) and for its six known impurities (88.5–98.9%). The test solution and related substances were found to be stable in the diluents for 24 h. The method was found to be stability indicating. The application of developed method was also verified by an assay of irbesartan and related substances in bulk drug samples.

Nagamani et al. [128] developed a reversed-phase high performance liquid chromatographic method for simultaneous determination of irbesartan and hydrochlorthiazide in tablet dosage form. The elution was performed with a mobile phase of methanol:0.05% orthophosphoric acid (90:10, v/v) on Intersil-BDS C18 column (250 × 4.6 mm, 5 μm). The detector wavelength was set at 226 nm. Retention times for irbesartan and hydrochlorthiazide were 2.869 and 3.942 min, respectively. The reliability and analytical performance of the procedure were statistically validated according to the respect of linearity, ranges, precision, accuracy, repeatability, reproducibility, detection and quantification limits. Linear ranges were established between 36–216 and 3–18 μg/mL for irbesartan for hydrochlorthiazide, respectively. The limits of detection and quantitation for irbesartan were 0.65 and 1.97 μg/mL, respectively. The method was successfully employed for the analysis of pharmaceutical formulations containing combined dosage form.

Srinivasu and Annapurna [129] developed and validated a stability-indicating liquid chromatographic method (gradient mode) for the simultaneous determination of irbesartan and hydrochlorothiazide in pharmaceutical dosage forms (tablets). Chromatographic separation was achieved on an X Bridge shield RP C18 column (150 mm × 3.0 mm, 3.5 μm) maintained at 45 °C and photodiode array detector. The mobile phase consisted of 0.1% v/v formic acid:acetonitrile (40:60, v/v), with a flow rate of 0.8 mL/min (ultraviolet detection at 235 nm). The linear ranges were 1.2–560 and 2–600 μg/mL for irbesartan and hydrochlorothiazide, respectively. The limits of detection and quantitation for irbesartan were 0.349 and 1.056 μg/mL, respectively, whereas those of hydrochlorothiazide were 0.654 and 1.980 μg/mL, respectively. Both irbesartan and hydrochlorothiazide were subjected to stress conditions of acidic, alkaline, and oxidation degradations. The forced degradation studies were performed using HCl, NaOH, and H_2O_2.

Zareh et al. [130] developed a gradient HPLC method for rapid simultaneous separation and determination of eight drugs of sartan and statin classes in their pure and dosage forms within 15 min. These drugs were irbesartan, losartan, valsartan, olmesartan, rosuvastatin, atorvastatin, lovastatin, and simvastatin. Separation was carried out on a Kinetex C_{18} 100A column (2.60 μm, 4.60 mm × 100 mm) using a gradient binary mobile phase of 0.05 M potassium dihydrogen phosphate buffer (pH 3.50 adjusted by orthophosphoric acid) and acetonitrile at room temperature. The flow rate was 1.00 mL/min and maximum absorption was measured using a diode array detector at 280 nm. The retention time of irbesartan was 4.72 min and LOD was 2.01 mM. The method was validated according to the ICH guidelines for the determination of the drugs in their dosage forms with highly precise recoveries. Also, a statistical comparison with reference methods was performed showing no significant differences between the proposed method and reported ones in terms of precision and accuracy.

Dagariya and Jat [131] described a validated reverse phase high performance liquid chromatographic method for simultaneous estimation of irbesartan, chlorthalidone and cilnidipine in tablet dosage form. Chromatographic separation was carried out on C18 column (250 × 4.6 mm, 5 μm) with a mobile phase comprised of potassium dihydrogen phosphate buffer 0.05 M (pH 3.5):acetonitrile:triethylamine (80:20:0.1%, v/v/v). The flow rate was 1.0 mL/min and the eluent was monitored at 222 nm. The chromatographic conditions separated irbesartan, chlorthalidone and cilnidipine at 3.807, 4.667 and 6.887 min, respectively. Linearity was found to be in the range of 30–90, 1.25–3.75 and 1–3 μg/mL for irbesartan, chlorthalidone and cilnidipine, respectively. The percentage recoveries of all the drugs were found to be 99.27–99.81%, 99.57–99.99% and 99.22–99.44% for irbesartan, chlorthalidone and cilnidipine, respectively. The method was found to be suitable for a simultaneous estimation of these drugs in a tablet.

Dedhiya et al. [132] develop and validated HPLC method for simultaneous estimation of irbesartan and atorvastatin in synthetic mixture. The chromatographic separation was performed on a Thermo C18 column (250 × 4.6 mm, 5 μm) using a mobile phase consisted of acetonitrile:methanol:0.1% formic acid (pH adjusted to 3.5 with triethylamine) in a ratio of 50:10:40 (%, v/v/v) at an isocratic flow rate of 1 mL/min. Detection was performed at 271 nm. The developed method was validated according to ICH Q2R1 guideline. The linearity was established over a concentration range of 300–1500 and 20–100 ng for irbesartan and atorvastatin, respectively. The retention times of irbesartan and atorvastatin were 4.01 ± 0.03 and 7.03 ± 0.03 min, respectively. Recovery of drug was achieved in the range of 99.84–99.98% and 99.90–100.67% for irbesartan and atorvastatin, respectively. Limit of detection was found to be 10.94 and 1.33 ng for irbesartan and atorvastatin, respectively. Limit of quantitation was found to be 33.14 and 4.03 ng for irbesartan and atorvastatin, respectively. The method was applied for simultaneous estimation of two drugs in their synthetic mixture and results were found to be in good agreement with the labeled claim.

Virani et al. [133] developed a reverse phase high performance liquid chromatographic method for simultaneous estimation of irbesartan and atorvastatin in synthetic mixture. Inertsil C18 column (150 mm × 4.6 mm, 5 μm) in a gradient mode with mobile phase of acetonitrile:0.1% formic acid (40:60, v/v) and pH adjusted to 3.5 ± 0.1 with orthophosphoric acid was used. The flow rate was 1.0 mL/min and absorbance of individual component was measured at 262 nm. The retention times of irbesartan and atorvastatin were found to be 3.993 and 7.733 min, respectively. Linearity for irbesartan and atorvastatin was in the range of 400–800 and 50–100 μg/mL with correlation coefficient values 0.9995 and 0.9994, the percentage recovery was 99.88% and 99.70%, respectively.

Annapurna et al. [134] developed a reverse phase liquid chromatographic technique for the separation and determination of irbesartan, candesartan and hydrochlorothiazide using QbD (Quality by Design). The method was optimized by experimental design approach to identify the chromatographic conditions where the adequate separation quality in minimal analysis duration. The relationship among independent variables and critical quality attributes was given by experimental design methodology. The segregation was consummated on Sunfire C18 type column (150 × 4.6 mm; 5 μm) as stationary phase; 0.1% acetic acid in water:acetonitrile (33:77%, v/v); flow rate was 0.7351 mL/min; detection was at 225 nm. The chromatographic efficiency was investigated for the composite effects of % organic phase and flow and finely optimized by employing central composite design. The method was validated and applied to the marketed formulations.

Prajapati et al. [135] developed a stability-indicating reverse phase high performance liquid chromatography (RP-HPLC) method for simultaneous estimation of irbesartan and cilnidipine in their combined dosage form. The separation was attained on Hypersil BDS C18 column (25 × 0.46 cm) using a mobile phase composed of phosphate buffer (pH 5.5):acetonitrile (40:60, v/v) at a flow rate of 1 mL/min and the detection wavelength was 240 nm. Retention times of irbesartan and cilnidipine were found to be 3.730 and 6.040 min, respectively. The method was validated for linearity, accuracy and precision. Linear ranges were 1–3 μg/mL for cilnidipine and 30–90 μg/mL for irbesartan. The percentage recovery obtained for cilnidipine and irbesartan were found to be in range of 100.61 ± 0.46%, 100.68 ± 0.80%, respectively.

Soni et al. [136] developed and validated a RP-HPLC method for the determination of irbesartan and cilnidipine in pharmaceutical dosage form. Chromatographic separation was performed on a Hypersil C18 BDS column (25 × 0.46 cm) using a mobile phase composed of phosphate buffer (pH 4.5):methanol (85:15, v/v) at a flow rate of 1 mL/min and the detection wavelength was 248 nm. The retention time of cilnidipine and irbesartan was obtained as 5.343 and 4.080 min, respectively. The obtained calibration curve was linear in the concentration range of cilnidipine is 0.5–1.5 μg/mL and for irbesartan is 15–45 μg/mL. For irbesartan, the limits of detection and quantitation were 1.167 and 3.538 μg/mL, respectively. For cilnidipine, the limits of detection and quantitation were 0.046 and 0.139 μg/mL, respectively. The method was successfully applied for the analysis of drugs in pharmaceutical formulation.

Marghany et al. [137] developed and validated HPLC method for the simultaneous determination of seven angiotensin II receptor blockers, namely, hydrochlorothiazide, chlorthalidone, eprosartan mesylate, valsartan, losartan potassium, irbesartan, and candesartan cilexetil. Different chromatographic parameters were tested and fully optimized. Best chromatographic separation was accomplished on a reversed-phase octadecylsilyl column

(250 × 4.6 mm, 5 μm) under gradient elution using methanol:sodium phosphate monobasic buffer (0.01 M, pH 6.5) as mobile phase. The detection of target analytes was obtained at 254 nm. The pH of the buffer has been selected according to Marvin® sketch software. The method was validated according to ICH guidelines and showed good precision (RSD < 1), good linearity ($r^2 \geq 0.999$), and high accuracy (between 98% and 102%) with LOD and LOQ of 40 and 160 ng/mL, respectively, for all the detected analytes.

Kumar et al. [138] developed and validated a reverse phase HPLC method with forced degradation studies for the simultaneous estimation of amlodipine besylate and irbesartan in the pharmaceutical formulation. The chromatographic separation of the two drugs were achieved using Enable C18 G column (250 × 4.6 mm; 5 μm) in isocratic mode with mobile phase consisting of sodium acetate buffer (pH 4.0) and acetonitrile (30:70, v/v) with a flow rate of 0.6 mL/min. UV detection was carried out at 238 nm. The method was validated for linearity, range, accuracy, precision, robustness, LOD and LOQ. The tablet formulation was subjected to stress conditions of degradation including acidic, alkaline, oxidative, thermal and photolysis. The retention time for amlodipine besylate and irbesartan were found to be 5.512 and 6.321 min, respectively. Linearity was observed over a concentration range 4–32 μg/mL for amlodipine besylate ($r^2 = 0.9999$) and 10–70 μg/mL for irbesartan ($r^2 = 0.9998$). The RSD for intra-day and inter-day precision was found to be 0.436% and 0.699% for amlodipine besylate and 0.435 and 0.30 for irbesartan. Amlodipine besylate shown stability toward acidic and thermal whereas in basic, oxidative and photolytic it shown less stability in which it degraded to more extent. Irbesartan shown stability toward thermal conditions whereas in remaining conditions it degrades to more extent especially in oxidative conditions. The method was also found to be simple, precise and sensitive for the simultaneous determination of amlodipine besylate and irbesartan in the tablet dosage form.

Kavya et al. [139] developed and validated HPLC method for the estimation of irbesartan in tablet dosage form. An Inertsil ODS C-18, 5 μm column having 250 × 4.6 mm internal diameter in isocratic mode with mobile phase containing acetonitrile:water (40:60, v/v) was used. The flow rate was 1.5 mL/min and effluents were monitored at 260 nm. The retention time for irbesartan was 4.5 min. The method was validated for linearity, accuracy, precision, specificity, LOD, LOQ and robustness. LOD and LOQ were found 10 and 30 ng respectively and recovery of irbesartan from tablet formulation was found 100.61%. The method was successfully applied for the quantitative determination of irbesartan in tablet formulation.

Ingale and Dalvi [140] developed and validated HPLC method for the estimation of irbesartan. A Microsorb-MV 100-5 C-18 (250 × 4.6 mm, 5 μm) column using UV detector with mobile phase containing methanol, 0.02% formic acid (70:30, v/v) was used. The flow rate was 1.0 mL/min and detection was carried out at 234 nm. The retention time for irbesartan was 8.18 min. The method was validated for linearity, accuracy, precision, LOD and LOQ and robustness. The results of all the validation parameters were well within their acceptance values. The proposed method was successfully applied for the quantitative determination of irbesartan.

Zheng et al. [141] developed a microemulsion liquid chromatographic method for the determination of irbesartan and related substances in tablets. Chromatographic separation was performed on an odyssil C18 column (4.6 × 250 mm, 5 μm) at a temperature of 30 °C. The mobile phase was SDS-butylalc-cyclohexane-water (3.0:6.0:0.8:92.2) at a flow rate of 1.0 mL/min. The detection wavelength was 230 nm. Related substances were completely separated from the major constituents. In the range of 50.01–175.04 μg/mL ($r = 0.9992$, $n = 6$), it had a good linear relationship for irbesartan with the peak area. The average recovery of irbesartan was 100.4% (RSD = 1.0%, $n = 9$) and its LOD was 0.02 μg/mL. While the related compound A was linear over the range of 0.0753–0.3518 μg/mL, with LOD of 0.025 μ/mL and its average recovery was 100.5% (RSD = 1.1%, $n = 9$). The contents of two specifications were 99.4% and 99.8%, respectively.

Shakeb and Puranik [156] developed a stability-indicating method for irbesartan and hydrochlorothiazide. Both drugs were subjected to different ICH prescribed stress conditions like acidic, alkaline, oxidation, reduction, thermal and photostability condition and found that degraded peaks did not interfere with the peaks of drug under the study. Separation was done on Enable C18 G column (250 × 4.6 mm, 5 μm) using a mobile phase consisted of a mixture of 50 mM ammonium acetate:acetonitrile (70:30, v/v) and the effluent was monitored at 235 nm using photodiode array detector. The method was linear over the concentration range of 150–350 μg/mL ($r^2 = 0.999$) with LOD and LOQ of 0.019 and 0.053 μg/mL for irbesartan, and range of 15–35 μg/mL ($r^2 = 0.999$) with LOD and LOQ of 0.023 and 0.070 μg/mL for hydrochlorothiazide. The method was validated for linearity, range, precision, accuracy, specificity, selectivity and intermediate precision.

Hashem et al. [142] developed HPLC method for simultaneous determination of four antihypertensive drugs (irbesartan, hydrochlorothiazide, losartan potassium and valsartan) in bulk powders and pharmaceutical preparations. The chromatographic separation of the four pharmaceuticals was achieved on a reversed-phase C18 Intersil ODS-3 column (5 μm, 4.6 × 250 mm) using a binary mobile phase of 45% acetonitrile and 55% 50 mM KH_2PO_4 (pH 4.5) at a flow rate of 1 mL/min. The detection wavelength was 210 nm. The total separation time was less than 12 min. The method was validated for system efficiency, linearity, accuracy, precision, LOD, LOQ, specificity, stability and robustness. The LOD values were 0.02, 0.12, 0.14 and 0.01 μg/mL for hydrochlorothiazide, losartan potassium, valsartan and irbesartan, respectively. Linearity ranges were 5–50 μg/mL for hydrochlorothiazide, 10–100 μg/mL for losartan potassium, irbesartan and valsartan. The recovery values of this method were between 97% and 103% and the reproducibility was within 1.67%. Statistical analysis proved that the method enabled reproducible and selective quantification of all four analytes as the bulk drug and in pharmaceutical preparations.

Anitha et al. [143] developed HPLC method with UV detection was developed for the estimation of irbesartan in tablet dosage form. Separation was achieved by Lithosphere 100 RP-18e, 5 μm column having 250 × 4.0 mm internal diameter with a mobile phase containing phosphate buffer and acetonitrile in 50:50 (v/v) with an ambient temp. The flow rate was 0.8 mL/min and eluent was monitored at 220 nm. The proposed method was found to be rectilinear over the concentration range of 20–60 μg/mL. This method was validated for system suitability, specificity, linearity, precision, accuracy, range, stability studies, ruggedness, robustness and filter validation. The results of all the validation parameters were well within their acceptance values. Therefore, the proposed method can be used for routine analysis of estimation of irbesartan in its tablet formulation.

Prabhu and Muralidhar [98] developed a reverse phase HPLC method for the determination of purity of irbesartan drug substance in bulk samples and its pharmaceutical dosage forms in the presence of its impurities. This method was capable of separation of related impurities along with irbesartan. This method could be also be used for the estimation of irbesartan in drug substance as well as in single tablet formulation. Two impurities were detected in drug sample by HPLC analysis. The chromatographic conditions were optimized using an impurity-spiked solution. Mass and infrared spectroscopic methods were used for the identification of impurities. The structure of the impurities were confirmed as 2-cyano-4′-bromomethyl biphenyl and 2-*n*-butyl-1,3-diazaspiro [4,4]-non-1-ene-4-one. The method was subsequently validated for the determination of irbesartan and its related compounds, as per ICH guidelines, for accuracy, precision, linearity and range, selectivity, LOD, LOQ and robustness. The LOD for irbesartan, impurity 1 and impurity 2 was found to be 18.51 μg/mL or ppm, 16.033 and 16.069 μg/mL or ppm, respectively. While LOQ was found to be 56.098, 48.587 and 48.69 μg/mL or ppm, respectively.

Miraz [144] described a reversed-phase liquid chromatographic method for the simultaneous determination of irbesartan, candesartan, gliquidone and pioglitazone. Separation was performed on Purospher® column with a dimension of (25 × 0.46 cm, 5 μm) RP-18 end-capped using a mobile phase consisted of methanol:water (90:10, v/v, pH was adjusted to 3.5 with phosphoric acid) with a flow rate of 1 mL/min and the detection was at 230 nm. The concentration-response relationship was found linear over a concentration range of 5–25 μg/mL for all of the analytes tested. The limits of detection and quantification were 0.83 and 2.78 for irbesartan, 0.30 and 1.01 for candesartan, 1.11 and 3.93 for gliquidone, and 0.41 and 1.41 μg/mL for pioglitazone, respectively. The method permitted the successful determination of these drugs in human serum.

Coudoré et al. [145] developed HPLC method with diode array detector for simultaneous determination of irbesartan and hydrochlorothiazide levels in the plasma of hypertensive patients given a fixed combination of 300 mg of irbesartan and 12.5 mg of hydrochlorothiazide. Compounds were extracted from acidified plasma samples with 3 mL ethyl acetate, and eluted at 6 and 19 min from a C_4 column by elution with an acetonitrile-phosphate buffer (pH 3.6) mobile phase gradient at a flow rate of 1 mL/min. The assay was linear over the ranges 20–4000 and 2.5–500 ng/mL for irbesartan and hydrochlorothiazide, respectively. Overall intra-assay and inter-assay variation were within acceptance limits. LOQ values were 20 and 2.5 ng/mL for irbesartan and hydrochlorothiazide, respectively. Plasma samples remained stable for 12 h at room temperature, through three thaw-freeze cycles, and for 2 and 7 months at −20 °C. In hypertensive patients, residual concentrations were 241.8 ± 39.0 and 22.3 ± 6.0 ng/mL for irbesartan and hydrochlorothiazide, respectively. There was no interference from other co-administered drugs. Despite the different physicochemical properties of both drugs, the method enabled accurate measurement of both drugs for assessment of compliance by patients treated by fixed-dose combination therapy with both drugs combination.

Maslarska et al. [146] developed HPLC method with UV detection for simultaneous analysis of three sartans (irbesartan, valsartan, and telmisartan) for quality control. Separation was done by isocratic elution on a LiChrospher C18 column (250 × 4 mm, particle size 5 μm) at the temperature 30 °C with a mobile phase consisting of 10 mM phosphate buffer:acetonitrile (65:35, v/v) at a flow rate of 1.0 mL/min. The column eluent was monitored with a UV detector at 225 nm. This allowed a rapid detection and identification as well as quantitation of the eluting peaks. Calibration curves for all drugs were in the range of 5–40 μg/mL and the linear regression coefficients were more than 0.995. Recovery rates for the sartans were in the range 96.5–103.1%. The LOD values were calculated between 0.04 and 0.06 μg/mL. Also, the LOQ values were 0.12–0.16 μg/mL. Within-day and between-day coefficient of variation for all sartans at all concentrations in the range of 0.83–3.79% was calculated.

Wani and Zargar [147] described an ultra-performance liquid chromatography-tandem mass spectrometry (UPLC-MS/MS) for quantification of irbesartan in human plasma. After a simple protein precipitation using methanol and acetonitrile, irbesartan and internal standard (telmisartan) were separated on Acquity UPLC BEH C18 column (50 × 2.1 mm, i.d. 1.7 μm) using a mobile phase consisted of acetonitrile:methanol:10 mM ammonium acetate (70:15:15, v/v/v) with a flow rate of 0.4 mL/min and detected MS/MS in negative ion mode. The ion transitions recorded in multiple reaction monitoring mode were *m/z* 427.2 → 193.08 for irbesartan and *m/z* 513.2 → 469.3 for telmisartan. The assay exhibited a linear dynamic range of 2–500 ng/mL with good correlation coefficient of (0.995) and with a lower limit of quantitation of 2 ng/mL. The intra- and inter-assay precisions were satisfactory; the relative standard deviations did not exceed 9.91%.

Patel et al. [148] described a liquid chromatographic-tandem mass spectrometric method for the simultaneous estimation of irbesartan and hydrochlorothiazide in human plasma using liquid-liquid extraction technique. This method used an electro spray ionization in a negative mode for both anlaytes using a triple quadrupole mass spectrometry where irbesartan D4 and hydrochlorothiazide 13C 15N2 D2 were used as an internal standard for irbesartan and hydrochlorothiazide, respectively. Analytes were recovered by diethyl ether:dichloromethane (70:30, v/v) and subsequently separated on an Ace 5 C18 column (100 × 4.6 mm, 5 μm) using methanol:0.1% formic acid in water (70:30, v/v) as a mobile phase, at a flow rate of 1.0 mL/min. Quantification of irbesartan and hydrochlorothiazide and irbesartan D4 and hydrochlorothiazide 13C 15N2 D2 was performed using multi-reaction monitoring mode (MRM) where transition of *m/z* 427.1 → 193.0 (irbesartan), 431.1 → 193.0 (irbesartan D4), 295.8 → 269.0 for hydrochlorothiazide and 300.9 → 271.1 for hydrochlorothiazide 13C 15N2 D2 in a negative mode. The calibration curve was linear over the concentration range of 10–5000 ng/mL for irbesartan and 1–500 ng/mL for hydrochlorothiazide. The intraday and inter-day precisions were less than 15% and the accuracy was all within ±15%. The method was fully validated for all the other parameters such as selectivity, matrix effect, recovery and stability as well.

Tutunji et al. [149] described an HPLC coupled with tandem mass spectrometry for the simultaneous determination of irbesartan and hydrochlorothiazide in human plasma. In this method, plasma samples were prepared using protein precipitation with acetonitrile. The two analytes and the internal standard (losartan) were separated on a reverse phase C18 column (50 × 4 mm, 3 μm) using water with 2.5% formic acid:methanol:acetonitrile (40:45:15, v/v/v) as a mobile phase (flow rate of 0.70 mL/min). Irbesartan and hydrochlorothiazide were ionized using electrospray ionization (ESI) source in a negative ion mode, prior to detection by multiple reaction monitoring (MRM) mode while monitoring at the following transitions: *m/z* 296 → 269 and *m/z* 296 → 205 for hydrochlorothiazide, 427 → 175 for irbesartan. Linearity was demonstrated over the concentration range 0.06–6.00 μg/mL for irbesartan and 1.00–112.00 ng/mL for hydrochlorothiazide. The calibration sensitivities were 0.2537 ± 0.0055 and 0.0129 ± 0.0002 for irbesartan and hydrochlorothiazide, respectively. The lower limit of quantitation of irbesartan and hydrochlorothiazide were found to be 0.06 μg/mL and 1.00 ng/mL, respectively. The limit of detection for irbesartan was 0.01 μg/mL and for hydrochlorothiazide was 0.51 ng/mL. The mean absolute recovery of irbesartan was 48.5%, 59.6% and 60.7% for concentration levels of 0.18, 3.00 and 4.80 μg/mL. The mean absolute recovery for hydrochlorothiazide was 28.8%, 33.0% and 34.1% for concentration levels of 3.00%, 55.00% and 88.00 ng/mL. Both irbesartan and hydrochlorothiazide were found to be stable under the mentioned test conditions. The method could be applied for the purpose of its application to measure concentration-time profiles for bioavailability, pharmacokinetic and bioequivalence decision after dosing with two tablet formulations containing 300 mg irbesartan and 12.5 mg hydrochlorothiazide tablet on healthy volunteers in a fasted state.

Nurmeilis et al. [150] developed and validated a liquid chromatography-tandem mass spectrometry (LC-MS/MS) method for the simultaneous determination of irbesartan and sinensetin, and application of the method to study a pharmacokinetic interaction of irbesartan and ethanol extract of *Orthosiphon stamineus* herba in rat plasma. Irbesartan and sinensetin were simultaneously extracted from plasma by protein precipitation with acetonitrile. Samples containing irbesartan and sinensetin were analyzed on C_{18} column Acquity® (100 mm × 2.1 mm, 1.7 μm) maintained at 40 °C. The gradient system of mobile phase composition was a mixture of acetonitrile and 0.1% formic acid (40:60, v/v), which was pumped at a flow rate of 0.3 mL/min. Mass detection was performed on Waters Xevo Triple Quadrupole equipped with an electro spray ionization (ESI) source at positive ion mode in the multiple reaction monitoring (MRM) mode. Irbesartan was detected at *m/z* 429.1 → 205.9, sinensetin was detected at *m/z* 373 → 342.9 and losartan as an internal standard was detected at *m/z* 423.05 → 404.9. The method was validated and showed good reproducibility and linearity of 0.99. The lower limits of quantification were 25 and 250 ng/mL for irbesartan and sinensatin, respectively. The precision values of within-run and between-run analysis were 9.3–5.25% and 1.52–5.47% (for irbesartan), 1.52–5.09% and 2.47–9.14% (for sinensetin) whereas the accuracy (% difference) of both irbesartan and sinensetin were less than 20%. Stability studies revealed that irbesartan and sinensetin were stable for 24 h at room temperature, 24 h in the autosampler, 3 freeze-thaw cycles, and at least 30 days at −20 °C. The validated method was applied to evaluate pharmacokinetic interactions of irbesartan and ethanol extract of *Orthosiphon stamineus* herba in rat plasma. The method was valid to evaluate irbesartan and sinensetin simultaneously *in vitro*.

Tan et al. [151] a liquid chromatography-tandem mass spectrometric method for the simultaneous quantification of irbesartan and candesartan in rabbit eye tissues to support a preclinical trial of eye-formulation in rabbits. Eye tissue samples (cornea, aqueous humor, vitreous body and retina/choroid) were first homogenized in H_2O-diluted rabbit plasma. Irbesartan and candesartan in the supernatants together with their respective internal standards (irbesartan-d_4 and candesartan-d_4) were extracted by solid-phase extraction. The extracted samples were injected onto a C_{18} column for gradient separation. The mass detection was in the positive electrospray ionization mode using the multiple reaction monitoring transitions of *m/z* 429 → 207, 433 → 211, 441 → 263, 445 → 267, for irbesartan and irbesartan-d_4, candesartan, candesartan-d_4, respectively. For the validated concentration ranges (5–5000 and 2–2000 ng/g for irbesartan and candesartan, respectively), the within-run and between-run accuracies (% bias) were within the range of −8.0 to 10.0. The percentage coefficient of variation ranged from 0.6 to 7.3. There was no significant matrix interference nor matrix effect from different eye tissues and different rabbits. The validated method was successfully used in the Good Laboratory Practice study of rabbits.

Nisa et al. [152] developed and validated HPLC method for determination of angiotensin receptor blocking agent irbesartan in human plasma. Liquid-liquid extraction of the drug was carried out. Chromatographic separation of irbesartan was conducted by the help of 4.0 mm × 25 cm column having L1 packing from plasma and mobile phase consisted of phosphate buffer and acetonitrile in a ratio of 67:33 (v/v). The flow rate was set at 1 mL/min and the detector at a wavelength of 220 nm. The resolution of irbesartan was well performed from plasma components. This method was validated and demonstrated linearity with a concentration range of 0.1 to 6 μg/mL of irbesartan in plasma. Intra-day, inter-day accuracy was found 89.33–96.37% while intra-day, inter-day precision was found within the limit of 0.02 and 2.15, respectively. The mean recovery of irbesartan was 97.28%. The efficacy of extraction was proved by above-mentioned results. In plasma, the 0.05 and 0.1 μg/mL dilutions were exhibited as the LOD and LOQ of irbesartan. Stability studies disclosed that irbesartan showed stability at −20 °C storage.

Zhou et al. [153] developed HPLC with fluorescence detection for detecting human plasma irbesartan. With telmisartan as the internal standard, the plasma was treated by ether:methylene chloride (7:3, v/v) extraction method. The column was Diamonsil C18, the mobile phase was water:acetonitrile (57:43, v/v), phosphoric acid adjusted to pH 3.00, fluorescence excitation wavelength was 254 nm, emission wavelength was 371 nm, a flow rate was 1.0 mL/min, the column temperature was room temperature and the injection volume 50 μL. Retention times of telmisartan and irbesartan were 6.1 and 4.3 min, respectively. In 0.05–5 μg/mL in the range of irbesartan detection was a good linear relationship ($r = 0.9996$), the LOD was 0.023 μg/mL. The average relative recoveries of high, medium and low concentrations (4, 0.8 and 0.08 μg/mL) were (101.18 ± 6.31%), (102.22 ± 5.32%), (98.45 ± 3.52%), recovery was (78.05 ± 6.66%), daytime, day RSD were less than 10%. Sample for three times freezing and long-term placement at −20 °C refrigerator had good stability. The method was accurate, high sensitivity and specificity, which can be used in pharmacokinetic interaction studies and drug available for irbesartan.

Elgawish et al. [154] developed a LC-MS/MS spectrometry method for the simultaneous determination of rosuvastatin and irbesartan in rat plasma:

insight into pharmacokinetic and drug-drug interaction studies. In the study, first chromatographic separation coupled tandem mass spectrometric was developed and fully validated for simultaneous measurement of rosuvastatin and irbesartan in rat plasma after oral administration. The two analytes were extracted from plasma sample using acetonitrile-induced protein precipitation then separated on an Agilent Eclipse Plus ODS (4.6 × 100 mm, 3.5 μm) column by gradient elution using 6 mM ammonium formate/0.1% formic acid and acetonitrile at a flow rate 0.4 mL/min. Multiple reaction monitoring in positive ion mode was used for quantification of precursors to production at m/z 492.1 → 206.9 for irbesartan, 482.1 → 258.1 for rosuvastatin, and 409.2 → 238.2 for the internal standard, amlodipine. Linearity was obeyed in the range of 1–10,000 and 1–5000 ng/mL with detection limits (S/N of 3) of 0.05 and 0.07 ng/mL for irbesartan and rosuvastatin, respectively. The method was validated in terms of selectivity, recovery, accuracy, precision, matrix effects, and stability as per US-FDA bioanalytical guidelines. The application of the method was applied to study pharmacokinetic interaction of irbesartan and rosuvastatin in rat plasma after a single oral dose. The area under the concentration-time curve (AUC), peak plasma concentrations (C_{max}), half-life time ($t_{1/2}$), and volume of distribution (V_d) of irbesartan and rosuvastatin were affected when the two drugs were co-administering. The method provided a valuable tool for studying drug-drug interaction and might be useful for therapeutic drug monitoring and bioequivalence studies.

Castro et al. [157] developed a method for the simultaneous determination of irbesartan and other sartan drugs (eprosartan, olmesartan, losartan, candesartan, telmisartan and valsartan), and the biodegradation product valsartan acid, in water samples (raw and treated wastewater, river and tap water). Solid-phase extraction and ultra-performance liquid chromatography-tandem mass spectrometry were employed as concentration and determination techniques, respectively. Different sorbents and elution solvents were tested for sample preparation. Under optimized conditions, samples at neutral pH (6–8 units) were concentrated using mixed-mode (reversed-phase and anionic exchange) cartridges. Thereafter, the sorbent was washed with 5 mL of a methanol: water (1:1) solution, dried under a nitrogen stream and compounds were eluted with 2 mL of methanol:NH_3 (98:2). The accuracy of the method (accounting for solid-phase extraction efficiency and matrix effects during electrospray ionization) was investigated using solvent-based calibration standards. Global recoveries, obtained for different water matrices (tap, river, treated and raw wastewater), ranged from 82% to 134%, with standard deviations between 2% and 18%. LOQs varied from 2 to 50 ng/L. Analysis of un-spiked samples confirmed: (1) the incomplete removal of sartans at sewage treatment plants, (2) the formation of valsartan acid during municipal water treatment, and (3) the presence of valsartan acid in the processed tap water samples. Additional findings of the study were the detection of hydroxylated derivatives of the sartan drugs irbesartan and losartan in wastewater, and the E-Z isomerization of eprosartan in environmental water samples.

Castro et al. [158] developed a single step sample preparation procedure for the simultaneous determination of five antihypertensive including irbesartan (propranolol, losartan, irbesartan, telmisartan, and valsartan), three antiarrhythmic drugs (flecainide, dronedarone, and amiodarone), and one of their metabolites (*N*-desethylamiodarone) in sludge from municipal sewage treatment plants. Matrix solid-phase dispersion and ultra-performance liquid chromatography with tandem mass spectrometry detection were selected as sample preparation and determination techniques, respectively. Under optimal conditions, matrix solid-phase dispersion extractions were carried out with freeze-dried samples (0.5 g) dispersed on 2 g of C_{18}. Exhaustive extraction of target compounds was achieved with 10 mL of a methanol/acetonitrile/formic acid (30:69:1) solution. The obtained extract was ready for analysis by the ultra-performance liquid chromatography with tandem mass spectrometry detection without any further treatment, except filtration. The overall recoveries of the method (calculated against solvent-based standards) varied from 82% to 124%, with standard deviations in the range from 2% to 16%. Thus, the method was free of matrix effects during electrospray ionization. The achieved LOQs stayed between 2 and 10 ng/g, and the linear response range extended to 5000 ng/g. The occurrence of target compounds was investigated in sludge from 14 different sewage treatment plants. High detection frequencies were observed for all compounds, with average concentrations above 100 ng/g for six cardiovascular drugs.

Klancar et al. [159] developed an analytical method based on solid-phase extraction and liquid chromatography coupled with tandem mass spectrometry for detection of a high number ($n = 44$) of selected pharmaceuticals (including irbesartan) in surface waters. The validated method showed suitable accuracy, precision, and linearity. The extraction efficiency for pharmaceuticals was mostly above 90% and the determined LOQs were generally below 5 ng/L and even down to 0.03 ng/L. The matrix effect was successfully minimized with isotopically labeled internal standards. Six river and lake samples from Slovene regions were afterward analyzed. Forty-two of the 44 target pharmaceuticals were detected in at least two water samples;

66% of them were detected in all samples. Three analytes were present above the LOQ in all samples: caffeine (1.1–3.1 ng/L), irbesartan (0.2–9.3 ng/L), and valsartan (0.8–47 ng/L). The highest concentration was measured for valsartan. Comprehensive comparison with the literature data showed that the found concentrations of pharmaceuticals were notably lower in our study and that a number of the pharmaceuticals had not been reported previously. Comparison of the measured and predicted (calculated) environmental concentrations of pharmaceuticals in surface waters showed high discrepancy between the two approaches, clearly indicating the need for comprehensive analytical methods for routine monitoring of these newly emerging pollutants. The results proved that the method was adequately sensitive to reliably quantify the majority of pharmaceutical contaminants even though they were found in notably lower concentrations in Slovenian surface waters than it was predicted from their consumption and from the related literature.

7. Stability and impurity profiling

Shah et al. [160] studied the stability of irbesartan toward different stress conditions. Irbesartan was subjected to hydrolytic, oxidative, photolytic and thermal stress, according to ICH guideline Q1A (R2). The drug showed degradation only in acidic, basic and photoacidic conditions, while it was stable to other stress conditions. A total of three degradation products were formed, which were separated on a C8 column employing a gradient HPLC method. Initially, a complete mass fragmentation pathway of the drug was established with the help of MS/TOF, MS^n and H/D exchange studies. Subsequently, the degradation products were subjected to LC-MS/TOF and on-line H/D exchange mass studies to obtain their accurate mass, fragment pattern and number of labile hydrogens. The MS results helped to assign tentative structures to degradation products, which were verified through ^{1}H and 2D COSY LC-NMR experiments. The products were identified as (2′-(1H-tetrazol-5-yl)biphenyl-4-yl) methanamine, 1-(1-((2′-(1H-tetrazol-5-yl)biphenyl-4-yl)methylamino) pentylideneamino)cyclopentane carboxylic acid and 2-butyl-3-(tetrazolo [1,5-f]phenanthridin-6-ylmethyl)-1,3-diazaspiro[4.4]non-1-en-4-one. The structures were justified by mechanisms of their formation (Fig. 9).

Saravanan and Ansari [161] studied the trace level impurities of irbesartan to its origin using LC-MS/MS and nMS2 technique. A simple and effective patented process was applied to get the impurity profile,

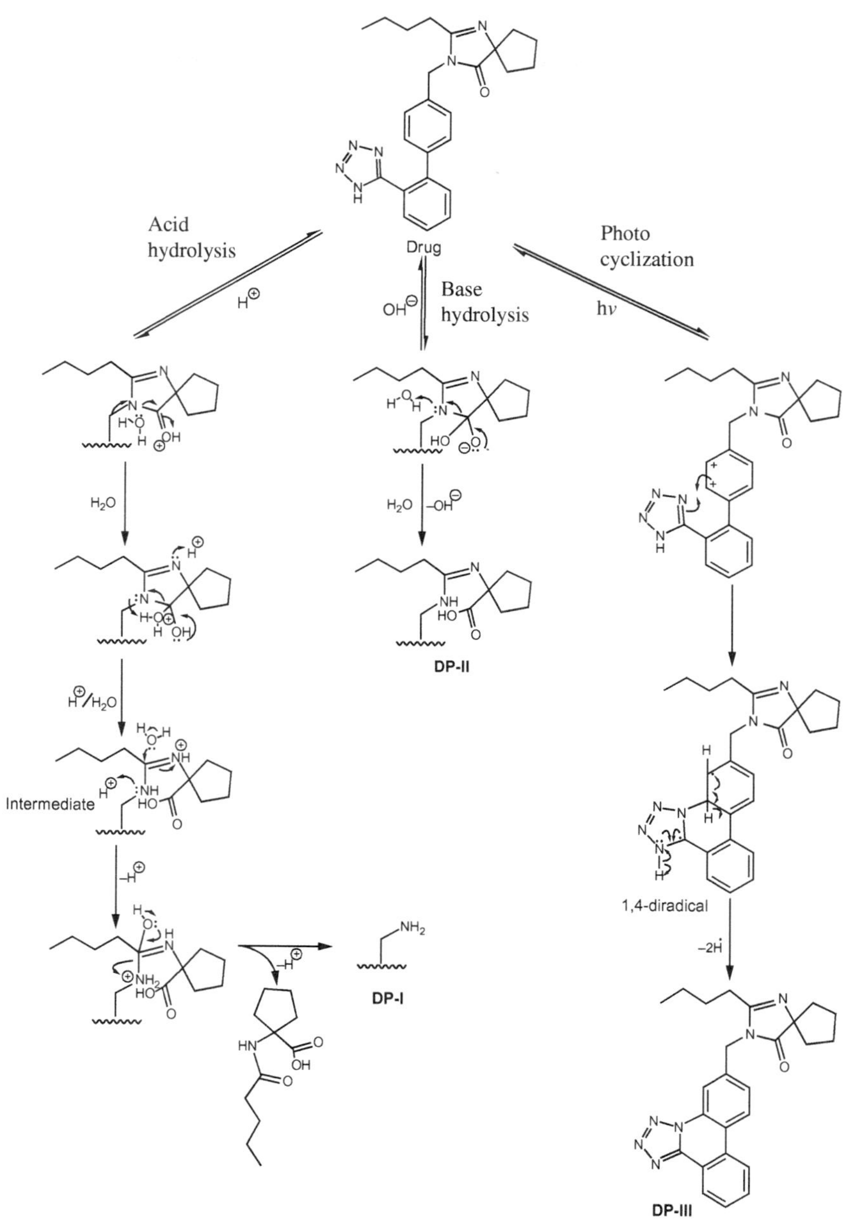

Fig. 9 Mechanism of degradation of irbesartan. *From R.P. Shah, A. Sahu, S. Singh. Identification and characterization of degradation products of irbesartan using LC-MS/TOF, MS(n), on-line H/D exchange and LC-NMR. J. Pharm. Biomed. Anal. 51 (2010) 1037–1346.*

and mass characterization was performed through the "nMS2" technique of triple quadrupole mass spectrometer analyzer. The molecular ion fragmentation occurred with multiple collision energies and provided meaningful MS/MS fragments for characterizing five trace level impurities less than 0.5% of irbesartan. The origin of the impurity formation in the synthetic process was successfully related to the spiro ketones. The results obtained in this research clearly indicated the approach of "nMS2" technique was very useful in the identification and structural prediction of trace level related impurities of irbesartan.

8. Pharmacological properties

8.1 Indications and usage

Irbesartan is indicated to treat hypertension and diabetic nephropathy in hypertensive patients with type 2 diabetes, elevated serum creatinine, and proteinuria [162]. A combination product with hydrochlorothiazide is indicated for hypertension in patients with uncontrolled hypertension with monotherapy or first line in patients not expected to be well controlled with monotherapy [163]. Irbesartan is also indicated for the treatment of diabetic nephropathy in patients with type 2 diabetes and hypertension, an elevated serum creatinine, and proteinuria (>300 mg/day). In this population, irbesartan reduces the rate of progression of nephropathy as measured by the occurrence of doubling of serum creatinine or end-stage renal disease (need for dialysis or renal transplantation) [162].

8.2 Pharmacodynamics and mechanism of action

Irbesartan is an angiotensin receptor blocker used to treat hypertension and diabetic nephropathy [160,163]. It has a long duration of action as it is usually taken once daily and a wide therapeutic index as doses may be as low as 150 mg daily, but doses of 900 mg/day were well tolerated in healthy human subjects [162,163].

Irbesartan prevents angiotensin II binding to the AT1 receptor in tissues like vascular smooth muscle and the adrenal gland. Irbesartan and its active metabolite bind the AT1 receptor with 8500 times more affinity than they bind to the AT2 receptor. Irbesartan's prevention of angiotensin II binding causes vascular smooth muscle relaxation and prevents the secretion of aldosterone, lowering blood pressure. Angiotensin II would otherwise bind to the AT1 receptor, inducing vasoconstriction and aldosterone secretion, raising blood pressure [162,163].

8.3 Pharmacokinetics

Irbesartan is 60–80% bioavailable with a T_{max} of 1.5–2h. Taking irbesartan with food does not affect the bioavailability [162,163]. In one study [125], healthy subjects were given single or multiple oral doses of 150, 300, 600, and 900mg of irbesartan. A single 150mg dose resulted in an AUC of 9.7 ± 3.0 μg/h/mL, a T_{max} of 1.5h, a half-life of 16 ± 7h, and a C_{max} of 1.9 ± 0.4 μg/mL. A single 300mg dose resulted in an AUC of 20.0 ± 5.2 μg/h/mL, a T_{max} of 1.5h, a half-life of 14 ± 7h, and a C_{max} of 2.9 ± 0.9 μg/mL. A single 600mg dose resulted in an AUC of 32.6 ± 11.9 μg/h/mL, a T_{max} of 1.5h, a half-life of 14 ± 8h, and a C_{max} of 4.9 ± 1.2 μg/mL. A single 900mg dose resulted in an AUC of 44.8 ± 20.0 μg/h/mL, a T_{max} of 1.5h, a half-life of 17 ± 7h, and a C_{max} of 5.3 ± 1.9 μg/mL.

Multiple 150mg doses resulted in an AUC of 9.3 ± 3.0 μg/h/mL, a T_{max} of 1.5h, a half-life of 11 ± 4h, and a C_{max} of 2.04 ± 0.4 μg/mL. Multiple 300mg doses resulted in an AUC of 19.8 ± 5.8 μg/h/mL, a T_{max} of 2.0h, a half-life of 11 ± 5h, and a C_{max} of 3.3 ± 0.8 μg/mL. Multiple 600mg doses resulted in an AUC of 31.9 ± 9.7 μg/mL, a T_{max} of 1.5h, a half-life of 15 ± 7h, and a C_{max} of 4.4 ± 0.7 μg/mL. Multiple 900mg doses resulted in an AUC of 34.2 ± 9.3 μg/h/mL, a T_{max} of 1.8h, a half-life of 14 ± 6h, and a C_{max} of 5.6 ± 2.1 μg/mL.

The volume of distribution of irbesartan is 53–93L and it is 90% protein bound in plasma, mainly to albumin and α1-acid glycoprotein [162,163]. Irbesartan is largely metabolized by glucuronidation and oxidation in the liver. The majority of metabolism occurs through the action of CYP2C9 with a negligible contribution from CYP3A4 [162,163]. Some hydroxylation also occurs in irbesartan metabolism [164]. Irbesartan can be glucuronidated by UGT1A3 to the M8 metabolite, oxidized to the M3 metabolite, or hydroxylated by CYP2C9 to one of the M4, M5, or M7 metabolites. The M4, M5, and M7 metabolites are all hydroxylated to become the M1 metabolite, which is then oxidized to the M2 metabolite. The M4 metabolite can also be oxidized to the M6 metabolite before hydroxylation to the M2 metabolite. Finally, the minor metabolite SR 49498 is generated from irbesartan by an unknown mechanism [165,166].

A 20% of a radiolabelled oral dose of irbesartan is recovered in urine, and the rest is recovered in the feces. Less than 2% of the dose is recovered in urine as the unchanged drug [167]. The terminal elimination half-life of irbesartan is 11–15h. Total plasma clearance of irbesartan is 157–176mL/min while renal clearance is 3.0–3.5mL/min [162,163].

9. Dosing information

9.1 Dosage and administration

Irbesartan is available as tablets (75, 150 and 300 mg). Irbesartan may be administered with other antihypertensive agents and with or without food. For treatment of hypertension, the recommended initial dose of irbesartan is 150 mg once daily. The dosage can be increased to a maximum dose of 300 mg once daily as needed to control blood pressure. For treatment of nephropathy in type 2 diabetic patients, the recommended dose is 300 mg once daily. In dose adjustment in volume and salt-depleted patients, the recommended initial dose is 75 mg once daily in patients with depletion of intravascular volume or salt (e.g., patients treated vigorously with diuretics or on hemodialysis) [162].

9.2 Use in specific populations

9.2.1 Pregnancy Category D

Use of drugs that act on the renin-angiotensin system during the second and third trimesters of pregnancy reduces fetal renal function and increases fetal and neonatal morbidity and death. Resulting oligohydramnios can be associated with fetal lung hypoplasia and skeletal deformations. Potential neonatal adverse effects include skull hypoplasia, anuria, hypotension, renal failure, and death. When pregnancy is detected, discontinue irbesartan as soon as possible. These adverse outcomes are usually associated with use of these drugs in the second and third trimesters of pregnancy. Most epidemiologic studies examining fetal abnormalities after exposure to antihypertensive use in the first trimester have not distinguished drugs affecting the renin-angiotensin system from other antihypertensive agents. Appropriate management of maternal hypertension during pregnancy is important to optimize outcomes for both mother and fetus. In the unusual case that there is no appropriate alternative to therapy with drugs affecting the renin-angiotensin system for a particular patient, apprise the mother of the potential risk to the fetus. Perform serial ultrasound examinations to assess the intra-amniotic environment. If oligohydramnios is observed, discontinue irbesartan, unless it is considered lifesaving for the mother. Fetal testing may be appropriate, based on the week of pregnancy. Patients and physicians should be aware, however, that oligohydramnios may not appear until after the fetus has sustained irreversible injury. Irbesartan crosses the placenta in rats and rabbits. In pregnant rats given irbesartan at doses greater than the maximum recommended human dose (MRHD), fetuses showed increased incidences of renal pelvic cavitation, hydroureter and/or absence of renal papilla. Subcutaneous edema also occurred in fetuses at doses about four times the MRHD (based on body surface area). These anomalies occurred when pregnant rats received irbesartan through day 20 of gestation but not when drug was stopped on gestation day 15. The observed effects are believed to be late gestational effects of the drug. Pregnant rabbits given oral doses of irbesartan equivalent to 1.5 times the MRHD experienced a high rate of maternal mortality and abortion. Surviving females had a slight increase in early resorptions and a corresponding decrease in live fetuses. Radioactivity was present in the rat and rabbit fetus during late gestation and in rat milk following oral doses of radiolabeled irbesartan [162].

9.2.2 Nursing mothers

It is not known whether irbesartan is excreted in human milk, but irbesartan or some of its metabolites is secreted at low concentration in the milk of lactating rats. Because of the potential for adverse effects on the nursing infant, discontinue nursing or discontinue irbesartan [162].

9.2.3 Pediatrics

In infants with histories of in utero exposure to an angiotensin II receptor antagonist, observe for hypotension, oliguria, and hyperkalemia. If oliguria occurs support blood pressure and renal perfusion. Exchange transfusion or dialysis may be required as means of reversing hypotension and/or substituting for disordered renal function. Irbesartan, in a study at a dose of up to 4.5 mg/kg/day, once daily, did not appear to lower blood pressure effectively in pediatric patients ages 6–16 years. Irbesartan has not been studied in pediatric patients less than 6 years old [162].

9.2.4 Geriatrics

In elderly subjects (age 65–80 years), irbesartan elimination half-life is not significantly altered, but AUC and C_{max} values are about 20–50% greater than those of young subjects (age 18–40 years). No dosage adjustment is necessary in the elderly. Race/Ethnicity In healthy black subjects, irbesartan AUC values are approximately 25% greater than whites; there is no difference in C_{max} values [162].

9.2.5 Patients with renal impairment

The pharmacokinetics of irbesartan is not altered in patients with renal impairment or in patients on hemodialysis. Irbesartan is not removed by hemodialysis. No dosage adjustment is necessary in patients with mild to severe renal impairment unless a patient with renal impairment is also volume depleted [162].

9.2.6 Patients with hepatic insufficiency

The pharmacokinetics of irbesartan following repeated oral administration are not significantly affected in patients with mild to moderate cirrhosis of the liver. No dosage adjustment is necessary in patients with hepatic insufficiency [162].

9.3 Drug-drug interactions

In vitro studies show significant inhibition of the formation of oxidized irbesartan metabolites with the known cytochrome CYP2C9 substrates/ inhibitors sulfenazole, tolbutamide and nifedipine. However, in clinical studies the consequences of concomitant irbesartan on the pharmacodynamics of warfarin were negligible. Based on in vitro data, no interaction would be expected with drugs whose metabolism is dependent upon cytochrome P450 isoenzymes 1A1, 1A2, 2A6, 2B6, 2D6, 2E1, or 3A4 [162].

In separate studies of patients receiving maintenance doses of warfarin, hydrochlorothiazide, or digoxin, irbesartan administration for 7 days has no effect on the pharmacodynamics of warfarin (prothrombin time) or pharmacokinetics of digoxin. The pharmacokinetics of irbesartan are not affected by coadministration of nifedipine or hydrochlorothiazide [162].

9.4 Contraindications

Irbesartan is contraindicated in patients who are hypersensitive to irbesartan and in diabetic patients taking aliskiren [162].

10. Toxicology

10.1 Warnings and precautions

10.1.1 Fetal toxicity

Use of drugs that act on the renin-angiotensin system during the second and third trimesters of pregnancy reduces fetal renal function and increases fetal and neonatal morbidity. Resulting oligohydramnios can be associated with fetal lung hypoplasia and skeletal deformations. Potential neonatal adverse effects include skull hypoplasia, anuria, hypotension, renal failure, and death. When pregnancy is detected, discontinue irbesartan as soon as possible [162].

10.1.2 Hypotension in volume or salt-depleted patients

In patients with an activated renin-angiotensin system, such as volume or salt-depleted patients (e.g., those being treated with high doses of diuretics), symptomatic hypotension may occur after initialization of treatment with irbesartan. Correct volume or salt depletion prior to administration of irbesartan or use a lower starting dose [162].

10.1.3 Impaired renal function

Changes in renal function including acute renal failure can be caused by drugs that inhibit the renin-angiotensin system. Patients whose renal function may depend in part on the activity of the renin-angiotensin system (e.g., patients with renal artery stenosis, chronic kidney disease, severe heart failure, or volume depletion) may be at particular risk of developing acute renal failure or death on irbesartan. Monitor renal function periodically in these patients. Consider withholding or discontinuing therapy in patients who develop a clinically significant decrease in renal function on irbesartan [162].

10.1.4 Carcinogenesis, mutagenesis, impairment of fertility

No evidence of carcinogenicity was observed when irbesartan was administered at dosages of up to 500/1000 mg/kg/day (males/females, respectively) in rats and 1000 mg/kg/day in mice for up to 2 years. For male and female rats, 500 mg/kg/day provided an average systemic exposure to irbesartan (AUC0–24 h, bound plus unbound) about 3 and 11 times, respectively, the average systemic exposure in humans receiving the maximum recommended dose (MRD) of 300 mg irbesartan/day, whereas 1000 mg/kg/day (administered to females only) provided an average systemic exposure about 21 times that reported for humans at the MRD. For male and female mice, 1000 mg/kg/day provided an exposure to irbesartan about three and five times, respectively, the human exposure at 300 mg/day [162].

Irbesartan was not mutagenic in a battery of in vitro tests (Ames microbial test, rat hepatocyte DNA repair test, V79 mammalian-cell forward gene-mutation assay). Irbesartan was negative in several tests for induction of chromosomal aberrations (in vitro-human lymphocyte assay; in vivo-mouse micronucleus study) [162].

Irbesartan had no adverse effects on fertility or mating of male or female rats at oral dosages ≤650 mg/kg/day, the highest dose providing a systemic exposure to irbesartan (AUC0–24 h, bound plus unbound) about five times that found in humans receiving the MRD of 300 mg/day [162].

10.2 Overdosage

No data are available in regard to overdosage in humans. However, daily doses of 900 mg for 8 weeks were well tolerated. The most likely manifestations of overdosage are expected to be hypotension and tachycardia; bradycardia might also occur from overdose. Irbesartan is not removed by hemodialysis. Acute oral toxicity studies with irbesartan in mice and rats indicated acute lethal doses were in excess of 2000 mg/kg, about 25- and 50-fold the MRHD (300 mg) on a mg/m^2 basis, respectively [162].

References

[1] S.C. Sweetman, Martindale: The Complete Drug Reference, 36th ed., Pharmaceutical Press, London, 2009, 1316.
[2] ChemSpider. 2020. Website: http://www.chemspider.com/Chemical-Structure.3618.html?rid=8d2b9c72-896d-4c6c-b31a-9a9fa16ef13c (Accessed April 6, 2020).
[3] PubChem. 2020 Website: https://pubchem.ncbi.nlm.nih.gov/compound/Irbesartan#section=MeSH-Entry-Terms (Accessed April 6, 2020).
[4] Drugs.Com. 2020 Website: https://www.drugs.com/international/irbesartan.html (Accessed April 6, 2020).
[5] The Merck Index. 2020 Website: https://www.rsc.org/Merck-Index/monograph/m6397/irbesartan?q=unauthorize (Accessed April 6, 2020).
[6] M.J. O'Neil (Ed.), The Merck Index: An Encyclopedia of Chemicals, Drugs, and Biologicals, Royal Society of Chemistry, Cambridge, UK, 2013, p. 946.
[7] B. Claude, B. Jean-Claude, C. Jacques, N. Dino, P. Tierre, M. Claude, M. Yvette, N-Substituted Heterocyclic Derivatives, Their Preparation and the Pharmaceutical Compositions in Which They Are Present, United States Patent: US5270317 (A) (issued on Dec. 14, 1993, 1993.
[8] C.A. Bernhart, B.P. Ferrari, Y.A. Muneaux, J.L.A. Assens, J. Clement, F. Haudricourt, C.F. Muneaux, J.E. Taillades, M-A. Vignal, J. Gougat, P.R. Guiraudou, C.A. Lacour, A. Roccon, C.F. Cazaubon, J-C. Breliere, G.L. Fur, D. Nisato, A new series of imidazolones: highly specific and potent nonpeptide AT1 angiotensin II receptor antagonists, J. Med. Chem. 36 (1993) 3371–3380.
[9] S.N. Rao, K.S. Babu, An improved and efficient synthesis of irbesartan, an antihypertensive drug, Org. Commun. 4 (2011) 105–111.
[10] O. Singh, H.K. Chopra, P. Loktongbam, Improved green method for synthesis of irbesartan, an antihypertensive drugs, IJSRP 7 (2017) 444–446.
[11] K.V.V.P. Rao, R. Dandala, V.K. Handa, I.V.S. Rao, A. Rani, A. Naidu, Improved methods for the synthesis of irbesartan, an antihypertensive active pharmaceutical ingredient, Synth. Commun. 37 (2007) 2897–2905.
[12] S.N. Rao, D. Sitaramaih, C.N. Rao, P.S. Rao, K.S. Babu, Commercial scalable process for the preparation of irbesartan intermediate, Rasayan J. Chem. 3 (2010) 681–689.

[13] M.C. DiPoto, J. Wu, Synthesis of 2-aminoimidazolones and imidazolones by (3 + 2) annulation of azaoxyallyl cations, Org. Lett. 20 (2018) 499–501.

[14] Z. Qin, Z. Pang, S. Hu, X. Dong, H. Huang, L. Lan, H. Qi, X. Huang, Q. Cai, L. Zhou, One-Pot Method for Preparing Irbesartan, Faming Zhuanli Shenqing CN 108840860 A (issued on Nov. 20, 2018), 2018.

[15] R. Qiao, J. Li, J. Wang, C. Li, Synthesis Method of Irbesartan From Glycine Methyl Ester Hydrochloride via Acylation, Aminolysis, Selfcondensation, Alkylation and Cyclization, Faming Zhuanli Shenqing CN 108276389 A(issued on July 13, 2018), 2018.

[16] Y. Zhang, Y. Zhang, R. Zhang, Method for Synthesizing High-Purity Antihypertensive Drug Irbesartan, Faming Zhuanli Shenqing CN 108218841 A (issued on June 29, 2018), 2018.

[17] S. Wang, W. Chen, P. Dong, F. Huang, Y. Bao, Method for Preparing Irbesartan Condensate, Faming Zhuanli Shenqing CN 106117146 A (issued on Nov. 16, 2016), 2016.

[18] DRUGBANK. 2020 Website: https://www.drugbank.ca/drugs/DB01029 (Accessed April 7, 2020).

[19] B.R. Ganapuram, M. Alle, R. Dadigala, G.M. Kotu, V. Guttena, Development, evaluation and characterization of surface solid dispersion for solubility and dispersion enhancement of irbesartan, J. Pharm. Res. 7 (2013) 472–477.

[20] A.K. Meka, S. Pola, K.R. Tupally, P.L. Abbaraju, Development, evaluation and characterization of surface solid dispersion for solubility and dissolution enhancement of irbesartan, Int. J. Drug Dev. Res. 4 (2012) 263–273.

[21] C.N. Nalini, M. Nivedhitha, A review on analytical methods of irbesartan and its combinations in pharmaceutical dosage forms, Curr. Pharm. Anal. 15 (2019) 1. https://doi.org/10.2174/1573412915666190802164428.

[22] C.S. Yogeesh, H.G. Sowmya, J.G.C. Babu, Analytical method development and validation of irbesartan: review, Eur. J. Pharm. Med. Res. 6 (2019) 235–238.

[23] United States Pharmacopeia, The United States Pharmacopeia 2019: USP 41; The national Formulary: NF 36, United States Pharmacopeial Convention, Rockville, MD, 2019, 2358.

[24] E. Pharmacopoeia, Council of Europe, F-67081 Strasbourg, France, 10th ed., 2010.

[25] V.K. Gupta, R. Jain, S. Agarwal, R. Mishra, A. Dwivedi, Electrochemical determination of antihypertensive drug irbesartan in pharmaceuticals, Anal. Biochem. 10 (2011) 266–271.

[26] B. Bozal, B. Doğan-Topal, B. Uslu, S.A. Özkan, H.Y. Aboul-Enein, Quantitative analysis of irbesartan in pharmaceuticals and human biological fluids by voltammetry, Anal. Lett. 42 (2009) 2322–2338.

[27] H.S. El-Desoky, M.M. Ghoneim, A.D. Habazy, Voltammetry of irbesartan drug in pharmaceutical formulations and human blood: quantification and pharmacokinetic studies, J. Braz. Chem. Sci. 22 (2011) 239–247.

[28] I.H.I. Habib, S.A. Weshahy, S. Toubar, M.M.A. El-Alamin, Adsorptive stripping voltammetric determination of irbesartan in bulk and pharmaceutical products, Pharm. Chem. J. 42 (2008) 438–442.

[29] P.M. Shah, J.S. Shah, D.G. Maheshwari, A review on analytical method for determination of II-receptor antagonists in different dosage forms, J. Global Trends Pharm. Sci. 6 (2015) 2805–2816.

[30] A. Asati, A. Shinde, S. Malik, K.C. Asati, Quantitative analysis method development and validation for Irbesartan in bulk drug by ultraviolet spectroscopy, J. Adv. Pharm. Edu. Res. 4 (2014) 101–105.

[31] K.K. Pradhan, U.S. Mishra, S. Pattnaik, D. Mishra, G. Panigrahi, K.C. Sahu, Method development, validation and stability study of irbesartan in bulk and pharmaceutical dosage form by UV-spectrophotometric method, Int. J. Pharm. Biol. Arch. 2 (2011) 1114–1122.

[32] S. Nissankararao, A.A. Kumar, R.D. Bhimavarapu, K.V. Prasanna, Estimation of irbesartan in bulk and dosage forms by new simple UV spectrophotometry using hydrotropic technique, Pharm. Anal. 4 (2013) 1–3.
[33] R. Biyani, K.S. Yadav, Development and validation of UV-spectrophotometric method for estimation of irbesartan by the hydrotrophy technique, J. Appl. Spectrosc. 86 (2019) 934–941.
[34] P. Virani, R. Sojitra, H. Raj, V. Jain, Development and validation of analytical method for irbesartan and atorvastatin by simultaneous equation spectroscopic method, Int. J. Pharm. Sci. Res. 6 (2015) 519–525.
[35] Y.M. Fayez, Simultaneous determination of some anti-hypertensive drugs in their binary mixture by novel spectrophotometric methods, Spectrochim. Acta A 132 (2014) 446–451.
[36] N. Rahman, M.R. Siddiqui, S.N.H. Azmi, Quantitative analysis of irbesartan in commercial dosage forms by kinetic spectrophotometry, Chem. Pharm. Bull. 54 (2006) 626–631.
[37] K. Divya, V. Sruthi, G.S. Kumar, S. Huidrom, M.A. Haque, V.V.L.N. Prasad, Simultaneous estimation of irbesartan and hydrochlorothiazide in combined pharmaceutical dosage form by UV spectroscopy, Int. J. Innov. Pharm. Sci. Res. 2 (2014) 1674–1680.
[38] A.R. Chabukswar, S.N. Shinde, B.S. Kuchekar, S.C. Jagdale, P.D. Lokhande, V.P. Choudhari, K.D. Ingale, Development and validation of novel UV methods for irbesartan and hydrochlorothiazide combination, Asian J. Res. Chem. 3 (2010) 728–731.
[39] I. Albero, V. Ródenas, S. Garcı´a, C. Sánchez-Pedreño, Determination of irbesartan in the presence of hydrochlorothiazide by derivative spectrophotometry, J. Pharm. Biomed. Anal. 29 (2002) 299–305.
[40] R. El-Bagary, H. Hashem, W.M. Abdelwahab, Septrofluorometric, spectrophotometric and LC determination of irbesartan, J. Chem. Pharm. Res. 3 (2011) 722–733.
[41] A.E. Fawzy, H.H. Hassan, G.M. Sonji, Analysis of eprosartan-hydrochlorothiazide and irbesartan-hydrochlorothiazide binary mixtures by derivative spectrophotometry, Int. J. Appl. Chem. 3 (2007) 1–12.
[42] N. Erk, Three new spectrophotometric methods applied to the simultaneous determination of hydrochlorothiazide and irbesartan, Pharmazie 58 (2003) 543–548.
[43] J. Joseph-Charles, S. Brault, C. Boyer, L. M-H, L. Cabrero, J.-P. Dubost, Simultaneous determination of irbesartan and hydrochlorothiazide in tablets by derivative spectrophotometry, Anal. Lett. 36 (2003) 2485–2495.
[44] M. Farouk, O. Abd Elaziz, M. Shehata, Simple novel spectrophotometric and spectrofluorimetric methods for determination of some antihypertensive drugs, J. Am. Sci. 7 (2011) 300–312.
[45] C. Vetuschi, A. Giannandrea, G. Carlucci, P. Mazzeo, Determination of hydrochlorothiazide and irbesartan in pharmaceuticals by fourth-order UV derivative spectrophotometry, IL Farmaco 60 (2005) 665–670.
[46] L. Sivasubramanian, K.S. Lakshmi, Spectrophotometric multicomponent analysis of irbesartan, hydrochlorothiazide and ramipril in pharmaceutical formulations by chemometric techniques, J. Anal. Pharm. Res. 2 (2016) 1–1000019.
[47] P. Shakya, P.K. Jain, S.P. Shrivastava, A. Gajbhiye, Simultaneous estimation of irbesartan and hydrochlorothiazide by UV spectroscopy, Int. J. Pharm. Pharm. Sci. 7 (2015) 389–391.
[48] M.R. Pattanaik, M.M. Annapurna, Simultaneous determination of irbesartan and hydrochlorothiazide by spectrophotometry, Int. J. Pharm. Technol. 8 (2016) 17545–17553.

[49] J.S.P. Kumar, M.M. Annapurna, New spectrophotometric methods for the simultaneous determination of irbesartan and hydrochlorothiazide in combined dosage forms, Pharm. Methods 6 (2015) 120–125.
[50] H. Syahputra, M.S.A. Muchlisyam, M.A. Masfria, Determination of simultaneous irbesartan and hydrochlorothiazide by ultraviolet spectrophotometry with dual wavelength method, Asian J. Pharm. Res. Dev. 7 (2019) 1–4.
[51] L. Sivasubramanian, K.S. Lakshmi, H-point standard addition method for simultaneous spectrophotometric determination of irbesartan, hydrochlorothiazide and telmisartan in tablets, Int. J. Res. Pharm. Chem. 4 (2014) 373–380.
[52] H. Abdellatef, Extractive-spectrophotometric determination of disopyramide and irbesartan in their pharmaceutical formulation, Spectrochim. Acta A 66 (2007) 1248–1254.
[53] M.A. Omar, O.H. Abdelmageed, A.A. Abdel-Gaber, A.M. Abdel-Megied, Spectrophotometric and spectrofluorimetric determination of certain angiotensin receptor blockers through complex formation, J. Pharm. Sci. Res. 3 (2011) 1499–1510.
[54] M.M. El-Sutohy, S.R. El-Shaboury, S.A. Hussein, N.A. Mohamed, Non-extractive spectrophotometric methods for determination of some angiotensin II receptor antagonists, Asian J. Pharm. Anal. 3 (2013) 3–8.
[55] A. Fathima, S. Rao, G. Venkateshwarlu, Spectrophotometric determination of drugs using 2,3 dichloro 5,6 dicyano-p-benzoquinone as analytical reagent, Int. J. ChemTech Res. 3 (2011) 1769–1780.
[56] I.A. Darwish, T.A. Wani, N.Y. Khalil, A.H. Bakheit, Novel 96-microwell spectrophotometric assays with high throughput for determination of irbesartan in its tablets, Dig. J. Nanomater. Biostruct. 7 (2012) 415–421.
[57] A.A. Abdelmonem, G.H. Ragab, H.A. Hashem, E.A. Bahgat, Spectrofluorimetric and spectrophotometric determination of irbesartan and bisoprolol hemifumarate independently in their tablets, UK J. Pharm. Biosci. 4 (2016) 43–52.
[58] M.M. Baraka, M.E. Elsadek, S.M. Ibrahim, M.A. El-didamoony, Spectrophotometric determination of irbesartan, losartan, atenolol and hydrochlorothiazide in bulk and dosage forms, Asian J. Pharm. Anal. Med. Chem. 4 (2016) 88–106.
[59] M.M. Ibrahim, M.A. Hegazy, A.E. El-Bayoumi, F.M. Abdel-Gawad, M.A.A. El-Ghani, Evaluating the reducing properties of some antihypertensive and antibacterial drugs, through their reaction with iron (III) ions, Anal. Chem. Ind. J. 2 (2013) 141–149.
[60] S. Ashour, R. Bayram, Selective and validated kinetic spectrophotometric method for the determination of irbesartan in pure and pharmaceutical formulations, Ann. Pharm. Fr. 77 (2019) 101–111.
[61] S.E.K. Tekkeli, A. Önal, New spectrophotometric and spectrofluorimetric methods for the determination of angiotensin II receptor antagonists in tablets and plasma, J. Anal. Chem. 68 (2013) 501–509.
[62] S. Ashour, R. Bayram, Sensitive extractional colorimetric analysis of fexofenadine hydrochloride and irbesartan bases through acid-dye complexation using naphthol blue black in pure form and pharmaceuticals, Modern Chem. (2017) 93–100.
[63] G.T. Rani, D.G. Sankar, L. Madhavi, B. Satyanarayana, Extractive visible spectrophotometric method for determination of telmisatan and irbesartan in bulk and pharmaceutical formulations, Asian J. Pharm. Clin. Res. 5 (2012) 41–44.
[64] K.A. Attia, M.W. Nassar, H.H. Abu Seada, A.M. Ahmed, A simple and green analytical method for determination of some angiotensin receptor antagonists by binary complex formation, Anal. Chem. Ind. J. 14 (2014) 167–176.
[65] S.R. El-Shaboury, S.A. Hussein, N.A. Mohamed, M.M. El-Sutohy, Spectrofluorimetric method for determination of some angiotensin II receptor antagonists, J. Pharm. Anal. 2 (2012) 12–18.

[66] H. Fu, H. Li, M. Shao, T. Yang, X. Zhang, R. Xu, Y. Wei, S. Chen, C. Ni, H. Wu, Simultaneous determination of repaglinide and irbesartan in biological plasmas using micellar enhanced excitation-emission matrix fluorescence coupled with ATLD method, Sci. China Chem. 59 (2016) 816–823.
[67] A.A. Khalid, H.H. AbuSeada, W.N. Mohammed, M.A. Adel, Development of regression analysis for determination of irbesartan in pharmaceutical tablets using fourier transform infrared spectrophotometry, Med. Anal. Chem. Int. J. 1 (2017) 1–10.
[68] A.G. Dal, S. Koyutürk, Simultaneous determination of irbesartan and hydrochlorothiazide in tablets by CE-DAD, J. Biol. Chem. 43 (2015) 145–152.
[69] S. Hillaert, W.V. Bossche, Simultaneous determination of hydrochlorothiazide and several angiotensin-II-receptor antagonists by capillary electrophoresis, J. Pharm. Biomed. Anal. 31 (2003) 329–339.
[70] Y.M. Fayez, M.A. Hegazy, Micellar Electrokinetic chromatography (MEKC) with multiresponse chemometric optimization for the determination of hydrochlorothiazide and coformulated antihypertensives in the presence of hydrochlorothiazide major impurity, Chromatogr. Sci. 54 (2016) 1050–1060.
[71] M. Zhang, F. Wei, Y.-F. Zhang, J. Nie, Y. Feng, Novel polymer monolith microextraction using a poly(methacrylic acid-ethylene glycol dimethacrylate) monolith and its application to simultaneous analysis of several angiotensin II receptor antagonists in human urine by capillary zone electrophoresis, J. Chromatogr. A 1102 (2006) 294–301.
[72] L. González, U. Akesolo, R.M. Jiménez, R.M. Alonso, Application of capillary zone electrophoresis to the screening of some angiotensin II receptor antagonists, Electrophoresis 23 (2002) 223–229.
[73] N.J. Shah, B.N. Suhagia, R.R. Shah, N.M. Patel, Development and validation of a HPTLC method for the simultaneous estimation of irbesartan and hydrochlorothiazide in tablet dosage form, Indian J. Pharm. Sci. 69 (2007) 240–243.
[74] S.A. Hussein, H.M. Abd El-Wadood, M.A.A. Abdallah, A.A. Khorshed, Application of high-performance thin-layer chromatography for screening and simultaneous determination of some angiotensin II receptor antagonists in dosage forms and plasma, J. Planar Chromatogr. 27 (2014) 192–198.
[75] A.S. Khodke, L.V. Potale, M.C. Damle, K.G. Bothara, A validated stability indicating HPTLC method for simultaneous estimation of irbesartan and hydrochlorothiazide, Pharm. Methods 1 (2010) 39–43.
[76] D.A. Shah, D.V. Patel, F.A. Mehta, U.K. Chhalotiya, K.K. Bhatt, High-performance thin-layer chromatography method for estimating the stability of a combination of irbesartan and amlodipine besylate, J. Taibah Univ. Sci. 9 (2015) 177–186.
[77] L. Sivasubramanian, K.S. Lakshmi, Simultaneous estimation of irbesartan, telmisartan, hydrochlorothiazide and ramipril in combined dosage forms by validated HPTLC method, J. Anal. Pharm. Res. 4 (2017) 2–7.
[78] B. Rosangluaia, P. Shanmugasundaram, M. Velraj, Validated HPTLC method for simultaneous estimation of irbesartan and hydrochlorthiazide in a tablet dosage form, Der Pharma Chem. 3 (2011) 310–317.
[79] B.H. Mehta, S.B. Morge, Simultaneous determination of irbesartan and hydrochlorothiazide by HPTLC method, Indian Drugs 47 (2010) 71–74.
[80] P. Kalaiselvi, K.G. Lalitha, HPTLC method development and validation for simultaneous estimation of chlorthalidone and irbesartan in combined tablet dosage form, Indian Drugs 51 (2014) 46–52.
[81] R.T. Sane, M. Francis, S. Pawar, Determination of irbesartan in pharmaceutical dosage forms by HPTLC, Indian Drugs 39 (2002) 32–35.
[82] B. Zeng, Y. Gu, K. Nguyen, J. Sherma, Development of quantitative HPTLC-densitometry methods following a model process for transfer of TLC screening

methods for pharmaceutical products containing moxifloxacin HCl, ofloxacin, amoxicillin trihydrate, acetylsalicylic acid + acetaminophen + caffeine, nimesulide, irbesartan, and pantoprazole, J. Liq. Chromatogr. Relat. Technol. 42 (2019) 324–329.

[83] S.H. Abd El-Rahman, H.M. Abd El-Wadood, M.A. Abdallah, A.A. Khorshed, Development of two high-performance thin-layer chromatographic methods for the determination of irbesartan in tablets and plasma, J. Planar Chromatogr. Mod. TLC 28 (2015) 83–89.

[84] S.H. Abd El-Rahman, H.M. Abd El-Wadood, M.A. Abdallah, A.A. Khorshed, HPTLC with fluorescence densitometry for simultaneous determination of some angiotensin II receptor blockers in tablets and plasma, J. AOAC Int. 98 (2015) 354–360.

[85] V. Khanvilkar, J. Shah, V. Kadam, Development and validation of HPLC assay for estimation of irbesartan in human plasma, Res. J. Pharm. Technol. 6 (2013) 292–295.

[86] K. Sujana, RP-HPLC-DAD method for determination of irbesartan in bulk and tablets exposed to forced conditions, Int. J. Pharm. Res. Dev. 3 (2011) 67–73.

[87] A. Yousheng, X. Chuhong, C. Huating, Content determination of irbesartan in serum by HPLC, China Pharm. (2004)04.

[88] R. Prashanthi, K. Raghavi, M. Sindhura, B. Anupama, B.N. Nalluri, Development and validation of a sensitive RP-HPLC-PDA method for assay of irbesartan in pure and pharmaceutical dosage forms, Int. J. Pharm. Bio Sci. 3 (2012) 397–407.

[89] R.R. Raju, N.B. Babu, Development and validation of HPLC method for the estimation of irbesartan in pharmaceutical dosage form, Pharmacophore 2 (2011) 108–112.

[90] S.K. Bae, M.J. Kim, E.J. Shim, D.Y. Cho, J.H. Shon, E.Y. Kim, J.G. Shin, HPLC determination of irbesartan in human plasma: its application to pharmacokinetic studies, Biomed. Chromatogr. 23 (2009) 568–572.

[91] M.B. Ubale, V.D. Dhakane, V.R. Chaudhari, Simultaneous high-performance liquid chromatographic determination of irbesartan and hydrochlorothiazide in pharmaceutical dosage form, J. Pharm. Sci. Innov. 1 (2012) 25–28.

[92] A.M. Alanazi, A.S. Abdelhameed, N.Y. Khalil, A.A. Khan, I.A. Darwish, HPLC method with monolithic column for simultaneous determination of irbesartan and hydrochlorothiazide in tablets, Acta Pharm. 64 (2014) 187–198.

[93] L.F. Tutunji, M.F. Tutunji, M.I. Alzoubi, M.H. Khabbas, A.I. Arida, Simultaneous determination of irbesartan and hydrochlorothiazide in human plasma using HPLC coupled with tandem mass spectrometry: Application to bioequivalence studies, J. Pharm. Biomed. Anal. 51 (2010) 985–990.

[94] Z. Vujić, N. Mulavdić, M. Smajić, J. Brborić, P. Stankovic, Simultaneous analysis of irbesartan and hydrochlorothiazide: an improved HPLC method with the aid of a chemometric protocol, Molecules 17 (2012) 3461–3474.

[95] S. Hemamrutha, R. Rambabu, S. Vidhyadhara, Development and validation of RP-HPLC method for simultaneous estimation of irbesartan and hydrochlorothiazide in bulk and pharmaceutical dosage forms, Int. J. Pharm. 3 (2013) 360–363.

[96] B. Raja, P. Himasri, B. Ramadevi, RP-HPLC method for the simultaneous estimation of irbesartan and hydrochlorthiazide in pharmaceutical dosage form, Int. Res. J. Pharm. Appl. Sci. 2 (2012) 29–38.

[97] A.R. Chabukswar, S.C. Jagdale, B.S. Kuchekar, P.D. Lokhande, S.N. Shinde, K.D. Ingale, A.K. Kolsure, Development and validation of a RP HPLC-PDA method for simultaneous estimation of hydrochlorothiazide and irbesartan, Der. Pharm. Chem. 2 (2010) 148–156.

[98] P. Prabhu, M. Muralidhar, Development and validation of a high performance liquid chromatography method for simultaneous determination of irbesartan and its related impurities in pharmaceutical tablets, Int. J. Pharm. Sci. Drug Res. 6 (2014) 145–153.

[99] S.K. Patel, M.S. Darji, Development and validation of reverse phase high performance liquid chromatographymethod for simultaneous estimation of amlodipine besylate and irbesartan in synthetic mixture, Int. J. Pharm. Drug Anal. 2 (2014) 94–99.
[100] H.M. Hafez, A.A. Elshanawane, L.M. Abdelaziz, M.M. Kamal, Quantitative determination of three angiotensin-II-receptor antagonists in presence of hydrochlorothiazide by RP-HPLC in their tablet preparations, Iranian J. Pharm. Res. 12 (2013) 635–643.
[101] H.K. Tiwari, T. Monif, P. Ranjan, P. Verma, S. Reyar, A.H. Khuroo, S. Mishra, Quantitative estimation of irbesartan in two different matrices and its application to human and dog bioavailability studies using LC-MS/MS, Asian J. Pharm. Sci. 8 (2013) 346–355.
[102] L. Gonzalez, J.A. Lopez, R.M. Alonso, R.M. Jiménez, Fast screening method for the determination of angiotensin II receptor antagonists in human plasma by high-performance liquid chromatography with fluorimetric detection, J. Chromatogr. A 949 (2002) 49–60.
[103] R. Youssef, A. Hbash, A. Hassan, Development and validation of RP-HPLC method for the estimation and separation of valsartan, losartan and irbesartan in bulk and pharmaceutical formulation, Int. J. Pharm. Sci. Rev. Res. 24 (2014) 311–314.
[104] K. Czerwińska, A.P. Mazurek, Identification and determination of selected angiotensin ii receptor antagonist group drugs by HPLC method, Acta Pol. Pharm. 68 (2011) 831–837.
[105] N. Goswami, A validated stability-indicating liquid chromatographic method for determination of process related impurities and degradation behavior of irbesartan in solid oral dosage, J. Adv. Pharm. Technol. Res. 5 (2014) 33–45.
[106] E. Caudrona, S. Laurenta, E.M. Billaud, P. Prognon, Simultaneous determination of the acid/base antihypertensive drugs celiprolol, bisoprolol and irbesartan in human plasma by liquid chromatography, J. Chromatogr. B Analyt. Technol. Biomed. Life Sci. 801 (2004) 339–345.
[107] R.A. Mhaske, S. Sahasrabudhe, A.A. Mhaske, RP-HPLC method for simultataneous determination of irbesartan, losartan, hydro-chlorothiazide and chlorthalidone-application to commercially available drug products, Int. J. Pharm. Sci. Res. 3 (2012) 1116–1123.
[108] X. Qiu, Z. Wang, B. Wang, H. Zhan, X. Pan, R. Xu, Simultaneous determination of irbesartan and hydrochlorothiazide in human plasma by ultra high performance liquid chromatography tandem mass spectrometry and its application to a bioequivalence study, J. Chromatogr. B Analyt. Technol. Biomed. Life Sci. 957 (2014) 110–115.
[109] L. Banjare, J.K. Chandra, P. Patel, Method development and validation for estimation of irbesartan in bulk drug and pharmaceutical dosage, J. Drug Deliv. Ther. 3 (2013) 87–90.
[110] A.R. Reddy, G.V.S. Kumar, S. Puranik, P. Giriprasad, K.A. Sridhar, Development and validation of stability indicating reverse phase HPLC method for simultaneous estimation of irbesartan and hydrochlorothiazide in bulk drug and tablet dosage form, Int. J. Pharm. Chem. Biol. Sci. 2 (2012) 696–703.
[111] S.V.S.G. Prasad, S. Shivakumar, T. Sudhir, R. Mital, G.D. Rao, LC/MS/MS method for the simultaneous estimation of Losartan potassium and irbesartan in rat plasma, Int. J. Pharm. Pharm. Sci. 1 (2009) 206–215.
[112] M. Farouk, O.A.A. Ghonim, A. Hemdan, M.A. Shehata, Novel validated chromatographic method for determination of some anti-hypertensive drugs, J. Am. Sci. 66 (2010) 476–486.
[113] Z. Li, F. Chen, X. Wang, C. Wang, Ionic liquids dispersive liquid-liquid microextraction and high-performance liquid chromatographic determination of irbesartan and valsartan in human urine, Biomed. Chromatogr. 27 (2013) 254–258.

[114] V.P. Rane, K.R. Patil, J.N. Sangshetti, R.D. Yeole, D.B. Shinde, Stability indicating LC method for simultaneous determination of irbesartan and hydrochlorothiazide in pharmaceutical preparations, J. Chromatogr. Sci. 48 (2010) 595–600.
[115] N. Erk, Simultaneous determination of irbesartan and hydrochlorothiazide in human plasma by liquid chromatography, J. Chromatogr. B Analyt. Technol. Biomed. Life Sci. 784 (2003) 195–201.
[116] A.K. Shakya, Y.M. Al-Hiari, O.M.O. Alhamami, Liquid chromatographic determination of irbesartan in human plasma, J. Chromatogr. B Analyt. Technol. Biomed. Life Sci. 848 (2007) 245–250.
[117] P. Jain, Y.R. Bhardwaj, D. Kishore, A liquid chromatography tandem mass spectrometry based method for the simultaneous determination of irbesartan and hydrochlorothiazide in human plasma, Int. J. Drug Dev. Res. 6 (2014) 18–31.
[118] H.W. Lee, H.Y. Ji, F.S. Park, K.C. Lee, H.S. Lee, Hydrophilic interaction chromatography-tandem mass spectrometric analysis of irbesartan in human plasma: application to pharmacokinetic study of irbesartan, Sep. Sci. 32 (2009) 2353–2358.
[119] M. Ganesan, S. Nanjundan, M. Gomathi, S. Muralidharan, Method development and validation of irbesartan using LCMS/MS: application to pharmacokinetic studies, J. Chem. Pharm. Res. 2 (2010) 740–746.
[120] G.V.R. Reddy, A.P. Kumar, B.V. Reddy, J. Sreeramulu, J.H. Park, Separation and quantification of key starting materials of irbesartan using liquid chromatography-mass spectrometry as a separation and identification tool, Anal. Lett. 42 (2009) 2087–2095.
[121] M.M. Eswarudu, T.N. Chary, S. Junapudi, M. Sushma, RP-HPLC method development and validation for simultaneous estimation of irbesartan and hydrochlorothiazide in pharmaceutical dosage form, Asian J. Res. Chem. 5 (2012) 472–476.
[122] V. Sathiyanarayanan, T.V.B. Reddy, M. Chandru, I.M. Bilal, S. Kuttirani, R. Prasanna, D. Eswaramoorthy, Stability indicating validated related substance UFLC method & alkali degradation impurities characterized by LCMS for irbesartan in bulk and pharmaceutical formulation product, J. Glob. Trends Pharm. Sci. 5 (2014) 1905–1911.
[123] P. Virani, S. Rajanit, R. Hasumati, V. Jain, Irbesartan: a review on analytical method and its determination in pharmaceuticals and biological matrix, Pharm. Anal. Qual. Assur. 2014 (2014) 1–8.
[124] H.A. Alhazmi, A.M. Alnami, M.A.A. Arishi, R.K. Alameer, M. Al Bratty, Z. Rehman, S.A. Javed, I.A. Arbab, A fast and validated reversed-phase HPLC method for simultaneous determination of simvastatin, atorvastatin, telmisartan and irbesartan in bulk drugs and tablet formulations, Sci. Pharm. 86 (2018) 1–13.
[125] S. Mondal, A. Pal, P. Mondal, D. Shit, S. Biswal, B.M. Babu, Determination of irbesartan using stability indicating reverse phase liquid chromatographic and UV spectrophotometric method, Int. J. Pharm. Invest. 10 (2020) 70–75.
[126] K.K. Pradhan, Development of validated stability indicating RP-HPLC method for estimation of irbesartan in bulk and pharmaceutical dosage forms, Int. J. Pharm. Sci. Res. 7 (2016) 4531–4539.
[127] K.T. Rao, L.V. Rao, V.V.K.M. Kandepi, Rapid stability-indicating RRLC method for simultaneous estimation of irbesartan and its related impurities, Indian J. Pharm. Sci. 78 (2016) 252–258.
[128] M. Nagamani, S.K. Sahoo, D. Kavitha, C. Bhagyalaxmi, L. Mahapatra, Validated RP-HPLC method for simultaneous estimation of irbesartan and hydrochlorothiazide in tablet dosage form, Int. J. Pharm. Res. Health Sci. 5 (2017) 1419–1423.
[129] T. Srinivasu, M.M. Annapurna, Development of a new validated stability indicating RP-HPLC method for the determination of irbesartan and hydrochlorothiazide, Asian J. Pharm. 12 (2018) S333–S339.

[130] M.M. Zareh, M.Z. Saad, W.S. Hassan, M.E. Elhennawy, M.K. Soltan, M.M. Sebaiy, Gradient HPLC method for simultaneous determination of eight sartan and statin drugs in their pure and dosage forms, Pharmaceuticals (Basel) 13 (2020) 32, 1–15.
[131] R.K. Dagariya, R.K. Jat, Method development and validation of irbesartan chlorthalidone and cilnidipine in their combined tablet dosage form by high performance liquid chromatography, J. Drug Deliv. Ther. 7 (2017) 88–96.
[132] P.P. Dedhiya, D.V. Jayswal, B.P. Marolia, S.A. Shah, Development and validation of RP-HPLC method for simultaneous estimation of irbesartan and atorvastatin in synthetic mixture, J. Pharm. Res. 11 (2017) 64–67.
[133] P. Virani, R. Sojitra, B. Savaj, H. Raj, V. Jain, K. Patel, V. Sutariya, Development and validation of RP-HPLC method for the simultaneous estimation of irbesartan and atorvastatin in synthetic mixture, Asian J. Pharm. Technol. 5 (2015) 175–181.
[134] M.M. Annapurna, B. Venkatesh, R.K. Chaitanya, Quality by design approach (QbD) for the simultaneous determination of anti-hypertensive drugs (candesartan, irbesartan and hydrochlorothiazide) by RP-HPLC, J. Anal. Pharm. Res. 4 (2017) 1–17.
[135] K.M. Prajapati, B. Patel, J. Patel, V. Darji, J.D. Raja, Stability indicating RP-HPLC method development and validation for simultaneous estimation of cilnidipine and irbesartan in its pharmaceutical dosage form, World J. Pharm. Pharm. Sci. 6 (2017) 1017–1026.
[136] C. Soni, C.J. Patel, M.M. Patel, Development and validation of stability indicating chromatographic method for simultaneous estimation of cilnidipine and irbesartan in pharmaceutical dosage form, World J. Pharm. Pharm. Sci. 7 (2018) 1124–1135.
[137] K.A. Marghany, R.A. Abdelsalam, G.M. Haddad, HPLC method transfer study for simultaneous determination of seven angiotensin II receptor blockers, J. Sep. Sci. 43 (2020) 1398–1405.
[138] T.H. Kumar, C. Asha, D.G. Sankar, Estimation of amlodipine besylate and irbesartan in pharmaceutical formulation by RP-HPLC with forced degradation studies, Int. J. App. Pharm. 11 (2019) 159–167.
[139] V. Kavya, K.G. Swamy, M.E. Uddin, T. Ramu, B. Navya, Development and validation of HPLC method for the estimation of irbesartan in pharmaceutical dosage form, J. Sci. Res. Pharm. 4 (2015) 66–68.
[140] P.L. Ingale, S.D. Dalvi, A.R. Salunkhe, Development and validation of HPLC method for the estimation of irbesartan, Indo Am. J. Pharm. Res. 6 (2016) 5242–5247.
[141] A. Zheng, L. Wang, N. Yu, N. Li, Determination of irbesartan tablets and related substances by microemulsion liquid chromatography, Guangdong Yaoxueyuan Xuebao 32 (2016) 444–448.
[142] H. Hashem, A.E. Ibrahim, M. Elhenawee, Simultaneous determination of the antihypertensives hydrochlorothiazide, losartan potassium, irbesartan and valsartan in bulk powders and pharmaceutical preparations by high performance liquid chromatography, Main Group Chem. 15 (2016) 335–346.
[143] N. Anitha, T. Mamatha, T.E. Rajan, G. Mrudul, P. Sultana, Analytical method development and method validation for the estimation of irbesartan in tablet dosage form by HPLC method, Am. J. PharmTech Res. 4 (2014) 469–482.
[144] A.Z. Miraz, HPLC-UV method for simultaneous determination of irbesartan, candesartan, gliquidone and pioglitazone in formulations and in human serum, J. Chin. Pharm. Sci. 27 (2018) 273–280.
[145] F. Coudoré, L. Harvard, S. Lefeuvre, E.M. Billaud, P. Beaune, G. Bobrie, M. Azizi, P. Prognon, S. Laurent, HPLC–DAD analysis of hydrochlorothiazide and irbesartan in hypertensive patients on fixed-dose combination therapy, Chromatographia 74 (2011) 559–565.

[146] V. Maslarska, V. Yankov, D. Obreshkova, S. Bozhanov, Simultaneous determination of sartans by high performance liquid chromatography with ultra violet detection, Indian J. Pharm. Edu. Res. 51 (2017) 343–348.
[147] T.A. Wani, S. Zargar, New ultra-performance liquid chromatography-tandem mass spectrometry method for the determination of irbesartan in human plasma, J. Food Drug Anal. 23 (2015) 569–576.
[148] C.D. Patel, G. Swati, B.H. Patel, Development and validation of bioanalytical method for simultaneous estimation of irbesartan and hydrochlorothiazide in human plasma using liquid chromatography-tandem mass spectrometry, Int. J. Pharm. Chem. Biol. Sci. 6 (2016) 414–423.
[149] L.F. Tutunji, M.F. Tutunji, M.I. Alzoubi, M.H. Khabbas, A.I. Arida, Simultaneous determination of irbesartan and hydrochlorothiazide in human plasma using HPLC coupled with tandem mass spectrometry: application to a bioequivalence study, J. Bioequiv. Availab. 7 (2015) 68.
[150] Nurmeilis, Y. Harahap, F.C. Saputri, A. Munim, R. Setiabudy, Determination of irbesartan and sinensetin simultaneously by liquid chromatography tandem mass spectrometry and their application to drug interaction study in rat plasma, Int. J. Pharm. Pharm. Sci. 8 (2015) 96–100.
[151] A. Tan, X. Gui, M. Wong, H. Deng, G. Gu, C. Fanaras, J.C. Fanaras, Simultaneous quantification of candesartan and irbesartan in rabbit eye tissues by liquid chromatography-tandem mass spectrometry, Biomed. Chromatogr. 34 (2020) e4808.
[152] Z. Nisa, S.I. Ali, M. Rizvi, M.A. Khan, R.A. Sultan, R. Fatima, N. Shaheen, F. Zafar, S.S. Kashif, N. Khatian, Development and validation of reverse phase HPLC method for determination of angiotensin receptor blocking agent irbesartan in plasma, Pak. J. Pharm. Sci. 32 (2019) 853–858.
[153] X. Zhou, H. Qi, J. Zhang, M. Wang, Q. Chang, Determination of plasma irbesartan by HPLC-fluorescence, Hebei Yiyao 36 (2014) 1733–1734.
[154] M.S. Elgawish, M.K. Soltan, M.M. Sebaiy, An LC-MS/MS spectrometry method for the simultaneous determination of rosuvastatin and irbesartan in rat plasma: Insight into pharmacokinetic and drug-drug interaction studies, J. Pharm. Biomed. Anal. 174 (2019) 226–234.
[155] A. Chavali, T. Jenkins, P. McConville, USP Method Transfer and Routine Use Analysis of Irbesartan Tablets From HPLC to UPLC, Waters Corporation, Printed/Produced in the U.S.A. 720004130EN VW-PDF. Available at: https://gimitec.com/file/720004130en.pdf, 2013.
[156] M. Shakeb, S.B. Puranik, Stability indicating method development and validation of irbesartan and hydrochlorothiazide with stress degradation, Int. J. Pharm. Pharm. Res. 3 (2015) 75–82.
[157] G. Castro, I. Rodriguez, M. Ramil, R. Cela, Selective determination of sartan drugs in environmental water samples by mixed-mode solid-phase extraction and liquid chromatography tandem mass spectrometry, Chemosphere 224 (2019) 562–571.
[158] G. Castro, I. Carpinteiro, I. Rodriguez, R. Cela, Determination of cardiovascular drugs in sewage sludge by matrix solid-phase dispersion and ultraperformance liquid chromatography tandem mass spectrometry, Anal. Bioanal. Chem. 410 (2018) 6807–6817.
[159] A. Klancar, J. Trontelj, R. Roskar, Development of a multi-residue method for monitoring 44 pharmaceuticals in slovene surface water by SPE-LC-MS/MS, Water Air Soil Pollut. 229 (2018) 1–18.
[160] R.P. Shah, A. Sahu, S. Singh, Identification and characterization of degradation products of irbesartan using LC-MS/TOF, MS(n), on-line H/D exchange and LC-NMR, J. Pharm. Biomed. Anal. 51 (2010) 1037–1346.

[161] S. Saravanan, A.T. Ansari, LC-MS/MS study of the trace level impurities of irbesartan an angiotensin ii receptor antagonist molecule to its origin through nMS^2 technique, Int. J. Pharm. Pharm. Sci. 11 (2019) 38–43.
[162] FDA, 2020. FDA Approved Drug Products: Irbesartan Oral Tablets, Highlights of PRESCRIBING INFORMATION, Available at: https://www.accessdata.fda.gov/drugsatfda_docs/label/2018/020757s069lbl.pdf (Accessed May 6, 2020).
[163] FDA, 2020. FDA Approved Drug Products: Irbesartan and Hydrochlorothiazide Oral Tablets, Highlights of Prescribing Information, Available at: https://www.accessdata.fda.gov/drugsatfda_docs/label/2018/020758s073lbl.pdf (Accessed May 6, 2020).
[164] M.R. Marino, K. Langenbacher, N.F. Ford, H.D. Uderman, Pharmacokinetics and pharmacodynamics of irbesartan in healthy subjects, J. Clin. Pharmacol. 38 (1998) 246–255.
[165] T.J. Chando, D.W. Everett, A.D. Kahle, A.M. Starrett, N. Vachharajani, W.C. Shyu, K.J. Kripalani, R.H. Barbhaiya, Biotransformation of irbesartan in man, Drug Metab. Dispos. 26 (1998) 408–417.
[166] S.F. Zhou, Z.W. Zhou, L.P. Yang, J.P. Cai, Substrates, inducers, inhibitors and structure-activity relationships of human Cytochrome P450 2C9 and implications in drug development, Curr. Med. Chem. 16 (2009) 3480–3675.
[167] Sandoz Canada: Irbesartan Product Monograph, 2020 Available at: https://www.sandoz.ca/sites/www.sandoz.ca/files/Irbesartan%20Product%20Monograph.pdf (Accessed May 6, 2020).

CHAPTER SIX

Validation of *in-vitro* bioassay methods: Application in herbal drug research

Gunawan Indrayanto*, Galih Satrio Putra, and Farida Suhud
Faculty of Pharmacy, Surabaya University, Surabaya, Indonesia
*Corresponding author: e-mail addresses: gunawanindrayanto@yahoo.com; gunawanindrayanto@staff.ubaya.ac.id

Contents

1. Introduction

Herbal drugs or traditional drugs have been used worldwide for centuries. Almost 80% of world populations, especially in the developing countries, use herbal drugs for their primary health care [1,2]. In Indonesia herbal drugs are very popular, called as *obat tradisional*, comprised of non-standardized herbal drugs (*jamu*), standardized herbal drugs and phyto-pharmaceuticals. *Ayurveda* has been used for a long time as traditional medicines in India. Also, Traditional Chinese Medicine (TCM) has also used herbs for almost thousand years. It is well reported that 50% Americans consume dietary supplements, and in many cases, dietary

Profiles of Drug Substances, Excipients, and Related Methodology, Volume 46
ISSN 1871-5125
https://doi.org/10.1016/bs.podrm.2020.07.005

supplement is comprised of plant materials referred to as *botanical natural product* [3]. Generally, an herbal drug consists of a single or a mixture of herbs or extracts prepared in a certain drug preparation *viz* tablet, capsule, syrup, sachet, herbs concoction *etc.* Due to its nature as a mixture of plant materials, the chemical composition of any herbal drug is very complex. It is well known that, the efficacy of any herbal drug depends on its *overall* chemicals or metabolite components, and not to *one* or *two* components only. There are countless scientific publications on herbal drugs, but unfortunately most of the herbal/traditional drugs are not yet scientifically fully established, regarding their safety and efficacy. This is the reason why the application of herbal drug(s) as modern drug(s) by medical doctors is very limited presently.

To develop a *modern drug* from herbal drug(s), Pan et al. [4] described three stages, which should be performed: (1) Pre-drug stage: collecting all information on the traditional herbal/folk medicine system and their practical experiences, (2) Quasidrug stage: preparation of extracts, phytochemical investigation, determination of lead compound, and testing bioactivities, and (3) Full drug stage: development of an isolated active compound into a modern drug, *e.g.,* artemisinin tablet as anti-malarial drug. Kellog et al. [3] recommended some steps for choosing *an ideal* botanical drug(s), which can be selected for research studies. The most important step is the authentication of the plant materials; this must be done before any chemical-, *in-vitro-,* and *in-vivo*-studies. Authentication can be carried out by using microscopical test, metabolomics and DNA bar coding. Another important step is the determination of the bioactive compound(s) of the herb(s). Two approaches have been applied for verifying the bioactive (or toxic) component(s) of any herbal drugs. The *first approach* is by using a bioassay guided isolation method as described by Rahman et al. [5]. *The second approach* is by utilizing a combination of chemical- and bioactivity-profiling (metabolomic), then evaluating by MVA; nice review articles on this 2nd approach was recently published Ory et al. [6] and Saesong et al. [7]. Fractionations and isolation of the active/toxic component(s) for confirming the bioactivity (1st approach), are not needed anymore; the active/toxic component(s) of the extract can be directly determined by using MVA, *i.e.,* PCA, PLS-DA, S-plot (2nd approach). These two approaches are briefly summarized in Fig. 1; detailed explanations and discussions of both methods can be referred in previous publications [5–7]. The chemical structure of the predicted bioactive or toxic compound can be determined using LC-HR MS/MS and many free online MS data bases (*e.g., Metfrag, Metlin,*

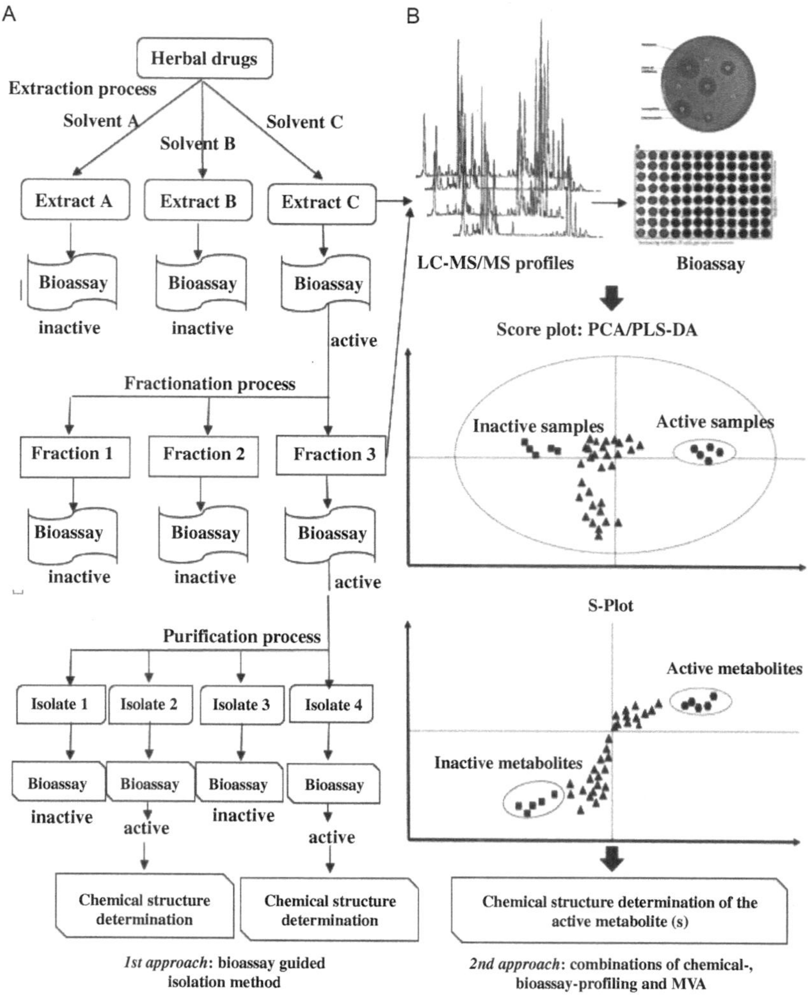

Fig. 1 The summarization of two approaches for the determination of the bioactive- or toxic component(s) of raw herbal drug or extracts. Solvents with different polarities (A, B, C) were used for performing extractions, after fractionations and isolation, the chemical structure of the active compound(s) can be determined (1st approach). By 2nd approach, extracts or fractions are directly analyzed using LC-MS/MS and their bioactivities were simultaneously evaluated; the active compound can be determined directly using MVA (PCA, PLS-DA, S-Plot). See text for more explanation.

GNPS, MzMine, CFMID, etc.) [1]. A recent review on the analytic techniques, that can be applied for metabolomic studies, has been published by Segers et al. [8]. Quantification of the bioactive- and/or toxic-marker(s) in any herbal drug(s) is very essential for performing a routine quality control

(QC) in Pharmaceutical industries; for this purpose, the authors recommended Marker's MaRS criteria [9] or Q-marker [10] as the method for the quality assessment. Dai et al. [11] applied 2nd approach for determining nine Q-markers of a TCM drug *Huang Qin* and applying it as an excellent routine QC method; those Q-markers showed bioactive equivalent to *Huang Qin* decoction.

As discussed above, for developing a modern drug from any herbal drug, chemical- and biological- assays are required, so the availability of *validated* methods is very crucial; without applying validated methods, all conclusions might be wrong or bias. Unfortunately, too many published works did not report their methods of validation (chemical and bioassay), this means the legitimacy of the conclusions cannot be verified. Review articles regarding the chemical method of validations, and its application for herbal drugs including QC, have been recently published by Indrayanto [1,2]. This present review will be focused and discussed on the method validation of the bioassay, especially on its application for herbal drug research, this included herbs, medicinal plants, extracts and (isolated) compounds.

2. Definition, principle and objective of the bioassay

Assay Guidance manual from Eli Lilly & Company [12] defined biological assay or bioassay as "a set of reagents that produces a detectable signal, allowing a biological process to be quantified". USP 42 ⟨1030⟩ [13] defined bioassay as "Analysis (as of a drug) to quantify the biological activity or activities of one or more components by determining its capacity for producing an expected biological activity on a culture of living cells (*in-vitro*) or on test organisms (*in-vivo*), expressed in terms of units". According to European Pharmacopeia 10 [14], "biological methods are described for the assay of certain substances and preparations whose potencies cannot be adequately assured by chemical or physical analysis. The principle applied wherever possible throughout these assays is that of comparison with a standard preparation to determine how much of the substance to be examined produces the same biological effect as a given quantity".

The aim of performing of biological assay or bioassay is to determine the effect of a substance on a certain type of living matter or a biological test system [15]. Bioassay can be performed either by direct or non-direct measurements. Direct bioassay means determining the concentration of a substance, which is required in order to elicit a specific response, *e.g.*, measuring the potency of powdered digitalis leaves by determining the concentration that can stop a cat's heart. Indirect bioassay can be determined by comparing the responses of equal concentrations of authentic standard and sample which are required to achieve a specified response.

According to USP 42 ⟨1030⟩ [13], ⟨1032⟩ [16] bioassay can be performed by using *in-vivo, ex-vivo and in-vitro* methods. In *in-vivo* bioassay, sets of dilutions of standards and test samples are administrated to animals to estimate the potency of test samples calculated from their curves of concentration *vs.* responses. Colony- and naïve-animals both are used in such studies, colony animals being preferable. Animals must be healthy and given time to stabilize in their environment to be suitable for use in a bioassay. *Ex-vivo* bioassay means evaluating the activity of test samples by using cells or tissues from humans and animal donors which are cultivated in the laboratory; bioassays that involve living tissues or cells from an animal require process of management like that of in *in-vivo* assay. *In-vitro* bioassay specified as quantification of biological activity or activities of one or more components by determining its capacity for producing an expected biological activity on a cultured cell lines, which can be derived from tumors, engineered cell lines, or non-transformed cell lines. The term *in-vitro* involves "the technique of performing a given procedure in an artificial environment outside the living organism" [17].

Bioassay can be also divided into various techniques, based on the living organisms used, *i.e.,* whole animals, isolated organ vertebrates, lower organisms (fungi, bacteria, insects, *etc.*), cultured cells, cultured tissue/organs, and isolated subcellular systems [5]. Bioassay can be applied for quality control (QC) purposes in a pharmaceutical (biotechnology) industry, but it can be utilized as well for evaluation the bioactivity of any herbal drugs and (isolated) chemical compound.

For studying the bioactivities of a certain sample(s), the bioassay study can be performed by four approaches: (1) Using a single bioassay technique for proving a single pharmacological activity (*e.g.,* antidiabetic), (2) Using a specific bioassay with different procedures to discover different types of bioactivities, (3) Using nonspecific bioassay for detecting multiple activities; *e.g.,* cytotoxicity can be applied for identifying of antibacterial-, anticancer-, insecticidal-activity, (4) Using combination of bioassay to find a specific or multiple activities [5].

Evaluation of the bioactivity of a certain sample, can be achieved by two stages. *Stage 1* is the optimization and validation of the dynamical signal range of responses of sample, and their variabilities in a certain biosystem. *Stage 2* is for the characterization and determination of the relative potency of the sample, by comparing to authentic reference compound [18]. General bioactivity primary screens that can be evaluated are anti-bacterial-, anti-diabetic-, anti-viral-, anti-fertility-, anti-cancer-, anti-inflammatory-, diuretic activity-, hepatotoxicity-, anti-emetic-assay. For many samples, bioassay can be carried out by using high throughput screening (HTS) using 96- or 384-well microtiter plate. There are two types of HTS namely cell-free or biochemical- and cell-based bioassays. Detailed method can be referred in the previous publication [5].

Some practical considerations should be considered for developing a HTS's bioassay. These are including plate format, plate type, final assay volume, assay homogeneity, reagent combination/addition, reagent stability and compound addition. Detail discussions regarding these parameters have been described in an e-book that published by *Aureliabo.com* [19]. Many factors can cause variations of the bioassay, *i.e.,* pH, temperature, ion concentration, reagent solubility, reagent stability, reagent aggregation, order of reagent addition and instrumentation, cell culture plastic, culture media, culture condition, serum, cell cycle and passage numbers [20]. Due too many variables of the bioassay, the availability of suitable authentic standard(s) is very essential, as requested by USP 42 ⟨1032⟩ [16]; without any suitable standard(s) the validity of the bioassay cannot be confirmed. In addition, each of the used biosystem should be verified regarding its performances.

3. Strategy for developing a bioassay and its validation

Steps that should be taken for performing a valid bioassay is summarized in Fig. 2. The first step is literature survey to find the most suitable and feasible methods in the laboratory, regarding the availability of the needed equipment's, organisms (bacteria, yeast, fungi, *etc.*), cultured cell lines, cultured tissues, cultured organs, reagents, the authentic standards (positive and negative) or commercially available bioassay kit(s). All reagents, standards and bioassay kits should be stored and used according to the manufacture's specifications; stability of all chemicals should be also evaluated [21].

The stability of test extract(s) or isolate compound(s) in a certain solvent (*e.g.,* water, DMSO, EtOH, *etc.*), which will be used in the bioassays, must be strictly evaluated. Extract or isolated compound must be stable during bioassay experiments. Stability evaluation can be performed using chromatographic- and spectroscopy-methods (UV/VIS, NIR). Chromatographic methods,

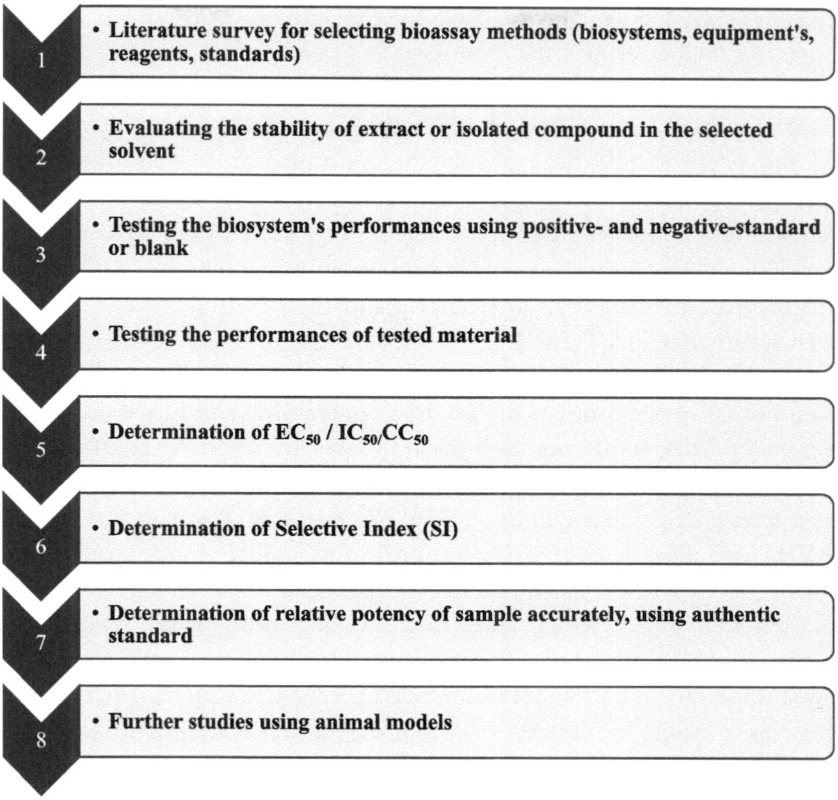

Fig. 2 Steps that should be taken for performing a valid *in-vitro* bioassay experiment.

i.e., (HP)TLC-densitometer, HPLC/UPLC-DAD/MS, GC–MS are preferred due to their selectivity. If the extract or isolated compound is not stable in the selected solvent, the bioassay cannot be performed. Another suitable solvent(s) must be selected and evaluated. Detailed stability testing method, prior to method validation, can be referred from the previous published article by the Indrayanto [1].

3.1 Evaluating bioassay's performances

All used cell lines should be first authenticated. Some test methods have been recommended by ATTC [22], *i.e.*, morphology check by microscope, growth curve analysis, species verification by isoenzymology, short tandem repeat (STR) analysis and mycoplasma detection. Comprehensive methods for cell line identification by previous studies can also be referred [23,24].

Commercially kits are also available for the authentication of the used cell lines, *e.g.*, GenePrint[R] 10 (Promega), IncuCyte[R] Live-Cell Analysis, Sartorius [24–26].

Due to many factors that can affect the biosystem(s) the performances of the biosystem should be evaluated before any data collections. For this purpose, the selected biosystem should be validated by using a *positive* standard and a *negative* standard or *blank*. Proposed panel of test organisms or cell lines, which can be used for primary *in-vitro* screening (antibacterial, anti-fungal, anti-parasitic, anti-cancer) and their related standard(s), have been summarized in Table 1.

If M_t and S_t were the mean and standard deviation (SD) of the maximum/top responses of the positive control, M_b and S_b were the mean and SD of the minimum/bottom responses of negative control/blank, respectively, the biosystem performances can be evaluated by using some parameters, *i.e.*, signal window (SW′), Z′ factor (Z′) and assay variability (AVR) (see Eqs. 1–3) [18,19]. According to guidance from OECD, Z′ can be defined as "A measure of the separation between solvent control and test item signal which takes into account the dynamic range of the *in-vitro* method, and the data variation associated with the signal and control measurements" [40]. The recommended acceptance criteria of the parameters are listed in Table 2 (modified from References 18,41,42). According to USP 42 ⟨111⟩ [43], an outlier test should be first performed prior any data evaluations.

$$\mathrm{SW}' = \frac{Mt - Mb - 3(St + Sb)}{St} \tag{1}$$

$$\mathrm{Z}' = \frac{Mt - Mb - 3(St + Sb)}{Mt - Mb} = 1 - \frac{3(St + Sb)}{Mt - Mb} \tag{2}$$

$$\mathrm{AVR} = \frac{3\,(St + Sb)}{Mt - Mb} \tag{3}$$

Bei Li [44] described an alternative method for evaluating the biosystem, *i.e.*, "Strictly standardized mean difference (SSMD)" using positive (t) and negative standard (b)

$$SSMD = \frac{Mt - Mb}{\sqrt{St^2 + Sb^2}} \tag{4}$$

If theoretically the response of strong positive control is greater than the negative control, excellent result is observed, if the value of SSMD is >4.7, good if 4.7 > SSMD > 3, inferior if 3 > SSMD > 2, and it is not

Table 1 Proposed organism/cell lines and standard for bioassay.

Bioassay	Organism/cell lines	Method	Medium	Reference standard	References
Antiparasitic					
Leishmaniasis	*e.g.*, *Leishmania amazonensis* ATCC®50,159™ IFLA/BR/67/PH8 *Leishmania chagasi* ATCC®50,133™ MHOM/BR/74/PP75 *Leishmania donovani* ATCC® 50212™ MHOM/IN/80/DD8 *etc.*	Promastigote assay	Tobie's medium	Amphotericin B, Fluconazole	[27]
	Leishmania mexicana (MHOM/MX/84/ISETGS)	Promastigote assay	Schneider's Drosophila medium	Pentamidine	[28]
Trypanosomiasis	*Trypanosoma cruzi* (MHOM/MX/1994/Ninoa)	Epimastigote assay	Schneider's Drosophila medium	Benznidazole	[28]
	Trypanosoma brucei strain GUTat 3.1	AlamarBlue assay	HMI-medium	Deferoxamine Diminazene-Aceturate	[29]
Malaria	*Plasmodium falciparum* 3D7 (ATCC® PRA-405D™) *Plasmodium falciparum* FCR3 (ATCC® 30932)	*e.g.*, FACS analysis Fluorescence Activated Cell Sorting	Hematocrit 5% of type O human red blood cells suspended in RPMI-1640 medium	Chloroquine	[30]

Continued

Table 1 Proposed organism/cell lines and standard for bioassay.—cont'd

Bioassay	Organism/cell lines	Method	Medium	Reference standard	References
Anti-fungal					
Yeast	*Candida parapsilosis* ATCC 22019 *Candida krusei* ATCC 6258 *Candida albicans* ATCC 64548 *Candida albicans* ATCC 64550	Broth dilution	RPMI 1640 with 2% Glucose	Amphotericin-B Flucytosine *etc.*[a]	[31]
Anti-bacterial					
Gram Negative	*Escherichia coli* ATCC 25922	Disk diffusion *etc.*	Mueller-Hinton agar (MHA)	Amikacin, Ampicillin, Aztreonam *etc.*[a]	[32]
	Pseudomonas aeruginosa ATCC 27853	Disk diffusion *etc.*	Mueller-Hinton agar (MHA)	Amikacin, Azlocillin, Aztreonam *etc.*[a]	[32]
	Hemophilus influenza ATCC 49766	Disk diffusion *etc.*	Hemophilus Test Medium (HTM)	Amoxicillin-clavulanate. Ampicillin, Ampicillin-Sulbactam *etc.*[a]	[32]
Gram Positive	*Staphylococcus aureus* ATCC 25923	Disk diffusion *etc.*	Mueller-Hinton agar (MHA)	Amikacin, Ampicillin, Azithromycin *etc.*[a]	[32]
	Streptococcus pneumoniae ATCC 49619	Disk diffusion *etc.*	Mueller-Hinton agar (MHA) with 5% sheep blood	Ampicillin, Azithromycin, Cefaclor, *etc.*[a]	[32]

Anaerobe	*Bacteroides fragilis* ATCC 25285	Agar dilution *etc.*	Brucella agar	Amoxicillin-clavulanate Ampicillin, Ampicillin-sulbactam, *etc.*[a]	[32]
	Bacteroides thetaiotaomicron ATCC 29741	Agar dilution *etc.*	Brucella agar	Amoxicillin-clavulanate Ampicillin, Ampicillin-sulbactam.*etc.*[a]	[32]
	Clostridium difficile ATCC 700057	Agar dilution *etc.*	Brucella agar	Amoxicillin-clavulanate Ampicillin, Ampicillin-sulbactam.*etc.* [a]	[32]
	Mycobacterium tuberculosis H37Rv	Löwenstein Jensen	Middlebrook 7H10, 7H11, 7H12	Isoniazid, Streptomycin, PAS (*p*-amino salicylic acid), Ethionamide *etc.*[a]	[33]
	Mycobacterium tuberculosis H37Rv (ATCC 27294)	Microplate Alamar blue assay	Difco Middlebrook 7H9 Broth	Rifampin, Isoniazid, Clofazimine	[34]
	Mycobacterium tuberculosis H37Rv (ATCC 27294)	Löwenstein Jensen	Middlebrook 7H9 medium	Isoniazid, Amikacin, Levofloxacin	[35]
Anticancer					
PANC-1 (ATCC® CRL-1469™) Pancreatic carcinoma cell lines		MTT assay	DMEM medium	Doxorubicin Cisplatin	[36]
Caco-2 (ATCC: HTB- 37™) Human colorectal adenocarcinoma cell lines		MTT assay	DMEM medium	Doxorubicin Cisplatin	[36]

Continued

Table 1 Proposed organism/cell lines and standard for bioassay.—cont'd

Bioassay	Organism/cell lines	Method	Medium	Reference standard	References
MCF-7 (ATCC: HTB-22™), Breast adenocarcinoma cell lines		MTT assay	DMEM medium	Doxorubicin Cisplatin	[36]
MCF7(ATCC HTB-22) Human estrogen-dependent breast carcinoma cell lines		MTT assay	RPMI-1640 medium	Doxorubicin Cisplatin	[37]
HeLa (ATCC CCL-2) Human cervix carcinoma cell lines		MTT assay	RPMI-1640 medium	Doxorubicin Cisplatin	[37]
PC-3 (ATCC® CRL-1435™) Human prostate cancer cell lines		XTT assay	RPMI-1640 medium	Gemcitabine	[38]
U-87 MG (ATCC® HTB-14™) Human primary glioblastoma cell lines		Alamar Blue Cell Viability Test	Eagle's minimum essential medium	Temozolomide	[39]

[a]Complete list of the standards can be referred Refs. [31,32].

Table 2 Acceptance criteria for SW′/SW, Z′/Z and AVR.

SW′/SW	Z′/Z	AVR
Recommended: $SW'/SW > 2$	Excellent: $Z'/Z > 0.5$	Recommended: $AVR < 0.6$
Acceptable: $SW'/SW > 1$	Moderate: $0 < Z'/Z < 0.5$	Unacceptable: $AVR > 0.6$
Unacceptable: $SW'/SW < 1$	Unacceptable: $Z'/Z < 0$	

acceptable if $SSMD < 2$. If theoretically the response of positive control is less than the negative control, excellent result is observed, if the value of SSMD is < -4.7, good if $-4.7 < SSMD < -3$, inferior if $-3 < SSMD < -2$, and it is not acceptable if $SSMD > -2$. Iversen et al. [18] described that Z' is more reliable parameter compared to SW'; the authors also recommended to use this parameter (Z'). An on-line calculator for Z' is available at http://www.screeningunit-fmp.net/tools/z-prime.php.

If Z', AVR values of standards were not yielded acceptable results, it can be concluded that the biosystem *was not optimal*, or the assay format *was not feasible* for generating the data as expected [41]; it could be also possible the used of positive and negative standards *were not suitable* for the intended bioassay application. The researcher must try selecting another biosystem and/or standard, which is more suitable for the desired application(s).

If the selected biosystems for specific application have been evaluated and showed acceptable values for standards, a preliminary experiment can be started for evaluating the performance of sample(s). A dose-response curve can be generated for sample(s), usually it forms a sigmoid curve. Max, mid and min responses must be determined. For evaluating the performance of a sample, Z factor of sample can be determined according to Eq. (5) [41]. Ms. was mean of responses of sample, Mc was mean of responses of control, Ss and Sc were SD of samples and controls, respectively. For agonist or activation assay, control was substituted with positive control (maximum activated signal), while for antagonist or inhibition assay control was substituted with negative control (minimum signal). The recommended acceptance criteria of Z and SW (signal window for sample) were identical to Z' and SW' (Table 2). Sample, which yielded Z out of acceptable value, is not recommended for further investigation.

$$Z = 1 - \frac{3(Ss + Sc)}{|Ms - Mc|} \tag{5}$$

To confirm and validating the results, it is recommended to do the experiment at least two to four times on different days [21,45], and calculate the average of SW/SW″, Z/Z′, and CV (Eqs. 6–8; n is number of replications). It will be nicer if the replications can be performed on at least three different days. If the bioassay was performed using HTS, additional validations should be performed, *i.e.,* evaluating the drift or edge effect, inter-plate and inter-days variations; the template of the plates lay out has been described in detail, the readers can refer to the guidance [21].

$$\mathrm{SW}'/\mathrm{SW} = \frac{\left(Mt - \frac{3St}{\sqrt{n}}\right) - \left(Mb + \frac{3Sb}{\sqrt{n}}\right)}{\frac{St}{\sqrt{n}}} \tag{6}$$

$$\mathrm{Z}'/\mathrm{Z} = \frac{\left(Mt - \frac{3St}{\sqrt{n}}\right) - \left(Mb + \frac{3Sb}{\sqrt{n}}\right)}{Mt - Mb} \tag{7}$$

$$\mathrm{CV} = \frac{\frac{S}{\sqrt{n}}}{M} \tag{8}$$

The recommended acceptance criteria were $\mathrm{SW}' \geq 2$ and $\mathrm{Z}' \geq 0.4$, CV of M_t and M_b are not more than 20%. Individual sample % activity (A_s) can be calculated [21].

$$\%\mathrm{As} = \left(\frac{Ri - M_b}{M_t - M_b}\right) 100 \tag{9}$$

where *Ri* is individual response of sample *i*; if the bioassay is an inhibition assay, % inhibition = 100 −% A_s.

A false positive or a false negative of the measured data can be anticipated using Z score (Zc) according to Martin et al. [42].

$$\mathrm{Zc} = \left|\frac{Ri - M_b}{S_b}\right| \tag{10}$$

where *Ri* is individual response of sample *i*, false positive, if |Zc| was >2 for solvent/blank sample, and false negative, if |Zc| <2 for positive control sample. "Hit of sample" can be determined also using |Zc| value (>2). Varbanov et al. [46] proposed another term, *i.e.,* "hit compound" which were defined as compound with normalized HTS score > Mb + 3 Sb. Normalization of the data can be performed according to Eq. (9).

3.2 Determination of EC_{50}/IC_{50}/CC_{50}

If the Z factor, SW and CV of sample showed acceptable values, half maximal effective concentration (EC_{50}), or half maximal inhibitory concentrations (IC_{50}) or 50% cytotoxic concentration (CC_{50}) or related terms of sample can be determined. Mostly it can be determined at the linier part of the sigmoid curve of dose respond relationship; but it could be also determined from the sigmoid curve directly using a software. For simplification only term of EC_{50} will be further mentioned. Transformation concentration data to logarithm ($\log_{10}$ or $\log_e$) scale usually yielded better fit of the linier regression curve. The acceptance criteria of the R^2 of linier regression for bioassay was ≥ 0.95. By mathematical calculation, it can be showed CV of the regression curve will be <9% if $R^2 \geq 0.95$ [47,48].

$$\%CV = 100\sqrt{e^{Sy} - 1} \quad (11)$$

$$Sy = \frac{\sum \left(Yi - \hat{Y}i\right)^2}{n - 2.}$$

where R^2 and CV were calculated using log transformed data, Yi was individual response of concentration Xi, and $\hat{Y}i = a + bXi$; a, b was intercept and slope of the regression curve, respectively.

The calculation of EC_{50} using four parameters logistic model (4PL), can be well applied, if there are *at least* two concentrations beyond both the lower (min) and upper (max) bend points, *i.e.*, beyond the linear portion of the curve [49]. See Fig. 3. Many free and commercially software's are available, that can be used for estimation of EC50. As examples, a free software from AAT Bioquest (https://www.aatbio.com/tools/ic50-calculator) can be applied for calculation EC_{50} using 4PL, linier, logarithm and semi log. Cheburator, a free software, designed to estimate the EC_{50} from the data for 96-well plates [50]. EC_{50}'s of antioxidants that calculated using some software's (*i.e.*, GraphPad Prism^R version 5.01, BLeSq^R 0.9.1, OriginProR 8.5.1, SigmaPlot^R 12, BioDataFit^R 1.02, and IBM SPSS Statistics^R Desktop 19.0), showed very similar results [51]. EC_{50} can be used as a parameter to define whether a sample is bioactive (see Table 3). It is urged to determine EC_{50} at least 3–6 times, and the RSD should be $\leq 10\%$. The variability of EC_{50} can be also estimated using MSR (minimum significant ratio) [52,53]. If S was standard deviation of log EC_{50} ($n = 6$), MSR can be calculated using Eq. (12) (for $\alpha = 0.05$). A robust assay has an MSR < 3. A weak assay has 3 < MSR < 5

$$MSR = 10^{2\sqrt{S}} \quad (12)$$

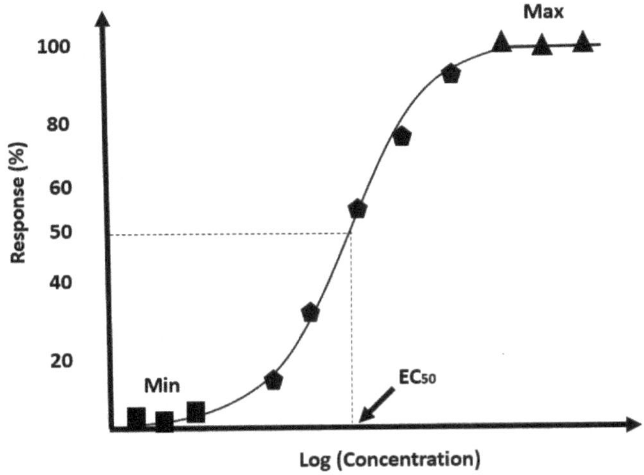

Fig. 3 A typical sigmoid dose-responses curve. Min: minimal response, Max: Maximal response, EC_{50}: 50% maximal effective concentration. At least two concentrations at Min-responses and at Max-responses are needed for determining EC_{50} using 4PL model.

Table 3 Classification of sample based on IC_{50} or CC_{50}.[a]

IC_{50} or CC_{50}		
Isolated compound	**Extract**	**Criteria**
<1 μM	–	Excellent or potent activity
1–20 μM	<10 μg/mL	Good activity or very strong cytotoxicity
20–100 μM	10–50 μg/mL	Moderate activity
–	10–100 μg/mL	Strong cytotoxicity
100–200 μM	50–100 μg/mL	Low activity
>200 μM	>100 μg/mL	Inactive
–	100–500 μg/mL	Moderate cytotoxicity

[a]According to Ref. [54] (for extract's cytotoxicity) and Ref. [55] (for extract's and compound's activity).

3.3 Selectivity index (SI)

Selectivity index (SI) can be defined as *the ratio* of the toxic concentration of a sample against its effective bioactive concentration [56]. The *ideal drug* should have *a relatively* high toxic concentration but with a *very low* active concentration. Evaluation of SI value (Eq. 13) for any research on herbal drug and/or isolated compound is very crucial for determining whether further works can be continued. For evaluating any anti-cancer activity of a sample, its cytotoxicity against non-malignant cell lines must be determined in order to calculate the SI value [57,58].

$$SI = \frac{IC_{50}^{no\ cancer\ cell}}{IC_{50}^{cancer\ cell}} \qquad (13)$$

A SI value ≥10 was assumed to belong of a selected potential sample that can be further investigated [57]. Weerapreeyakul et al. [54] proposed a lower SI value (≥3) for classifying of a prospective anti-cancer sample. Evaluating anti-cancer activity of a sample using malignant cell lines or animal *only* but without determining of SI, is a poor predictor for further (clinical) study [59]. Ideally the drug can kill the cancer cells, but it should not affect the normal cells [58]. A similar SI equation used for evaluating anti-viral, anti-microbial and antioxidant of sample was described: [60,61].

$$SI = \frac{CC_{50}}{IC_{50}} = \frac{LC_{50}}{MIC} \qquad (14)$$

where CC_{50}/LC_{50} was 50% cytotoxic concentrations; IC_{50} was 50% inhibition concentrations or effective concentration, MIC was minimum inhibitory or effective concentrations. Awouafack et al. [61] also recommended an acceptance criterion of SI ≥ 10 for a selective bioactive sample.

Countless scientific publications have reported the bioactivity of many medicinal plants, herbal drugs and isolated compound(s), but unfortunately most of them without mentioning the SI data. Cos et al. [62] stated, without the SI data of the reported bioactive sample(s), the described publication was *very limited value* only. Adamu et al. [63,64] reported the anti-bacterial, antioxidant activities and their SI of South African plant leaves' extracts, and the most active extract showed SI of 5.2. Recently Famuyide et al. [65] described SI (0.24–21.07) of extracts of *Eugenia* and *Syzygium* sp., which showed anti-bacterial activity against some gram-positive and gram-negative bacteria; according to these authors the extracts could be assumed bioactive *and non-toxic* if SI was >1, but according to Nogueira and Estólio do Rosário SI should not be lower than 2 [66]. Due to many variations of the acceptance criterium of SI by some publications as described above, further works need to be accomplished for determining the *minimum* acceptable value of SI. See also section 4.

Relatively low of SI (<1) means the sample could be toxic and cannot be used as a (herbal) drug. If the calculated SI value is between 1 and 10, re-evaluating using other biosystem(s) is advocated for confirmation. Sample that showed a relatively low data of SI (< 1) must be contained toxic component(s). A combination of chemical profiling (LC HR MS/MS) and MVA can be applied to determine the toxic compounds (Fig. 1, 2nd approach). The SI of the sample(s) can be increased by removing the toxic compound(s) using certain chemical method(s). Ratih et al. [67] applied acidified water to remove the toxic alkaloids from the leaves of *Justicia gendarussa,* before performing the extraction; using this method free toxic alkaloids extracts can be prepared.

4. *In-vitro* toxicity testing

Presently many toxicity studies were performed by using *in-vitro* methods, instead of using animals. This is due to the application of the *3R principle* that described by WMS. Russel and RL Burch in 1959 as cited by Singh et al. [68]. In principle, 3R meant reduction of the number animals which used for toxicity testing, reducing of suffering and pain of the animal, and using of non-animal techniques, *i.e.,* established cell line, cell- and tissue-cultures [69].

Many alternative methods for various toxicity testing methods (*in-vivo* and *in-vitro* methods), and their method validations were already described, discussed and edited by Eskes and Wheland [70]. Krihsna and Gopalakrishnan [71] reported some *in-vitro* biosystems for evaluating the safety and toxicity of nutraceutical, *i.e.,* immortalized cell line, primary hepatocytes suspensions, primary hepatocytes cell cultures, and *in-vitro* chip technology. Method assessments for toxicity evaluations of herbal drugs have been reviewed and discussed [72]. Some newest methods for specific toxicity testing, *e.g.,* carcinogenicity and reproductive toxicity have been reported by Corvi et al. [73]. Another common method for evaluating the toxicity of herbal drugs is by using zebrafish method [74]. Toxicity of chemical compounds can be predicted also by using *in-silico* approaches [75]. *In-vitro* methods were recommended to be used as primary cytotoxicity screening instead of classical methods using animals (rat, mice) or brine shrimp lethality assay [66].

MTT assay was generally used for the evaluation *in-vitro* toxicity of herbal material or isolated compound. Detailed MTT assay has been described by previous publication [76,77]; accordingly, the CC_{50} of positive standard should be included in its specification range ($\pm$2.5 SD); the variability of intra- and inter assay should be <15%. Other *in-vitro* methods for cell viability assays have been described by the published guidance [12]. Classification of the cytotoxicity [54] or activity [55] of crude extracts or isolated compounds were summarized by Table 3.

Choosing the most suitable cell lines for toxicity evaluation, is the most important step for determining SI value of sample; as example, if anti-cancer activity of a sample of herbal drug was evaluated using certain cancer cell lines, the toxicity test should be performed on its related non-malignant cell lines. According to the author's opinion, it will be nicer, if it is at least two different toxicity methods could be performed (*e.g.*, using different cell lines, and or zebrafish methods), this because the exact mechanisms of the toxicity of the sample might be not yet fully elucidated or unknown. A Proposed panel of cell lines for primary *in-vitro* cytotoxicity evaluations were presented in Table 4; these described cell lines can be applied for calculating SI of sample(s). A nice completed multitiered system for toxicity test was recommended by Sachana and Hargreaves [78]; these multitiered tests applying combination test using of monoculture of specific cell type, coculture, organotypic tissue slices, and at last animal model. By this combination testing's, an accurate SI can be determined, but it might be also very expensive.

5. Determination the relative potency of sample using authentic standard

At stage 2 the relative potency of tested sample will be accurately determined by comparing to the authentic standard. The accurate potency of a sample (extract or isolated compound) is needed to be established, before any *in-vivo* experiments especially for clinical trial. If the sigmoid curves of dose-response relationship of sample and standard showed very similar pattern, it can be assumed that mechanism of action is similar. If the curves are not similar, another suitable standard must be selected. Complete methods of calculations including the mathematical equations were already described in detail by European Pharmacopeia (EP)10, 5.3 [14], USP 42 general chapters ⟨111⟩ [43], ⟨1032⟩ [16], ⟨1033⟩ [48] and ⟨1034⟩ [85]. These official methods were intended for quality control purposes in the Pharmaceutical Industry, but these methods can be as well applied for determining the accurate relative potency of sample(s). The principle of bioassays which described in those Pharmacopeias was *dilution* assay; it means that the sample to be assayed contained the identical bioactive compound to standard, but in different concentrations. For herbal drug research the

Table 4 Proposed cell lines for evaluation the cytotoxicity and SI.

Cell lines	Method	Medium	Application (SI)	References
Hep-G2 (ATCC HB-8065)	MTT assay	RPMI supplemented with 10% fetal bovine serum, 1% L-glutamine (200 mM) and penicillin (100 U/mL)/ streptomycin (100 μg/mL) incubated at 37 °C in a humidified 6% CO_2 with 95% air atmosphere.	Anti-plasmodial activity	[79]
WI-38 human normal fetal lung fibroblasts	MTT assay	Dulbecco's modified Eagle medium (DMEM) supplemented with 10% heat-inactivated FBS, 1% L-glutamine (200 mM) and penicillin (100 UI/ml)–Streptomycin (100 μg/mL) incubated overnight at 37 °C in a 5% CO_2 atmosphere	Anti-malarial activity	[80]
Vero monkey kidney cell lines	MTT assay	MEM, supplemented with 0.1% gentamicin and 5% fetal bovine serum. Incubated overnight at 37 °C in a 5% CO_2 atmosphere	Anti-bacterial and antifungal activity	[81]
CRFK cells				
Bovine dermis cells				
Vero (ATCC CCL-81) monkey kidney normal cell lines	MTT assay	Dulbecco's modified Eagle medium (DMEM) supplemented with 5% Fetal Bovine Serum (FBS), 100 U/mL penicillin, and 100 μg/mL streptomycin incubated 48 h at 37 °C in a 5% CO_2 atmosphere	Anti-cancer activity	[37]
MDCK cells Madin-Darby Canine Kidney	MTT Assay	DMEM supplemented with 10% fetal bovine serum (FBS) incubated 48 h at 37 °C in a 5% CO_2 atmosphere.	Anti-malarial activity	[30]

Table 4 Proposed cell lines for evaluation the cytotoxicity and SI.—cont'd

Cell lines	Method	Medium	Application (SI)	References
MRC-5 Human diploid embryonic cell lines	MTT assay	MEM medium supplemented with 10% FCS incubated at 37 °C under 5% CO_2–95% air for 7 days	Anti-plasmodial and Anti-trypanosomal activity	[82]
HBL100 (Cell Line Services code 300178) Human epithelial normal cell lines	MTT assay	RPMI culture medium supplemented with 10% heat-inactivated fetal bovine serum. Cell culture media were supplemented with 4 mM glutamine 100 μg/mL gentamicin and penicillin. Streptomycin (200 units/mL and 200 μg/mL).	Anti-cancer activity	[83]
HaCat (Cell Line Services code 330493) Human keratinocyte normal cell lines				
NHDF (PromoCell code c-12300). Human dermal fibroblasts normal cell lines				
CCL-1 (NCTC clone 929) noncancerous normal mouse adipose areolar	Crystal-violet method	RPMI-1640 medium, supplemented with 10% equine serum, L-glutamine (2 mM), and 100 U/mL penicillin and 100 μg/mL streptomycin incubated 37 °C in a 5% CO_2 atmosphere.	Anticancer activity	[84]

chemical structure of the active compound in sample might be different to standard, but their bioactive mechanisms should be similar. The methods of calculations for bioassays, which were described in EP 10 [14] and USP 42 ⟨1034⟩ [85]. are almost identical, but it seemed EP 10 [14] depicted more detailed and completed mathematical equations. The described equations of EP 10 [14] can be applied manually for determining the potency of the samples; the readers can simply use Excel to perform any needed calculations. Some free and commercial software's also are available for the purpose of calculations. USP 42 ⟨1033⟩ [48] described relatively more detailed method validation compared to the EP 10. The summarization of USP 42 ⟨1033⟩ [48] and EP 10 [14] methods will be described below.

5.1 Validation method for the determination of relative potency

According to Bower [86] the possible sources of variability of bioassay are analyst, day, reagent lot variations, day-to-day variations, inter- and intra-assay variation; the recommended acceptance criteria of the CV is $<20\%$. For validation purposes of biological assay in a Pharmaceutical Industry, USP 42 ⟨1033⟩ [48] recommended applying at least two analyst, two media lots, and two runs for each of concentrations level.

$$\%\mathrm{CV} = 100\left(\sqrt{e^{SD^2} - 1}\right) \quad (15)$$

$$\%\mathrm{GCV} = 100\left(e^{SD} - 1\right) \quad (16)$$

Eq. (15) was used for original data, while Eq. (16) was used for log transformed data; SD was calculated using log transformed data. Confidence interval (CI) of the results of analysis can be calculated using Eqs. (17) and (19) [48].

$$\mathrm{CI} = \mathrm{Mean} \pm \mathrm{tab}.\frac{SD}{\sqrt{n}} \quad (17)$$

$$\text{Relative bias (RB)} = 100\left(\frac{e^{Mean}}{Target\ cocentration} - 1\right)\% \quad (18)$$

$$\text{Confidence interval of RB} = \mathrm{CIRB} = 100\left(\frac{e^{CI_L}}{1.00}\right)\%, \left(\frac{e^{CI_U}}{1.00}\right)\% \quad (19)$$

where e^{Mean} was geometric mean, e^{CI_L} and e^{CI_U} were low and upper confidence interval (Eq. 19), Mean was calculated using log transformed data. If USL was upper specification limit, LSL was low specification limit,

δ_p was variance of product and δ_{RA} was release assay variance, process capability index (Cp) can be calculated using Eq. (20). The minimum acceptance criteria of Cp should be >1

$$Cp = \frac{USL - LSL}{\sqrt{\delta_p^2 + RB^2 + \delta_{RA}^2}} \tag{20}$$

For determining the Cp that will be used for herbal drug research, the authors recommended a simpler modification of Eq. (20) (*i.e.*, Eq. 21), as described previously [87,88]; USL and LSL should be specified first for each of the tested sample (if any). If it is not yet specified, the author recommend 95–105% of target concentrations.

$$\text{Cp} = \min\left\{\frac{USL - Mean}{3SD}, \frac{Mean - LSL}{3SD}\right\} \tag{21}$$

If the analysis was performed in different conditions SD should be calculated as intermediate precision as described by USP 42 ⟨1033⟩ [48] and USP 42 ⟨1010⟩ [89]. The lower confidence interval of Cp (Cp(L)) can be estimated using Eq. (22) [88].

$$\text{Cp(L)} = \text{Cp}\left[1 - Z_{\alpha/2}\sqrt{\left(\frac{1}{9n(Cp^2} + \frac{1}{2(n-1)}\right)}\right] \tag{22}$$

for $\alpha = 0.05$, Z-distribution score ($Z_{\alpha/2}$) will be 1.96. Table 5 shows the required estimates of Cp(L) such that the criterion of the lower bound of an approximate 95% confidence interval on Cp of at least 1, is met for a given replication; the more replications (n), the smaller the value of Cp(L).

Table 5 Required Cp (L) values.

n	Cp(L)
5	≥3.32
10	≥1.91
15	≥1.63
20	≥1.50

Modified from T. Coffey, K.M. Bower. A statistical approach to assess and justify potential product specifications. BioProcess Int. 15(2) (2017). https://bioprocessintl.com/analytical/qa-qc/statistical-approach-assess-justify-potential-product-specifications/.

5.2 Summarization of calculation of relative potency according to EP 10.0, 5.3 [14]

Quantitative assay can be applied by using quantitative- and quantal-responses. The applied statistical methods were based on the assumptions, that the responses are normally distributed, homogenous and independent. If homogeneity of variants is not satisfactory, transformation responses y to ln y or log y is recommended; if observation follows a Poison distribution, transformation y to $\sqrt{y}$ is useful; if the dose is proportional to area instead of measured diameter transformation y to y^2 can be applied. For quantitative purposes, assays can be performed using parallel-method, slope-ratio method and sigmoid dose relationship.

Three restrictions were normally imposed on the assay design, *i.e.*, (1) each sample/preparation must be evaluated with identical number of dilutions, (2) in the parallel-line model, the ratio of adjacent doses must be constant for all treatments in the assay; in the slope-ratio model, the interval between adjacent doses must be constant for all treatments in the assay, and (3) there should be an identically number of experimental units to each treatment

5.2.1 Parallel-line method

Assay design that can be applied for parallel-methods are: Completely randomized design, Randomized block design, Latin square design and Cross-over design. The requirements for validity test using this parallel method are: (1) linier regression term is significant ($P < 0.05$), (2) the term of non-parallelism is not significant ($P > 0.05$), and (3) the term of non-linearity is not significant ($P > 0.05$). Completed ANOVA equations for calculations of the validity of the method, estimation of potency of sample, and the confidence limit have been described in detail. Examples of calculations that were given are (1) Two-dose multiple assay with completely randomized design, (2) Three-dose Latin square design, (3) Four-dose randomized block design, (4) Five-dose multiple assay with completely randomized design, (5) Twin cross over design.

5.2.2 Slope-ratio method

The slope-ratio model, this model generally used some microbiological assays when the independent variable is an essential growth factor below the optimal concentration of the medium. The described assays designs are (*hd* +*1*)- and (*hd*)-design, where h is number of preparations including standard preparations, d is number of dose level in each preparation (excluding the blank). The requirements for validity test using this slope-ratio method are: (1) the variance due to blanks in ($hd+1$)-design is not significant ($P>0.05$); This indicates that the responses of the blanks do not significantly differ from the common intercept and the linear relationship is valid down to zero dose; (2) variation due to intersection is not significant ($P>0.05$) and (3) variation due to non-linearity is not significant ($P>0.05$) for assay including at least three preparations. A significant variation due to blanks means that the hypothesis of linearity is not valid near zero dose. If this is probable to be systematic rather than incidental for the type of assay, the (hd)-design is more suitable. Completed ANOVA equations for validity evaluations, calculation potency of sample and its confidence limit have been described in detail. Examples of calculations that were given are (1) A completely randomized (0, 3, 3) design, (2) A completely randomized (0, 4, 4, 4) design; (0, 3, 3) means three equivalent dilutions both standard and test preparations, in addition blank is also prepared; (0, 4, 4, 4) means four duplicate concentrations of standards and two test preparations and blank.

5.2.3 Sigmoid dose-responses curve

Sigmoid dose-responses curve model is generally for immunoassays; for a valid assay the curves of samples and standard should have same slope-factor, maximum and minimum responses. The horizontal location could be located at different location. And this related to the potency of sample. If a software was not available, and if upper limit (α), and lower limit (τ) are available, then select mean of responses (μ) between 20% and 80% of the limit, then transform the responses (y) of the selected dose to:

$$y = \ln\left(\frac{\mu-\tau}{\alpha-\mu}\right) \tag{23}$$

and used parallel line to calculate. An example of a serological assay of tetanus sera using four parameters logistic curve (4PL) was described; methods for evaluating the validity of parallelism and linearity of sample and standard were explained in detail.

5.2.4 Assay using quantal responses

For quantal assays the responses were dichotomous or binary outcome, animal dead or alive, certain effect might be observable or not. In quantitative we get n separate responses for each treatment, but by quantal the fraction of unit in each treatment group showing a response. When the fractions plotted against logarithms of the doses, a sigmoid curve usually observed instead of linier curve (the *Probit* method). If the responses are depended on process of growth, the *logistic* distribution model is preferred. Three conditions should be fulfilled: (1) the relation between log dose and responses shows a cumulative distribution curve, (2) curves of standard and sample must be parallel and identically shaped, only differ in horizontal location, and (3) there is no natural response to extreme low dose, and to extremely high dose. Completed equations in tabulated format are presented in detail both for Probit and Logit methods. Method of calculation of median dose (ED_{50}) is also described. This section is completed with detail examples of calculations, so researchers can easily apply for their measured data.

6. Summary and recommendations

This present review described the validation method for the *in-vitro* bioassay and its application in herbal drug research including herbs, medicinal plants, extracts and isolated compounds; this validation method can be applied as well for any synthetic compounds. There was no different method for performing bioassay of isolated compound and synthetic compound.

Prior doing any bioassay experiment(s), the authentication of the used organisms or cell lines using standardized methods is required. It is recommended to use defined cell lines or organisms for certain bioactivity- and/or toxicity-testing (*e.g.*, from ATCC, see Tables 1 and 4).

Stability studies of the tested extract(s) or isolated compound(s) in the selected solvent must be evaluated; chromatographic methods, *i.e.*, TLC, HPLC/UPLC, GC are preferred due to their selectivity. If the extracts or isolated compound is not stable in the required analyses time, new solvent(s) must be selected and evaluated. The selection of extraction method and solvent extractor is very crucial in order preventing the degradation of the biochemical components in herbs. If the traditional application of herbal drug is by decoction, it is very advisable using water as extractor instead of using organic solvent.

It is also advised to use authentic analytical grade chemical standard(s) as listed by Table 1. Due to too many variables of a biosystem, its performances must be evaluated using a positive- and a negative-standard/blank. Sample can only be tested regarding its bioactivity, if the performances of the used biosystem showed acceptable values as described by Table 2.

The reliability of any EC_{50}/CC_{50} value should be verified using minimum significant ratio (MSR). The application of two or more biosystems for bioactivity- and toxicity-study is encouraged for confirming the results. It is very important to note that determining the selective index for every tested sample is very essential; although an extract or a compound showed (very) high activity (Table 3), but if it has a relatively low SI (<1–2), the sample needs not to be further investigated.

Resolve an accurate relative potency of extract or isolated compound must be performed before continuing the work with animal and/or clinical trial. For this purpose, the author recommends applying the calculation's methods which were described by EP 10.0, 5.3 instead of USP 42. EP 10.0, 5.3 described complete mathematical equations (ANOVA) for estimating of the relative potency of sample so the readers can do calculation manually and/or using Excel.

Reliable results and conclusions only can be achieved, if all used chemical- and bioassay methods have been validated before any data collections. Without any method validations, the conclusions may be not correct or wrong; results cannot be repeated by other researches, so the reported data have very limited scientific value.

Acknowledgments

The authors wish to acknowledge Ms. Febry Ardiana (PT Bernofarm, Gedangan, Surabaya, Indonesia) for providing Reference Nos. [13,16,43,48,85,89], and Ms. Nurliya Irfiani (PT Merck Sharp Dohme Pharma Tbk, Pandaan, Pasuruan, Indonesia) for supplying Reference No. [14]. Mr. James Kuzba and Mrs. Nuryanti Basuni (Medical device Company. Vesta a Lubrizol Company at Franklin WI, USA) for checking the English usage.

Abbreviation

4PL	four parameters logistic model
ANOVA	analysis of variance
ATCC	American type culture collection
AVR	assay variability
CC_{50}	cytotoxic concentration 50%
CI	confidence interval
Cp	capability index
Cp(L)	lower confidence interval of CP
CRFK	Crandell feline kidney cells
CV	coefficient variance
DAD	diode-array detection
DMEM	Dulbecco's modified eagle medium
DMSO	dimethyl sulfoxide

DNA	deoxyribonucleic acid
EC_{50}	effective concentration 50%
EP	European Pharmacopeia
FACS	fluorescence activated cell sorting
FBS	fetal bovine serum
GC	gas chromatography
GC–MS	gas chromatography mass spectrometry
GCV	geometric coefficient variance
HPLC	high performance liquid chromatography
HPTLC	high performance-thin layer chromatography
HTM	Haemophilus test medium
HTS	high throughput screening
IC_{50}	inhibitory concentrations 50%
LC-HRMS	liquid chromatography-high resolution mass spectrometry
LD_{50}	lethal dose 50%
Mb	mean of minimum bottom/blank/negative control's responses
Mc	mean of control's responses
MDCK	Madin-Darby canine kidney
MEM	minimal essential medium
MHA	Mueller-Hinton agar
Ms.	mean of sample's responses
MS/MS	tandem mass spectrometry
MSR	minimum significant ratio
Mt.	mean of top/maximum/positive control's responses
MTT	3-(4,5-dimethylthiazol-2-yl)-2,5-diphenyltetrazolium bromide
MVA	multivariate analysis
n	number of replications
NCTC	National Collection of Type Cultures
NIR	near infrared spectrophotometer
OECD	Organization for Economic Co-operation and Development
PCA	principal component analysis
pH	power of hydrogen
PLS-DA	partial least squares-discriminant analysis
QC	quality control
R^2	coefficient of determination of regression curve
RPMI	Roswell Park Memorial Institute
Sb	SD of the minimum/bottom responses of negative control/blank
Sc	SD of sample responses
SD	standard deviation
SI	selectivity index
Ss	SD of sample responses
SSMD	strictly standardized mean difference
St	SD of the maximum/top responses of the positive control
STR	short tandem repeat
SW'	signal window
TCM	traditional Chinese medicines
TLC	thin layer chromatography
UPLC	ultra performance liquid chromatography

USP	United States Pharmacopeia
UV/VIS	ultraviolet/visible spectrophotometer
XTT	2,3-bis-(2-methoxy-4-nitro-5-sulfophenyl)-2H-tetrazolium-5-carboxanilide
Z′	Z factor

References

[1] G. Indrayanto, Validation of chromatographic methods of analysis: application for drugs that derived from herbs, in: H.G. Brittain (Ed.), Profiles of Drug Substances, Excipients, and Related Methodology, vol. 43, Elsevier Academic Press, San Diego, New York, Boston, London, Sydney, Tokyo, Toronto, 2018, pp. 243–258. (ISSN 1871-5125).

[2] G. Indrayanto, Recent development of quality control methods for herbal derived drug preparations, Nat. Prod. Commun. 13 (12) (2018) 1599–1603.

[3] J.J. Kellogg, M.F. Paine, J.S. McCune, N.H. Oberlies, N.B. Cech, Selection and characterization of botanical natural products for research studies: a NaPDI center recommended approach. Natural Product Report 36 (2018) 1196–1221. https://doi.org/10.1039/c8np00065d.

[4] S.Y. Pan, S.F. Zhou, S.H. Gao, Z.L. Yu, S.F. Zhang, M.K. Tang, J.N. Sun, D.L. Ma, Y.F. Han, W.F. Fong, K.M. Ko, New perspectives on how to discover drugs from herbal medicines: CAM's outstanding contribution to modern therapeutics, Evid. Based Complement. Alternat. Med. 2013 (2013) 627375. 25 pages. https://doi.org/10.1155/2013/627375.

[5] A. Rahman, M.I. Choudhary, W.J. Thomson, Bioassay Techniques for Drug Development, Taylor & Francis e-Library, Singapore, 2001, pp. 3–4. ISBN 0-203-30453-5.

[6] L. Ory, E.H. Nazih, S. Daoud, J. Mocquard, M. Bourjot, L. Margueritte, M.A. Delsuc, J.M. Bard, Y.F. Pouchus, S. Bertrand, C. Roullier, Targeting bioactive compounds in natural extracts—development of a comprehensive workflow combining chemical and biological data. Anal. Chim. Acta 1070 (2019) 29–42. https://doi.org/10.1016/j.aca.2019.04.038.

[7] T. Saesong, P.-M. Allard, E.F. Queiroz, L. Marcourt, N. Nuengchamnong, P. Temkitthawon, N. Khorana, J.-L. Wolfender, K. Ingkaninan, Discovery of lipid peroxidation inhibitors from Bacopa species prioritized through multivariate data analysis and multi-informative molecular networking. Molecules 24 (2019) 2989. https://doi.org/10.3390/molecules24162989.

[8] K. Segers, S. Declerck, D. Mangelings, Y.V. Heyden, A.V. Eeckhaut, Analytical techniques for metabolomics studies: a review. Bioanalysis 11 (24) (2019) 2297–2318. https://doi.org/10.4155/bio-2019-0014.

[9] A. Bensoussan, S. Lee, C. Murray, S. Bourchier, F. van der Kooy, J.L. Pearson, J. Liu, D. Chang, C.S. Khoo, Choosing chemical markers for quality assurance of complex herbal medicines: development and application of the herb MaRS criteria. Clin. Pharmacol. Ther. 97 (6) (2015) 628–640. https://doi.org/10.1002/cpt.100.

[10] J. Xie, A.-H. Zhang, H. Sun, G.-L. Yan, X.-J. Wang, Recent advances and effective strategies in the discovery and applications of natural products. RSC Adv. 8 (8) (2018) 812–824. https://doi.org/10.1039/c7ra09475b.

[11] X.-M. Dai, D.-N. Cui, J. Wang, W. Zhang, Z.-J. Zhang, F.-G. Xu, Systems pharmacology based strategy for Q-markers discovery of HuangQin decoction to attenuate intestinal damage. Front. Pharmacol. 9 (3) (2018) 236. https://doi.org/10.3389/fphar.2018.00236.

[12] Guidance for Assay Development & HTS, Version 5, Introduction, Eli Lilly and Company and the National Institutes of Health Chemical, 2007, p 1, 6–167. https://hts.ku.edu/sites/hts.drupal.ku.edu/files/docs/NCGC_Assay_Guidance_Manual.pdf.
[13] USP 42 -NF 37 ⟨1030⟩ Biological assay Chapters-Overview and Glossary, USPC 2019. https://online.uspnf.com/uspnf/document/GUID-53143C8C-1146-44D4-9A45-3A51469F86FC_1_en-US?highlight=1030%20. (Thu Jun 27 2019).
[14] European Pharmacopoeia 10.0, 5.3 Statistical Analysis of Results of Biological Assay and Test, European Pharmacopoeia ONLINE, EDQM Council of Europa, Strasbourg, France, 2020. https://www.edqm.eu/en/european_pharmacopoeia_10th_edition.
[15] A. Rosso, Statistical Analysis of Experimental Designs Applied to Biological Assays, Master thesis. Lund University, Lund, Sweden, 2010. https://lup.lub.lu.se/student-papers/search/publication/1620399.
[16] USP 42-NF 37 < 1032 > Design and Development of Biological Assay, USPC 2019. https://online.uspnf.com/uspnf/document/GUID-5005D5F9-7475-4169-9E10-A13AAD3189D8_1_en-US. (Mon May 27 2019).
[17] A.K. Jain, D. Singh, K. Dubey, R. Maurya, S. Mittal, A.K. Pandey, In vitro toxicology. Models and methods for in vitro toxicity. in: A. Dhawan, S. Kwon (Eds.), In Vitro Toxicology, Academic Press, Elsevier Inc, 2018, pp. 47–64. https://doi.org/10.1016/B978-0-12-804667-8.00003-1.
[18] P.W. Iversen, B.J. Eastwood, G.S. Sittampalam, K.L. Cox, A comparison of assay performance measures in screening assays: signal window, Z′ factor, and assay variability ratio. J. Biomol. Screen. 11 (3) (2006) 247–252. https://doi.org/10.1177/1087057105285610.
[19] Aurelia Bioscience, How to Develop Effective In Vitro Assays for Early Drug Discovery. A Comprehensive Guide to Assay Design and Development, Aureliabio.com. http://www.aureliabio.com/wp-content/uploads/Aurelia-eBook_How-to-develop-bioassays-that-enhance-your-early-drug-discovery-research.pdf, Sat August 1, 2020.
[20] I. Padmalayam, Assay Development in Drug Discovery, Drug Discovery Division, Southern Research Institute, Birmingham, Alabama 35205, Sat August 1, 2020. https://www.uab.edu/medicine/adda/images/PadmalayamAssayDevelopment.pdf.
[21] P.W. Iversen, B. Beck, Y.F. Chen, W. Dere, V. Devanarayan, B.J. Eastwood, M.W. Farmen, S.J. Iturria, C. Montrose, R.A. Moore, J.R. Weidner, G.S. Sittampalam, HTS assay validation 2012 (updated), in: G.S. Sittampalam, A. Grossman, K. Brimacombe (Eds.), Assay Guidance Manual, Eli Lilly & Company and the National Center for Advancing Translational Sciences, Bethesda, 2004. https://www.ncbi.nlm.nih.gov/books/NBK83783/.
[22] American Type Culture Collection (ATCC), Cell Line Authentication Test Recommendations, Tech Bulletin no 8. https://www.atcc.org/~/media/PDFs/Technical%20Bulletins/tb08.ashx, 2010.
[23] V. Marx, Cell-line authentication demystified, Nat. Methods 11 (5) (2014) 483–487. https://www.nature.com/articles/nmeth.2932.pdf?origin=ppub.
[24] J.L. Almeida, K.D. Cole, A.L. Plant, Standards for cell line authentication and beyond. PLoS Biol. 14 (6) (2016) e1002476. https://doi.org/10.1371/journal.pbio.1002476.
[25] T. Nelson, E. Endsley, D. McEwen, L. Oupika, B. Neagle, V. Groppi, Cell Culture Quality Control Assay, Application Note, IncuCyte, Sartorius, 2019. https://www.sartorius.com/resource/blob/63378/449a19d7aeab75fd483719dc769fbdc6/8000-0023-c00-cell-culture-quality-control-application-note-data.pdf.
[26] Sponsored Paper, The importance of cell line authentication, Biotechniques 54 (April 2013) 222. https://www.future-science.com/doi/pdf/10.2144/000114014.
[27] R. Kariyawasam, P. Challa, R. Lau, A.K. Boggild, Susceptibility testing of Leishmania spp. against amphotericin B and fluconazole using the Sensititre™ YeastOne™ YO9 platform. BMC Infect. Dis. 19 (2019) 593. https://doi.org/10.1186/s12879-019-4237-3.

[28] M.J. Chan-Bacab, E. Hernandes-Nunez, G. Navarrete-Vazquez, Nitazoxanide, tizoxanide and a new analogue [4-nitro-N-(5-nitro-1,3-thiazol-2-yl)benzamide; NTB] inhibit the growth of kinetoplastid parasites (*Trypanosoma cruzi* and *Leishmania mexicana*) *in vitro*. J. Antimicrob. Chemother. 63 (6) (2009) 1292–1293. https://doi.org/10.1093/jac/dkp117.
[29] C.M. Amisigo, C.A. Antwi, J.P. Adjimani, T.M. Gwira, In vitro anti-trypanosomal effects of selected phenolic acids on *Trypanosoma brucei*. PLoS One 14 (5) (2019) e0216078. https://doi.org/10.1371/journal.pone.0216078.
[30] S. Yeo, D.-X. Liu, H.S. Kim, H. Park, Anti-malarial effect of novel chloroquine derivatives as agents for the treatment of malaria. Malar. J. 16 (2017) 80. https://doi.org/10.1186/s12936-017-1725-z.
[31] European Committee on AntiMicrobial Susceptibility Testing (EUCAST), Method for the determination of broth dilution minimum inhibitory concentrations of antifungal agents for yeasts, in: EUCAST Definitive Document E.DEF 7.3.1, 2017, pp. 1–21. http://www.eucast.org/fileadmin/src/media/PDFs/EUCAST_files/AFST/Files/EUCAST_E_Def_7_3_1_Yeast_testing__definitive.pdf.
[32] Clinical and Laboratory Standards Institute (CLSI), Performance Standards for Antimicrobial Susceptibility Testing; Thirty Ed. CLSI Standard M100-S30, Clinical and Laboratory Standards Institute, Wayne, PA, 2020, pp. 68–226. http://em100.edaptivedocs.net/GetDoc.aspx?doc=CLSI%20M100%20ED30:2020&format=SPDF.
[33] World Health Organization (WHO), Guidelines for Drug Susceptibility Testing for Second-Line Anti-Tuberculosis Drugs for Dots-Plus. https://apps.who.int/iris/bitstream/handle/10665/83807/WHO_CDS_TB_2001.288_eng.pdf;jsessionid=269EEADFA659E327B04FA7E31B0D0E1E?sequence=1, 2001.
[34] Y. Lu, M. Zheng, B. Wang, L. Fu, W. Zhao, P. Li, J. Xu, H. Zhu, H. Jin, D. Yin, H. Huang, A.M. Upton, Z. Ma, Clofazimine analogs with efficacy against experimental tuberculosis and reduced potential for accumulation. Antimicrob. Agents Chemother. 55 (11) (2011) 5185–5193. https://doi.org/10.1128/AAC.00699-11.
[35] European Committee on Anti-Microbial Susceptibility Testing (EUCAST), Reference Protocol for MIC Determination of Anti-Tuberculosis Agents against Isolates of the *Mycobacterium tuberculosis* Complex in Middlebrook 7H9 Broth, Version 6.1. 4th. http://www.eucast.org/fileadmin/src/media/PDFs/EUCAST_files/Mycobacteria/Methods_in_AMST/Technical_protocol_AMST_MIC_reference_method_190719.pdf, 2019.
[36] N. Alhourani, V. Kasabri, Y. Bustanji, R. Abbassi, M. Hudaib, Potential antiproliferative activity and evaluation of essential oil composition of the aerial parts of *Tamarix aphylla* (L.) H. Karst: a wild grown medicinal plant in Jordan. Evid. Based Complement. Alternat. Med. 2018 (2018) 9363868. https://doi.org/10.1155/2018/9363868.
[37] P.W. Prasetyaningrum, A. Bahtiar, H. Hayun, Synthesis and cytotoxicity evaluation of novel asymmetrical mono-carbonyl analogs of curcumin (AMACs) against Vero, HeLa, and MCF7 cell lines. Sci. Pharm. 86 (2018) 25. https://doi.org/10.3390/scipharm86020025.
[38] C. Eroglu, E. Avci, M. Secme, Y. Dodurga, H. Vural, E. Kurar, The combination effect of ferulic acid and gemcitabine on expression of genes related apoptosis and metastasis in PC-3 prostate Cancer cells. Eur J Biol. 77 (1) (2018) 32–37. https://doi.org/10.26650/EurJBiol.2018.0003.
[39] S. Chiu, B. Batsaikhan, H.M. Huang, J.Y. Wang, Application of electric cell-substrate impedance sensing to investigate the cytotoxic effects of andrographolide on U-87 MG Glioblastoma cell migration and apoptosis. Sensors 19 (2019) 2275. https://doi.org/10.3390/s19102275.
[40] Organisation for Economic Co-operation and Development (OECD), Guidance Document on Good in Vitro Method Practices (GIVIMP), ENV/JM/MONO.(2018) 19. http://www.oecd.org/officialdocuments/publicdisplaydocumentpdf/?cote=ENV/JM/MONO(2018)19&doclanguage=en.

[41] J. Zhang, T.D.Y. Chung, K.R. Oldenburg, A simple statistical parameter for use in evaluation and validation of high throughput screening assays. J. Biomol. Screen. 4 (2) (1999) 67–73. https://doi.org/10.1177/108705719900400206.
[42] H.L. Martin, M. Adams, J. Higgins, J. Bond, E.E. Morrison, S.M. Bell, S. Warriner, A. Nelson, D.C. Tomlinson, High-content, high-throughput screening for the identification of cytotoxic compounds based on cell morphology and cell proliferation markers. PlOS one 9 (2) (2014) 1–8. https://doi.org/10.1371/journal.pone.0088338.
[43] USP 42-NF 37 ⟨111⟩ Design and Analysis of Biological Assay, USPC 2019. https://online.uspnf.com/uspnf/document/GUID-E40D050D-B163-43F4-8D82-891AF3907355_1_en-US?highlight=111. (Wed Jul 10 2019).
[44] B. Li, Data Analysis Approaches in High Throughput Screening, Technical Journal Club. https://www.rna.uzh.ch/dam/jcr:ffffffff-b34e-2810-0000-00003717f5fb/BeiLi_JC_08042014.pdf, 2014.
[45] M.A. Bray, A. Carpenter, Advanced assay development guidelines for image-based high content screening and analysis, 2017 Jul 8. in: G.S. Sittampalam, A. Grossman, K. Brimacombe (Eds.), Assay Guidance Manual [Internet], Bethesda (MD), Eli Lilly & Company and the National Center for Advancing Translational Sciences, 2004. https://www.ncbi.nlm.nih.gov/books/NBK126174/.
[46] H.P. Varbanov, F. Kuttler, D. Banfi, G. Turcatti, P.J. Dyson, Repositioning approved drugs for the treatment of problematic cancers using a screening approach. PLoS One 12 (2) (2017) e0171052. https://doi.org/10.1371/journal.pone.0171052.
[47] K.M. Bower, The relationship between R^2 and precision in bioassay validation, BioProcess International 16 (4) (2018) 26. https://bioprocessintl.com/upstream-processing/assays/the-relationship-between-r2-and-precision-in-bioassay-validation/.
[48] USP 42-NF 37 ⟨1033⟩ Biological Assay Validation, USPC 2019. https://online.uspnf.com/uspnf/document/GUID-952E8C3B-738B-40F8-A552-E14026AC78A9_1_en-US. (Mon May 27, 2019).
[49] J.L. Sebaugh, Guidelines for accurate EC50/IC50 estimation. Pharm. Stat. 10 (2011) 128–134. https://doi.org/10.1002/pst.426.
[50] D. Nevozhay, Cheburator software for automatically calculating drug inhibitory concentrations from in vitro screening assays. PLoS One 9 (9) (2014) e106186. https://doi.org/10.1371/journal.pone.0106186.
[51] Z. Chen, R. Bertin, G. Froldi, EC50 estimation of antioxidant activity in DPPH assay using several statistical programs. Food Chem. 138 (2013) 414–420. https://doi.org/10.1016/j.foodchem.2012.11.001.
[52] J.V. Haas, B.J. Eastwood, P.W. Iversen, V. Devanarayan, J.R. Weidner, Minimum significant ratio—a statistic to assess assay variability, 2013 Nov 1 [Updated 2017 Nov 20]. in: G.S. Sittampalam, A. Grossman, K. Brimacombe (Eds.), Assay Guidance Manual [Internet], Eli Lilly & Company and the National Center for Advancing Translational Sciences, Bethesda (MD). 2004. https://www.ncbi.nlm.nih.gov/books/.
[53] N.P. Coussens, G.S. Sittampalam, R. Guha, K. Brimacombe, A. Grossman, T.D.Y. Chung, J.R. Weidner, T. Riss, O.J. Trask, D. Auld, J.L. Dahlin, V. Devanaryan, T.L. Foley, J. McGee, S.D. Kahl, S.C. Kales, M. Arkin, J. Baell, B. Bejcek, N. Gal-Edd, M. Glicksman, J.V. Haas, P.W. Iversen, M. Hoeppner, S. Lathrop, E. Sayers, H. Liu, B. Trawick, J. McVey, V.P. Lemmon, Z. Li, O. McManus, L. Minor, A. Napper, M.J. Wildey, R. Pacifici, W.W. Chin, M. Xia, X. Xu, M. Lal-Nag, M.D. Hall, S.M.J. Inglese, A. Simeonov, C.P. Austin, Assay guidance manual: quantitative biology and pharmacology in preclinical drug discovery. Clin. Transl. Sci. 11 (2018) 461–470. https://doi.org/10.1111/cts.12570.
[54] N. Weerapreeyakul, A. Nonpunya, S. Barusrux, T. Thitimetharoch, B. Sripanidkulchai, Evaluation of the anticancer potential of six herbs against a hepatoma cell line. Chin. Med. 7 (2012) 15. https://doi.org/10.1186/1749-8546-7-15.

[55] R. Batista, A.J.S. Júnior, A. Braga de Oliveira, Plant derived antimalarial agents: new leads and efficient phytomedicines. Part II. Non-alkaloidal natural products. Molecules 14 (2009) 3037–3072. https://doi.org/10.3390/molecules14083037.
[56] J.C. Pritchetta, L. Naesens, J. Montoya, Treating HHV-6 infections, the laboratory efficacy and clinical use of anti-HHV-6 agents. in: L. Flamand, I. Lautenschlager, G. Krueger, D. Ablashi (Eds.), Human Herpes Viruses HHV-6A, HHV-6B, and HHV-7. Diagnosis and Clinical Management, third ed., Elsevier, Amsterdam, 2014, pp. 311–331. https://doi.org/10.1016/B978-0-444-62703-2.00019-7.
[57] O.A. Peña-Morán, M.L. Villarreal, L. Álvarez-Berber, A. Meneses-Acosta, V. Rodríguez-Lóp, Cytotoxicity, post-treatment recovery, and selectivity analysis of naturally occurring podophyllotoxins from *Bursera fagaroides* var. *fagaroides* on breast Cancer cell lines. Molecules 21 (2016) 1013. https://doi.org/10.3390/molecules21081013.
[58] M. López-Lázaro, Editorial: how many times should we screen a chemical library to discover an anticancer drug? Drug Discov. Today 20 (2015) 167–169. https://doi.org/10.1016/j.drudis.2014.12.006.
[59] M. López-Lázaro, Two preclinical tests to evaluate anticancer activity and to help validate drug candidates for clinical trials. Onco. Targets. Ther. 2 (2) (2015) 91–98. https://doi.org/10.18632/oncoscience.132.
[60] S.A. Fedoreyev, N.V. Krylova, N.P. Mishchenko, E.A. Vasileva, E.A. Pislyagin, O.V. Iunikhina, V.F. Lavrov, O.A. Svitich, L.K. Ebralidze, G.N. Leonova, Antiviral and antioxidant properties of echinochrome A. Mar. Drugs 16 (2018) 509. https://doi.org/10.3390/md16120509.
[61] M.D. Awouafack, L.J. McGaw, S. Gottfried, R. Mbouangouere, P. Tane, M. Spiteller, J.N. Eloff, Antimicrobial activity and cytotoxicity of the ethanol extract, fractions and eight compounds isolated from *Eriosema robustum* (Fabaceae). BMC Complement. Altern. Med. 13 (2013) 289. https://doi.org/10.1186/1472-6882-13-289.
[62] P. Cos, A.J. Vlietinck, D.V. Berghe, L. Maesa, Anti-infective potential of natural products: how to develop a stronger in vitro 'proof-of-concept'. J. Ethnopharmacol. 106 (2006) 290–302. https://doi.org/10.1016/j.jep.2006.04.003.
[63] M. Adamu, V. Naidoo, J.N. Eloff, The antibacterial activity, antioxidant activity and selectivity index of leaf extracts of thirteen South African tree species used in ethnoveterinary medicine to treat helminth infections. BMC Vet. Res. 10 (2014) 52. https://doi.org/10.1186/1746-6148-10-52.
[64] M. Adamu, V. Naidoo, J.N. Eloff, Efficacy and toxicity of thirteen plant leaf acetone extracts used in ethnoveterinary medicine in South Africa on egg hatching and larval development of Haemonchus contortus. BMC Vet. Res. 9 (2013) 38. https://doi.org/10.1186/1746-6148-9-38.
[65] I.M. Famuyide, A.O. Aro, F.O. Fasina, J.N. Eloff, L.J. McGaw, Antibacterial and antibiofilm activity of acetone leaf extracts of nine under-investigated south African *Eugenia* and *Syzygium* (Myrtaceae) species and their selectivity indices. BMC Complement. Altern. Med. 19 (2019) 141. https://doi.org/10.1186/s12906-019-2547-z.
[66] F. Nogueira, V.E. do Rosário, Methods for assessment of antimalarial activity in the different phases of the Plasmodiumlife cycle. Rev. Pan-Amaz Saude 1 (3) (2010) 109–124. https://doi.org/10.5123/S2176-62232010000300015.
[67] G.A.M. Ratih, M.F. Imawati, R.R. Nugroho, D.I. Purwanti, S. Wongso, B. Prajogo, G. Indrayanto, Phytochemicals of Gandarusa (*Justicia gendarussa*) and its Preparations. Natural Product Communications, June 2019. https://doi.org/10.1177/1934578X19851406.
[68] S. Singh, V.K. Khanna, A.B. Pant, Development of in vitro toxicology: a historic story. in: A. Dhawan, S. Kwon (Eds.), In Vitro Toxicology, Academic Press, Elsevier Inc, London, San Diego, Cambridge, Oxford, 2018, pp. 1–17. https://doi.org/10.1016/B978-0-12-804667-8.00001-8.

[69] S. Srivastava, S. Mishra, J. Dewangan, A. Divakar, P.K. Pandey, S.K. Rath, Principles for in vitro toxicology. in: A. Dhawan, S. Kwon (Eds.), In Vitro Toxicology, Academic Press, Elsevier Inc, London, San Diego, Cambridge, Oxford, 2018, pp. 21–39. https://doi.org/10.1016/B978-0-12-804667-8.00002-X.
[70] Validation of alternative methods for toxicity testing, C. Eskes, M. Whelan (Eds.), Advances in Experimental Medicine and Biology, In: vol. 856, Springer International Publishing, Switzerland, 2016. (ISBN 978-3-319-33826-2).
[71] G. Krishna, G. Gopalakrishnan, Alternative in vitro models for safety and toxicity evaluation of nutraceuticals, in: R.C. Gupta (Ed.), Nutraceuticals Efficacy, Safety and Toxicity, Academic Press, Elsevier, London, San Diego, Waltham, Oxford, 2016, pp. 355–385. ISBN: 978-0-12-802147-7.
[72] A. Aydın, G. Aktay, E. Yesilada, A guidance manual for the toxicity assessment of traditional herbal medicines, Nat. Prod. Commun. 11 (11) (2016) 1763–1773.
[73] R. Corvi, H. Spielmann, T. Hartung, Alternative approaches for carcinogenicity and reproductive toxicity. in: M. Balls, R. Combes, A. Worth (Eds.), The History of Alternative Test Methods in Toxicology, Academic Press, Elsevier, London, San Diego, Waltham, Oxford, 2019, pp. 209–217. https://doi.org/10.1016/B978-0-12-813697-3.00024-X.
[74] C.D. Jayasinghe, U.A. Jayawardena, Toxicity assessment of herbal medicine using Zebrafish embryos: a systematic review, Evid. Based Complement. Alternat. Med. 2019 (2019) 7272808. 17 pages. https://doi.org/10.1155/2019/7272808.
[75] R. Parthasarathi, A. Dhawan, In Silico approaches for predictive toxicology. in: A. Dhawan, S. Kwon (Eds.), In Vitro Toxicology, Academic Press, Elsevier Inc, London, San Diego, Cambridge, Oxford, 2018, pp. 91–108. https://doi.org/10.1016/B978-0-12-804667-8.00005-5.
[76] L. Tolosa, M.T. Donato, M.J.G. Lechon, General cytotoxicity assessment by means of the MTT assay. in: M. Vinken, V. Rogiers (Eds.), Protocols in in Vitro Hepatocyte Research, Methods in Molecular Biology, vol. 1250, Springer Science Business Media, New York, 2015, pp. 333–348. https://doi.org/10.1007/978-1-4939-2074-7_26.
[77] T.L. Riss, R.A. Moravec, A.L. Niles, S. Duellman, H.A. Benink, T.J. Worzella, L. Minor, Cell viability assays, 2013 May 1 [Updated 2016 Jul 1]. in: G.S. Sittampalam, A. Grossman, K. Brimacombe (Eds.), Assay Guidance Manual, Eli Lilly & Company and the National Center for Advancing Translational Sciences, Bethesda (MD), 2004. https://www.ncbi.nlm.nih.gov/books/NBK144065/.
[78] M. Sachana, A.J. Hargreaves, Toxicological testing: in vivo and in vitro models. in: R.C. Gupta (Ed.), Veterinary Toxicology, Academic Press, Elsevier Inc, San Diego, London, Cambridge, Oxford, 2018, pp. 143–160. https://doi.org/10.1016/B978-0-12-811410-0.00009-X.
[79] P. Verhaeghe, A. Dumètre, C. Castera-Ducros, S. Hutter, M. Laget, C. Fersing, M. Prieri, J. Yzombard, F. Sifredi, S. Rault, P.R.P. Vanella, N. Azas, 4-Thiophenoxy-2-trichloromethyquinazolines display in vitro selective antiplasmodial activity against the human malaria parasite *Plasmodium falciparum*. Bioorg. Med. Chem. Lett. 21 (2011) 6003–6006. https://doi.org/10.1016/j.bmcl.2011.06.113.
[80] M.C. Jonville, H. Kodjab, L. Humeaub, J. Fournelb, P. De Molc, M. Caoa, L. Angenot, M. Frédéricha, Screening of medicinal plants from Reunion Island for antimalarial and cytotoxic activity. J. Ethnopharmacol. 120 (2008) 382–386. https://doi.org/10.1016/j.jep.2008.09.005.
[81] V.P. Bagla, L.G. McGaw, E.E. Elgorashi, J.N. Eloff, Antimicrobial activity, toxicity and selectivity index of two biflavonoids and a flavone isolated from Podocarpus henkelii (Podocarpaceae) leaves. BMC Complement. Altern. Med. 14 (2014) 383. https://doi.org/10.1186/1472-6882-14-383.

[82] E. Abdel-Sattar, F.M. Harraz, S.M.A. Al-Ansari, S. El-Mekkawy, C. Ichino, H. Kiyohara, K. Otoguro, S. Omura, H. Yamada, Antiplasmodial and antitrypanosomal activity of plants from the kingdom of Saudi Arabia. J. Nat. Med. 63 (2) (2008) 232–239. https://doi.org/10.1007/s11418-008-0305-5. Natural Resource Letter.

[83] F. Lefranc, G. Nuzzo, N.A. Hamdy, I. Fakhr, L.M.Y. Banuls, Goietsenoven, G. Guido Villani, V. Mathieu, R. Soest, R. Kiss, M.L. Ciavatta, In vitro pharmacological and toxicological effects of norterpene peroxides isolated from the Red Sea sponge *Diacarnus erythraeanus* on normal and cancer cells. J. Nat. Prod. 76 (2013) 1541–1547. https://doi.org/10.1021/np400107t.

[84] R.B. Badisa, L.T. Ayuk-Takem, C.O. Ikediobi, E.H. Walker, Selective anticancer activity of pure licamichauxiioic-B acid in cultured cell lines. Pharm. Biol. 44 (2) (2006) 141–145. https://doi.org/10.1080/13880200600592301.

[85] \USP 42-NF 37 <1034> Analysis of Biological Assays, USPC 2019. https://online.uspnf.com/uspnf/document/GUID-CC133EE7-6ACF-4565-8740-AE4EECF3AF58_1_en-US (Mon May 27 2019).

[86] K.M. Bower, Certain approaches to understanding sources of bioassay variability, BioProcess Int. 16 (2018) 6–8. https://bioprocessintl.com/upstream-processing/assays/certain-approaches-to-understanding-sources-of-bioassay-variability/.

[87] S. Kromidas, Validierung in der Analytik, in: Zweite Auflage, Wiley-VCH, Weinheim, Germany, 2011, pp. 202–211.

[88] T. Coffey, K.M. Bower, A statistical approach to assess and justify potential product specifications, BioProcess Int. 15 (2) (2017). https://bioprocessintl.com/analytical/qa-qc/statistical-approach-assess-justify-potential-product-specifications/.

[89] USP 42-NF 37⟨1010⟩ Analytical data- interpretation and treatment, USPC 2019. https://online.uspnf.com/uspnf/document/GUID-5C0818CD-E76F-44B8-B504-C202CA762F2A_3_en-US (Fri March 1 2019).

Cumulative index

Bold numerals refer to volume numbers.

D

E

F

G

M

N

O

P

Q

R

S

T

U

V

W

X

Y

Z

Kolibri 360 GmbH
Ferdinand-Jühlke-Straße 7
99095 Erfurt, Deutschland
produktsicherheit@kolibri360.de